COMPARATIVE VERTEBRATE COGNITION: ARE PRIMATES SUPERIOR TO NON-PRIMATES?

DEVELOPMENTS IN PRIMATOLOGY: PROGRESS AND PROSPECTS

Series Editor:

Russell H. Tuttle
University of Chicago, Chicago, Illinois

This peer-reviewed book series will meld the facts of organic diversity with the continuity of the evolutionary process. The volumes in this series will exemplify the diversity of theoretical perspectives and methodological approaches currently employed by primatologists and physical anthropologists. Specific coverage includes: primate behavior in natural habitats and captive settings: primate ecology and conservation; functional morphology and developmental biology of primates; primate systematics; genetic and phenotypic differences among living primates; and paleoprimatology.

ALL APES GREAT AND SMALL
VOLUME 1: AFRICAN APES
Edited by Birute M. F. Galdikas, Nancy Erickson Briggs, Lori K. Sheeran, Gary L. Shapiro and Jane Goodall

THE GUENONS: DIVERSITY AND ADAPTATION IN AFRICAN MONKEYS
Edited by Mary E. Glenn and Marina Cords

ANIMAL BODIES, HUMAN MINDS: APE, DOLPHIN, AND PARROT LANGUAGE SKILLS
William A. Hillix and Duane M. Rumbaugh

COMPARATIVE VERTEBRATE COGNITION: ARE PRIMATES SUPERIOR TO NON-PRIMATES?
Lesley J. Rogers and Gisela Kaplan

COMPARATIVE VERTEBRATE COGNITION: ARE PRIMATES SUPERIOR TO NON-PRIMATES?

Edited by

Lesley J. Rogers and Gisela Kaplan

Centre for Neuroscience and Animal Behaviour
School of Biological, Biomedical and Molecular Sciences
University of New England
Armidale, NSW 2351, Australia

KLUWER ACADEMIC / PLENUM PUBLISHERS
New York, Boston, Dordrecht, London, Moscow

Library of Congress Cataloging-in-Publication Data

Comparative vertebrate cognition: are primates superior to non-primates? / edited by Lesley J. Rogers and Gisela Kaplan.
p. cm.
Includes bibliographical references and index.
ISBN 0-306-47727-0
1. Cognition in animals. 2. Primates—Psychology. 3. Vertebrates—Psychology.
I. Rogers, Lesley J. II. Kaplan, Gisela T.

QL785.C537 2003
596.1513—dc22

2003060171

ISBN 0-306-47727-0

233 Spring Street, New York, New York 10013

http://www.wkap.nl/

10 9 8 7 6 5 4 3 2 1

A C.I.P record for this book is available from the Library of Congress

Printed in the United States of America

PREFACE

Primates have featured in human cultural history for a long time. Adventurers, hunters, and travelers described them in their diaries. Even before Darwin, there was some recognition that they were different and fascinating, although not necessarily likeable. Rennie, in his observation of orangutans in 1838 condemned orangutans as slovenly and useless creatures: "Their deportment is grave and melancholy, their disposition apathetic, their motions slow and heavy, and their habits so sluggish and lazy, that it is only the cravings of appetite, or the approach of imminent danger, that can rouse them from their habitual lethargy, or force them to active exertion." (cit. in Yerkes, R. M., and Yerkes, A. W. (1929) *The Great Apes. A Study of Anthropoid Life*, Yale University Press, New Haven). Starting with Darwin, interest in primates became scientific and, tragically, led to the first major holocaust of apes, who were killed by their thousands in short spans of time to supply museums with specimens. Gradually zoos also captured them in increasing numbers for "exhibits." Because of their relatedness to humans, primates were then also regarded as good test subjects in biomedical research and the beginning of the trade and supply of monkeys commenced, taking them from the wild in uncontrolled and ever growing numbers for use in medical laboratories throughout the best part of the 20th century.

Against this background, the efforts that were made in the 1960s and throughout the 1970s to say that primates and, in particular great apes, were "special" was a much needed and radical departure from the poor practice and derogatory attitudes that had gone before. Attitudes began to change along with the sudden rise to fame of the great apes, thanks to the extremely devoted research work done by a number of researchers, many of whom are now household names. These changes resulted in questioning the ethics of how we treated primates.

Export and import of many primates is now banned in most countries; hence, taking them from the wild is often no longer an option, no matter how justifiable the purpose may be. Poaching is usually punishable by law (although not always enforceable) and use of primates for frivolous research purposes has been curbed over the years.

Overall, primates are now better protected in the wild and are generally treated better in captivity than they were in the 1950s. Such new appreciation can be attributed in no small measure to primatology's own hard work in disseminating the persistent take-home message that primates are special, deserving of our interest, protection and moral support. As a result of these sustained efforts in the last 50 years or so, the image of great apes gradually changed from that of brutes like King Kong (gorilla), the silly clown (chimpanzee), or the slovenly imbecile (orangutan) to that of creatures worthy of appreciation or, less needed, we believe, of some kind of new "noble savage" perhaps even with a mind of its own.

So what are we doing here in Primatology by publishing a book seemingly questioning the very essence on which we have finally achieved some change in moral status for primates? There are two points to be made in reply. The appalling past treatment of primates and their depletion in the wild had to be fought by political means and discourses in the political realms of human societies. The claim that primates are special was an important political point of argument; and so was the evolving concept of great apes as flagship species.

This book, voicing the opinions of some leading primatologists, ethologists, psychobiologists, neuroscientists and anthropologists, is not speaking from the standpoint of a political engagement with primates but of a scientific engagement with primates in relation to all other species. The extremely lively and seminal primatological debates of the 1970s, especially on the cognitive skills of apes, have carried us almost through to this day. Thirty years later, it is time to take stock and revise what we know, casting the net wider than the primate order to see what we have achieved theoretically and what is sustainable scientifically. Such stock taking exercises help us to adjust our research directions, reflect on our practices, and perhaps also give rise to new and different questions.

Often inadvertently, but sometimes deliberately, we make statements implying superiority of the behavior or cognitive abilities of apes compared to monkeys, or primates compared to nonprimates. Yet recent evidence shows that at least some of the characteristics that we thought made primates unique and superior are also seen in some species of non-primates. Thus we are beginning to become aware of the risks of moving in one intellectual milieu and choosing to work on a species instead of choosing a problem and then testing it on a range of species. With this recognition has come another motivation for editing this book. Each contributor has addressed the question "Are the cognitive abilities of primates simply different from other mammals or are they an evolutionary advance on them?" or, more generally, "Are primates special?" Each contributor has approached the question from a different perspective and considered a range of different species. We have covered in this book research findings about a wide span of vertebrate species, all relevant to understanding primates. Of course, we recognize that there are differences in cognition between all species, including the various primates, but we have examined whether there is any substance behind the often stated or hinted assumption that primates are some sort of improved cognitive design and that, amongst primates, the apes are a further improvement that foregrounded the evolution of the human brain and mind. We have approached this topic fully aware of the difficulties in defining what we mean by higher cognition, or "intelligence," but by our various ways of dealing with this uncertainty as we compare primate and nonprimate species we believe we will lead to a change in thinking.

The aim of the book is to explore afresh the long-standing interest, and emphasis on, the "special" capacities of primates. Some of the recent discoveries of the higher cognitive abilities of other mammals and also birds challenge the concept that primates are special and even the view that the cognitive ability of apes is more advanced than that of non-primate mammals and birds. It is therefore timely to ask whether primates are, in fact, superior to non-primates, and to do so from a broad range of perspectives. It will become apparent to the reader that researchers are only at the beginning of the search to find out whether primates are special and, of course, by "special" is meant not just different, which applies to all biological categories, but "better" in the ephemeral sense of being more like us and being cognitively superior to all other species.

Some of the contributors conclude that all primates (or some primate species) do, indeed, have abilities that separate them from other mammals and that these abilities are clearly special. Others argue that primates as a group are no more special than a range of other species, or they are special only because they are seen in relation to human primates. Our aim has not been to downplay studies on primates, or to simply criticize them, in comparison to studies on nonprimates, but to present the results of similar studies on primates and nonprimates to see whether any or many of the known cognitive abilities of primates are shared by some of the nonprimate taxa.

Divided into five sections this book deals with topics about higher cognition and how it is manifested in different species, and also considers aspects of brain structure that might be associated with complex behavior.

One important theme of this book is, of course, the evolution of higher cognition and another is the applicability of different testing methods to different species. We would fall short if we attempted here to summarize the breadth of areas covered by the chapters, since each one is a complete essay in itself. We hope they will be read to enlighten scientifically, rather than in any sense to lessen our respect for primates politically. At the very least, or perhaps the most, we hope that this volume will stimulate comparative research on behavior and so ensure that future dividing lines between species are erected only when their foundations are in science and not the result of either cultural distortions or political gain.

We would like to express our gratitude to Russell Tuttle, editor of the series, and Andrea Macaluso, editor with Kluwer, whose invitation to us to proceed with this book gave us the opportunity to explore our ideas further, and to thank Richard Andrew, Dario Maestripieri, and Michael Beran for their valuable comments on an earlier draft of the manuscript. We thank Leanne Stewart for assistance in preparing the index. We are most grateful to all of the contributors for adding so many ideas and giving us so many enjoyable pages to read.

Lesley J. Rogers and Gisela Kaplan
April, 2003

CONTENTS

Contributors xv

PART I: COMPLEX COGNITION

1. **Comparing the Complex Cognition of Birds and Primates** 3
Nathan J. Emery and Nicola S. Clayton
Introduction 3
Why Might Primates be Superior to Non-Primates? 4
Comparing Birds and Primates 5
Primates have a Neocortex Larger than Predicted for their Body Size 5
Primates have an Expanded Prefrontal Cortex 8
Primates Demonstrate Social Learning and Imitation 9
Primates Understand Others' Mental States 14
Primates Display Insight, Innovation, and they Construct and Use Tools 23
Insight and Innovation 23
Manufacture and Use of Tools 25
Primates Utilize Symbolic and Referential Communication 27
Primates Demonstrate Elements of Mental Time Travel 29
The Retrospective Component—Episodic Memory 30
Do Animals have Episodic-like Memory? 31
Episodic-like Memory in Scrub-Jays 32
The Prospective Component—Future Planning 33
Is there any Evidence of Future Planning in Animals? 35
Food Caching by Scrub-Jays: A Candidate for Future Planning in Animals? 36
The Perils of Primatocentrism and "Scala Naturae" 36
Uses and Abuses of the Ecological/Ethological Approach to Cognition 38

Species Differences in Ecology and Cognition 39
Ethologically Relevant Stimuli are Difficult to Control 40
How Far can the Natural Behavior of an Animal be Translated to the Laboratory? 41
The Great Divide: Awareness of "Self" 41
Is there a Case for Convergent Cognitive Evolution and Divergent Neurological Evolution? 45
Acknowledgments 46
References 46

2. Visual Cognition and Representation in Birds and Primates **57**
Giorgio Vallortigara
Introduction 57
Integration and Interpolation of Visual Information in the Spatial Domain 58
Integration and Interpolation of Visual Information in the Temporal Domain 69
Representing Objects 72
Objects in Space: Use of Geometric and Nongeometric Information 77
Conclusions 84
Acknowledgments 85
References 85

PART II: SOCIAL LEARNING

3. Socially Mediated Learning among Monkeys and Apes: Some Comparative Perspectives **97**
Hilary O. Box and Anne E. Russon
Introduction 97
Socially Mediated Learning 98
Imitation 106
Imitation in Monkeys 106
Imitation in Great Apes 110
Primate Imitation in Broader Perspective 115
Culture 118
Behavioral Traditions among Monkeys 119
Behavioral Traditions among the Great Apes 121

Facilitating Influences on Behavioral Traditions among Great Apes 123
References 127

4. Social Learning, Innovation, and Intelligence in Fish **141**
Yfke van Bergen, Kevin N. Laland, and William Hoppitt
Introduction 141
Traditions and Social Learning in Guppies 144
Innovation in Guppies—Is Necessity the Mother of Invention? 148
Conformity and Social Release 152
Primate Supremacy Reconsidered 156
Conclusions 163
Acknowledgments 164
References 164

PART III: COMMUNICATION

5. The Primate Isolation Call: A Comparison with Precocial Birds and Non-primate Mammals **171**
John D. Newman
Introduction 171
The Mammalian Isolation Call 172
The Primate Isolation Call 174
Isolation Call Development 175
A Well-Studied Primate 176
Neurochemical (Pharmacological) Control of Isolation Call Production 177
Neural Mechanisms of Isolation Call Production 177
Neural Mechanisms of Isolation Call Perception 179
Conclusions 181
References 181

6. Meaningful Communication in Primates, Birds, and Other Animals **189**
Gisela Kaplan
Introduction 189
Communication from the Point of View of the Receiver 191

Referential Signaling 191
Vocal Signaling in General 191
Motivational versus Referential Signals 192
Attributing Meaning in Alarm and Food Calling 194
Deception in Vocal Signaling 197
Nonvocal Communication 199
Human Language and Animal Studies 201
Co-evolutionary Events 204
Complex Communication, Social Organization, and the Hunt 205
Advantages of Living Together 207
Hierarchy, Group Complexity, and Feeding 212
Conclusion 214
References 215

PART IV: THEORY OF MIND

7. Theory of Mind and Insight in Chimpanzees, Elephants, and Other Animals? 227
Moti Nissani
Elephant Cognition 228
Do Elephants and Chimpanzees know that People See? 231
Experiment 1: Do Elephants know that People See? 240
Experiment 2: Do Chimpanzees know that People See? 245
Insight in Animals? 248
Retractable Cord-Pulling in Elephants 251
Do Elephants know when to Suck or Blow? 254
Conclusion 257
Acknowledgments 257
References 258

8. The Use of Social Information in Chimpanzees and Dogs 263
Josep Call
Reading Attention 265
What can Others See 266
What Organ is Responsible for Vision 269
Following Attention 271
Attention Following into Distant Space 272

Attention Following in Object Choice 272
Directing Attention 275
Discussion 278
References 283

PART V: BRAIN, EVOLUTION, AND HEMISPHERIC SPECIALIZATION

9. Increasing the Brain's Capacity: Neocortex, New Neurons, and Hemispheric Specialization 289
Lesley Rogers
Introduction 289
Brain Size Relative to Body Weight 290
Neocortex/Isocortex 296
Frontal Lobes 299
Relative Differences in the Size of Different Regions of the Brain 299
Coordinated Size Change 300
Mosaic Evolution 302
Linking the Size of Brain Regions to Specific Behavior 304
Correlations between Brain Size and Behavior 305
Foraging for Food 305
Social Intelligence 306
Social Learning, Innovation, and Tool Use 308
Hemispheric Specialization 310
Corpus Callosum 312
Experience and Brain Size 314
Assumptions/New Neurons 316
Conclusion 317
References 318

10. The Evolution of Lateralized Motor Functions 325
Michelle A. Hook
Whole-body Turning 327
Lower Vertebrates: Fish, Amphibians, and Reptiles 327
Birds 329
Non-Primate Mammals: Rodents, Dolphins, Cats, and Dogs 329
Non-Human Primates 331

Summary of Turning Biases 331
Hand Preferences for Simple Actions 332
Lower Vertebrates 332
Birds 333
Non-Primate Mammals: Rodents 333
Non-Primate Mammals: Cats and Dogs 334
Non-Human Primates 335
Summary of Hand Preferences for Simple Actions 338
Complex Visuospatial Tasks 339
Non-Primate Mammals: Cats 339
Non-Human Primates 340
Summary of Complex Visuospatial Tasks 342
Manipulation and Tool Use 343
Birds 343
Non-Human Primates 344
Summary of Manipulation and Tool Use 345
Foot Preferences in Locomotion 346
Birds 346
Non-Human Primates 347
Summary of Foot Preferences in Locomotion 348
Production of Emotional Responses and Vocalizations 348
Lower Vertebrates: Fish, Amphibians, and Reptiles 348
Birds 349
Non-Primate Mammals: Rodents 351
Non-Human Primates 352
Summary 354
Are Primates Special? 355
Acknowledgments 359
References 359

Epilogue **371**

About the Editors **375**

Index **377**

CONTRIBUTORS

Josep Call, Max Planck Institute for Evolutionary Anthropology, Leipzig, Germany

Nicola S. Clayton, Department of Experimental Psychology, University of Cambridge, Cambridge, UK

Nathan J. Emery, Sub-department of Animal Behaviour, University of Cambridge, Cambridge, UK

Michelle A. Hook, Department of Psychology, Texas A&M University, College Station, TX, USA

William Hoppitt, Sub-Department of Animal Behaviour, University of Cambridge, Madingley, UK

Gisela Kaplan, Centre for Neuroscience and Animal Behaviour, School of Biological, Biomedical and Molecular Sciences, University of New England, Armidale, NSW, Australia

Kevin N. Laland, Sub-Department of Animal Behaviour, University of Cambridge, Madingley, UK

John D. Newman, Laboratory of Comparative Ethology, NICHD, National Institutes of Health, Department of Health and Human Services, Poolesville, Maryland USA

Moti Nissani, Department of Interdisciplinary Studies, Wayne State University, Detroit, MI, USA

Hilary O. Box, Department of Psychology, University of Reading, Reading, UK

Lesley Rogers, Centre for Neuroscience and Animal Behaviour School of Biological, Biomedical and Molecular Sciences, University of New England, Armidale, NSW, Australia

Anne E. Russon, Department of Psychology, Glendon College of York University, Toronto Ontario, Canada

Giorgio Vallortigara, Department of Psychology and B.R.A.I.N. Centre for Neuroscience, University of Trieste, Trieste, Italy

Yfke van Bergen, Sub-Department of Animal Behaviour, University of Cambridge, Madingley, UK

COMPARATIVE VERTEBRATE COGNITION: ARE PRIMATES SUPERIOR TO NON-PRIMATES?

PART ONE

Complex Cognition

CHAPTER ONE

Comparing the Complex Cognition of Birds and Primates

Nathan J. Emery and Nicola S. Clayton

INTRODUCTION

At first glance, birds and non-human primates (hereafter primates) seem very different. Birds have beaks, feathers, produce offspring that gestate in shells, and can fly. Primates are covered in hair, have forward facing eyes and grasping hands, and while some are arboreal, none of them can fly. Although there are vast morphological differences between the two groups, there is growing evidence of strong similarities in their mental abilities, particularly in the realm of advanced cognitive processing. This suggestion is initially surprising considering the vast difference between the size and structure of avian and primate brains. Even more surprising is the fact that a number of recent experiments have failed to demonstrate some complex cognitive abilities in primates (including chimpanzees, gorillas, and orangutans) that have been convincingly

---

Nathan J. Emery • Sub-Department of Animal Behaviour, University of Cambridge, Cambridge, UK. **Nicola S. Clayton** • Department of Experimental Psychology, University of Cambridge, Cambridge, UK.

Comparative Vertebrate Cognition, edited by Lesley J. Rogers and Gisela Kaplan. Kluwer Academic/Plenum Publishers, 2004.

demonstrated in some species of birds, such as the corvids (jays, crows, ravens, and magpies) and parrots.

In this chapter, we will review evidence that some species of birds (focussing primarily on the corvids and parrots), with their much smaller brains and relative absence of cortical structures, have been shown to possess abilities that have either not been tested in primates or that primates have not displayed. This is unlikely to be due to a direct deficit in the mental cognition of the great apes per se (certainly primates' evolutionary relatedness to humans would dispute this idea), but to a lack of ethologically based tasks for testing primate cognition. Of course, some primate studies have begun to incorporate a high degree of ethological validity (such as gaze following in rhesus monkeys and knowledge attribution in chimpanzees (Emery et al., 1997; Hare et al., 2000, 2001; Tomasello et al., 1998)) and it is promising to note that these studies have shown a number of feats that the less naturalistic paradigms have not. Nonetheless, we throw down the gauntlet to primate researchers to design experiments with the same ecological validity as in experiments on corvids and parrots. Although avian and primate brains differ significantly in size and structure, similar principles of organisation are evident. We suggest that birds and primates reflect a case of divergent evolution in relation to neuroanatomy, but convergent evolution in relation to mental processes.

WHY MIGHT PRIMATES BE SUPERIOR TO NON-PRIMATES?

As the sub-title of this book is "Are Primates Superior to Non-Primates?" we begin this chapter by presenting seven claims as to why this question may warrant asking in the first place (but we admit that there are likely to be more). All the reasons are based on a largely anthropocentric view, namely that primates have been suggested to form a special group of animals because they demonstrate cognitive abilities that are similar to our own. The seven claims for the special status of primates are:

1. Primates have a neocortex larger than predicted for their body size.
2. Primates have an expanded prefrontal cortex.
3. Primates demonstrate social learning and imitation.
4. Primates understand others' mental states.
5. Primates display insight and innovation, and they construct and use tools.

6. Primates utilize symbolic and referential communication.
7. Primates demonstrate elements of "mental time travel" (i.e., episodic memory and future planning).

The next section will discuss these claims in detail, and for each case provide counter evidence that some avian species demonstrate some, or all, of the above traits. In some domains, namely social cognition and mental time travel, the evidence presented for birds is surprisingly more convincing than that shown for non-human primates. At the end of the chapter, we will evaluate why we think this intuitive discrepancy may have occurred, through a discussion of the perils of primatocentrism, the still prevailing influence of the concept of "scala naturae" in comparative psychology, and the use and abuse of the ethological approach to animal cognition. We close the chapter with a discussion of why corvids and parrots may present a case for convergent mental evolution when compared to primates, but also provide evidence for a complex psychological trait that has so far only been demonstrated in chimpanzees (*Pan troglodytes*), orangutans (*Pongo pygmaeus*), and bottle-nosed dolphins (*Tursiops truncatus*), but not in birds.

COMPARING BIRDS AND PRIMATES

Primates have a Neocortex Larger than Predicted for their Body Size

The great apes, including humans, have brains larger than would be predicted for their body size. Jerison (1971) called this the encephalization quotient (K). A closer examination of the regression lines on Jerison's classic figure reveals that the carrion crow (*Corvus corone*) also appears to be greatly encephalized, with a K value much higher than the regression line and higher than that of the chimpanzee. Jerison's data, however, were based on comparisons between whole brain size and body size. Body size has been criticized as an inaccurate scaling measure (as the brain stops developing sooner than the rest of the body and external, ecological variables may have an undue influence on body size; Deacon, 1990). Specifically, a larger body may reflect possession of a larger gut that would require no additional computational power (Byrne, 1995). Therefore, a better indication of the influence of the body on brain size may be the size of the brainstem (Passingham, 1982). The primary function of the brainstem is to regulate visceral and somatic functions, such as breathing rate and the control of blood pressure. These functions are largely dependent on body size, such as the force required to pump blood around a large body compared to a small body.

Overall brain size has also been criticized as a poor measure of cognitive capacity (Byrne, 1996; Deacon, 1990; see also Chapter 9 by Rogers). Many brain areas that control primary sensory and motor functions are not associated with "intelligence" or cognition. Other subcortical structures, such as the amygdala, hypothalamus, and hippocampus function in the control of species-specific behaviors, such as aggression, sex, and parenting (Panksepp, 1998) and behaviors dependent on spatial memory, such as homing, brood parasitism, migration, and food-caching (Clayton, 1998).

In mammals, the neocortex is the brain area most associated with cognitive processing (memory, reasoning, concept formation, and social intelligence). Although, many areas of the neocortex are also important for basic sensory and motor functions, the available comparative data for neocortex size in a number of primate species is limited to the neocortex as a whole (Stephan et al., 1981). Birds do not have a neocortex, but certain areas of the forebrain have been suggested to represent functionally equivalent (analogous) structures to the mammalian neocortex (see next section), ones that may correlate with measures of higher cognition, such as feeding innovation (Lefebrve et al., 1997; Nicolakakis and Lefebvre, 2000; Timmermans et al., 2000) and tool use (Lefebvre et al., 2002).

A comparative analysis was performed to investigate whether the neocortex of primates represents a special case for dramatic expansion, or whether a similar pattern is seen in the forebrain of some bird species. The only comprehensive set of avian neuroanatomical data available (Portmann, 1946, 1947) presents forebrain and brainstem weights for 140 avian species (including 42 Passeriforme species, 29 Coraciomorphae species, 28 Pelargomorphae species, and 41 Alectoromorphae species). Portmann's data was transformed from brain indices (using his basic unit; the brainstem of a Galliforme of comparable size to the chosen bird) to the actual weight of the forebrain (hemispheres) and brainstem. A forebrain : brainstem ratio was therefore calculated for each species that was independent of phylogenetic anomalies. For the primates (19 Prosimian species and 26 Simian species, excluding *Homo sapiens*) and insectivores (18 species; basal mammals), a neocortex : brainstem ratio was calculated using volumetric data from Stephan et al. (1981).

Interestingly, when the forebrain : brainstem ratio of corvids (Corvidae) and parrots (Psittaciformes) is compared to other avian species (Figure 1A), the pattern of difference closely resembles the difference of neocortex: brainstem ratio between the great apes and other primates and insectivores (Figure 1B).

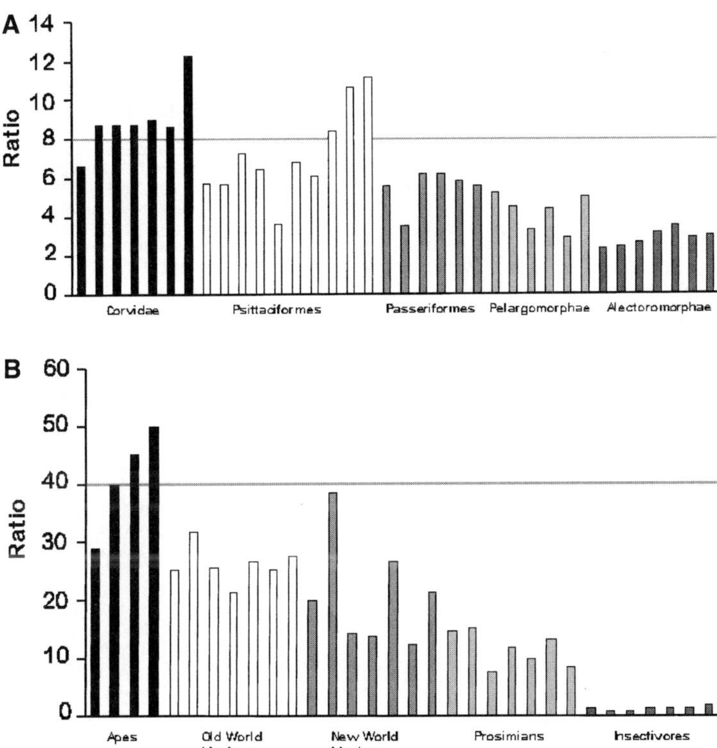

Figure 1. (A.) Forebrain:brainstem ratios in some representative species of different groups of birds; Corvidae (corvids), Psittaciformes (parrots), Passeriformes (perching songbirds, not including corvids), Pelargomorphae (inc. birds of prey), Alectoromorphae (chickens, pigeons and doves). Original data from Portmann (1947), in which forebrain was represented as Hemisphere index, and brainstem as Brainstem index. The indices were transformed to raw data by multiplying values with a basal number (an individual value calculated from the predicted value of the brainstem of another species from the same taxonomic group of comparable body size, but less developed). (B.) Neocortex size: brainstem size rations in representative species of different groups of primates (Apes, not including humans, Old World monkeys, New World monkeys and prosimians) and insectivores. Original data from Stephan et al. (1981) and Zilles and Rehkamper (1988).

Primates have an Expanded Prefrontal Cortex

Although birds do not have a six-layered neocortex, Karten (1969, 1991) has suggested that the nuclear arrangement in the avian dorsal ventricular ridge (DVR) processes stimuli in similar ways to the mammalian neocortex. All mammals have a neocortex; however, an expanded prefrontal cortex has been suggested to be special to the primates and the seat of their advanced cognitive abilities. The avian equivalent of the primate prefrontal cortex may be either the hyperstriatum ventrale (HV) or the neostriatum caudolaterale (NC). Evidence for this has been derived from lesion studies on tasks comparable to those used to test mammalian prefrontal cortex function, such as delayed alternation, reversal learning, and go/no go tasks (Güntürkün, 1997; Hartmann and Güntürkün, 1998). Güntürkün and colleagues lesioned only the NC, however, and the tasks used were those traditionally used to examine dorsolateral prefrontal cortex function in mammals, so-called executive functions (Stuss and Alexander, 2000). Although the comparable studies have yet to be performed, the avian HV may represent a functional equivalent to other prefrontal cortical areas, such as the orbitofrontal cortex. Lesions to these avian brain areas cause deficits in filial imprinting or selective attachment to the mother (Horn, 1985), and lesions of the orbitofrontal cortex in primates cause dramatic disturbances in social attachment and affiliative behavior (Raleigh and Steklis, 1981).

Additional evidence for the functional equivalence of the avian neostriatum and mammalian prefrontal cortex comes from studies of the distribution of dopamine in the avian telencephalon (Durstewitz et al., 1999), which is also widely distributed in the mammalian prefrontal cortex (Goldman-Rakic et al., 1992).

If these areas do constitute equivalent structures to the primate prefrontal cortex, and we hypothesize that corvids and parrots have cognitive abilities compatible to primates, then we may expect that these areas are larger in the corvids and parrots compared to other birds (a more precise localization to areas of the forebrain in those species demonstrated in the previous section). We therefore compared the volumes of four regions of the forebrain (the hyperstriatum accessorium [HA], hyperstriatum dorsale [HD], HV, and neostriatum) of six available avian species (*Coturnix coturnix, Perdix perdix, Phasianus colchicus, Passer domesticus, Garrulus glanddarius,* and *Corvus corone* (Rehkamper et al., 1991)) and calculated ratios of these brain regions to brainstem (as previously), and as a percentage of telencephalon. Percentage of telencephalon did not differ between these species for all the examined brain structures, thereby suggesting that overall body size was an irrelevant factor. Brain area : brainstem ratio, however, revealed that the neostriatum is larger than predicted in the carrion crow (compared to the other birds; Figure 2A).

An additional data set comparing different forebrain regions (HA, HD, HV, neostriatum, paleostriatum augmentatum, paleostriatum primitivum, and archistriatum) in four corvids (Carrion crow, *Corvus corone*; magpie, *Pica pica*; rook, *Corvus frugilegus*; and jackdaw, *Corvus monedula*) confirmed that the neostriatum displays the greatest expansion within the corvid forebrain (Voronov et al., 1994; see Figure 2B); however, a somewhat surprising result is the greater neostriatal expansion in the magpie and jackdaw brain (relative to body size as brainstem size within the same birds was unavailable for these species). These patterns mirror that for the prefrontal cortex of the great apes, which are approximately the same relative size in chimpanzees, orangutans, gorillas (*Gorilla gorilla*), and humans, but larger than in gibbons (*Hylobytes lar*) and macaques (*Macaca* sp.; Semendeferi et al., 2002).

Primates Demonstrate Social Learning and Imitation

The ability to learn information about objects, individuals or locations, or the precise methods or actions required to achieve a particular goal from another individual is called social learning and it can have many forms. Space does not permit us to describe the different forms of social learning in much detail, but the significant categories include stimulus and local enhancement, contagion, social facilitation, observational conditioning, copying, goal emulation, and imitation (see reviews in Tomasello and Call, 1997; Whiten and Ham, 1992; and papers in Heyes and Galef, 1996; see also Chapter 3 by Box and Russon).

For food-caching birds, one of the problems of living in social groups concerns the pilfering of food (kleptoparasitism; Brockmann and Barnard, 1979). Hiding food caches in the presence of others is risky because an observer may subsequently steal those caches when the storer is out of sight. For storers, there are many potential counter-strategies available to reduce the potential for pilfering (see next section). For pilferers, the ability to locate caches made by others quickly and efficiently may be the important difference between successful pilfering and potential aggression from the storer. Therefore, pilfering birds may require a sophisticated observational spatial memory for learning about the precise location of another individual's caches.

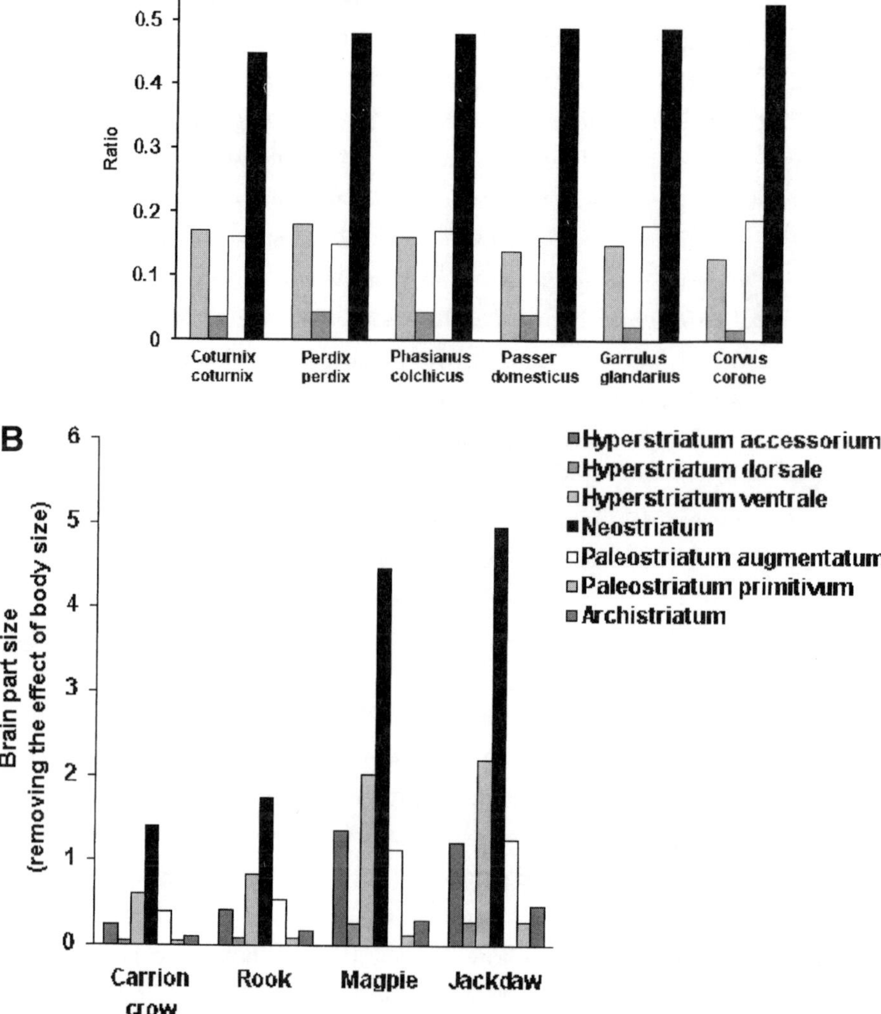

Figure 2. (A.) Ratios of subdivisions of the avian forebrain (Hyperstriatum accessorium, H. dorsale, H. ventrale, and Neostriatum) to the brainstem in representative species of Galliformes; *Coturnix coturnix, Perdix perdix*, and *Phasianus colchicus* and Passeriformes; *Passer domesticus, Garrulus glandarius* and *Corvus corone*. Data from Rehkamper et al. (1991). (B.) Ratios of subdivisions of the avian forebrain (H. accessorium, H. dorsale, H. ventrale, neostriatum, Paleostriatum augmentatum, P. primitivum and Archistriatum) in representative species of Corvidae; *Corvus corone* (carrion crow), *Corvus frugilegus* (rook), *Pica pica* (magpie) and *Corvus monedula* (jackdaw) to body size (brainstem values were unavailable). Original neural data from Voronov et al. (1994), and body sizes from Madge and Burn (1999).

This behavior has been studied in closely-related species of New World corvids (jays and nutcrackers). Bednekoff and Balda (1996a, 1996b) tested the ability of pinyon jays (*Gymnorhinus cyanocephalus*), Clark's nutcrackers (*Nucifraga columbiana*), and Mexican jays (*Aphelocoma ultramarina*) to remember where another bird had cached, by examining their cache retrieval efficiency. The birds were allowed to observe another bird caching, and were then given the opportunity to recover those caches either 1 day or 2 days later. Bednekoff and Balda (1996a) found that highly-social pinyon jays could remember the location of caches made by another bird, in specific locations at 1- and 2-day retention intervals and in general locations at a 7-day retention interval.

At the 1-day retention interval, Clark's nutcrackers (a relatively asocial species, but one that relies heavily on cached food for survival over the winter) performed more accurately than chance, as both storers and observers, and there was no difference between the two groups (i.e., observers could locate other's caches as well as the birds that made the caches). At the 2-day interval, Clark's nutcrackers accurately recovered their own caches, but not those they had observed. At the 1-day retention interval, Mexican jays (a very social species that lives in large flocks) behaved the same as the nutcrackers, and were more accurate than expected by chance when recovering caches they had made and those made by another bird. As with the pinyon jays, there was no difference between recovering their own caches and another's caches at the 2-day retention interval (Bednekoff and Balda, 1996b). Balda et al. (1997) have suggested that as pinyon jays and Mexican jays are social species, and display sophisticated social learning capabilities; but Clark's nutcrackers are an asocial species and display constrained social learning, there may be an adaptive specialization within social corvids to learn information from others. This argument stems from their work on the adaptive specialization of spatial memory in these species (see following arguments), comparing social versus non-social species on a social learning task and a non-social learning task (Templeton et al., 1999).

The western scrub-jay (*Aphelocoma californica*) is a territorial species, which forms pairs only during the breeding season, and thus might be described as asocial. If the adaptive specialization for social learning hypothesis is correct, then we would predict that closely-related asocial jays would fail to locate another's caches during similar tests for observational spatial memory. Griffiths, Duarte, and Clayton (unpublished observations; see also Clayton et al., 2001b), examined three groups of western scrub-jays; Storers, Observers, and Controls (a group that could hear another bird caching in an adjacent cage, but not see it). (The act of caching can be heard as a series of bill taps when the bird buries the food in the sand. This can be easily heard by other individuals in the vicinity, including the experimenter.) The birds were given a 3-hr retention interval to enhance the possible retrieval accuracy, and to mimic natural behavior; caches are very unlikely to remain after 1–7 days. This retention is considerably shorter than in the Bednekoff and Balda (1996a, 1996b) experiments.

Each group was compared with the other two groups. Observers made significantly fewer looks than Controls before searching in a location where food had been cached by the Storer during the caching phase, but the Observers made significantly more looks than Storers to search in a location where food had been cached compared to Observers. The Storers also made significantly fewer looks to search for the food than Controls (Figure 3).

Imitation is said to represent a form of social learning that may require complex representations, not only of another's perspective, but also their intentions, and how they relate to the individual observing the actions (see reviews in Heyes and Galef, 1996; Whiten and Ham, 1992; and Chapter 3). True imitation has been defined as "the copying of a novel or otherwise improbable act or utterance, or some act for which there is clearly no instinctive tendency" (Thorpe, 1963, p. 135) and as such the copied actions/vocalizations should not already be within the animal's natural repertoire (Clayton, 1978).

In birds, two forms of imitative learning have been investigated, vocal mimicry and motor imitation. Male songbirds not only copy the song of their fathers (Catchpole and Slater, 1995), but some species, such as mynah birds (*Gracula religiosa*) and parrots (Psittaciformes) can imitate the vocalizations of other birds, human speech, and general noises (such as doors closing, and walking; Baylis, 1982). As there is little evidence for vocal imitation in non-human primates, the case of motor imitation may be more appropriate for comparison between birds and primates. To date, only one study on an African gray parrot (*Psittacus erythacus*) has examined the ability to imitate novel motor patterns (Moore, 1992), with each action associated with a verbal label.

The two-action method of motor imitation has been proposed as the most appropriate method for examining imitative behavior in animals (Heyes, 1996). The technique was initially used to test whether budgerigar observers learned to remove a red cardboard square from a white pot demonstrated by a conspecific, either using the beak or the foot (Dawson and Foss, 1965). Similar experiments

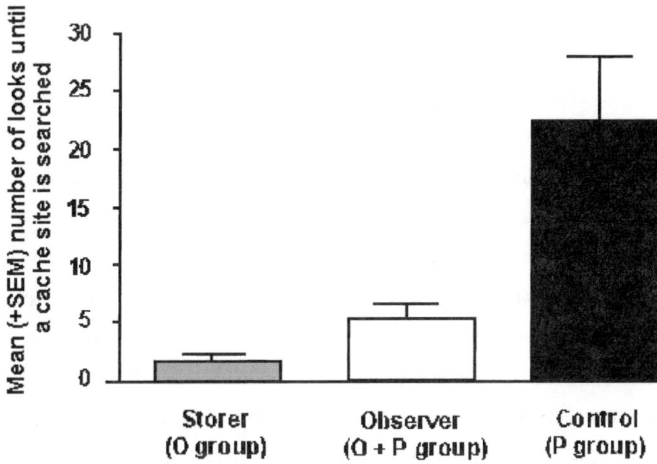

Figure 3. Mean (+SEM) number of looks made until a cache site is searched in three groups of western scrub-jays performing an observational spatial memory test. The Storer group (O group) made the caches, the Observer (O + P group) watched the Storer make the caches, and were then allowed to pilfer those caches, and the Control group (P group) were located in an adjacent cage, and so could hear the Storer caching, but could not see them being made. Unpublished data by Griffiths, Duarte and Clayton.

where two demonstrated actions can achieve a goal have been performed in rats (*Rattus norvegicus*; Heyes and Dawson, 1990), starlings (*Sturnus vulgaris*; Campbell et al., 1999), Japanese quail (*Coturnix japonica*; Atkins and Zentall, 1996), and marmosets (*Callithrix jacchus*; Bugynar and Huber, 1997). An example of the two-action method in corvids was provided in ravens (*Corvus corax*); two groups of dyads, either controls or observer-demonstrator pairs, were presented with boxes that could be opened by levering the lids (Fritz and Kotrschal, 1999). The demonstrator was trained to open the box using a different method from levering (pulling up). Control birds only levered open the box, whereas the observers opened the box both ways (levering and the demonstrated pull and open technique).

Perhaps, the most convincing demonstration of true imitation in non-human primates using a form of the two-action method has been the use of an "artificial fruit," a clear Perspex box that contained food, and that could be opened using two-actions (poke a bolt or turn a pin), and with the actions performed in a particular sequence (turn then poke). A non-functional action (twist) was also included in demonstrations to determine whether the observer would copy an action that did not lead to the goal. The artificial fruit has provided evidence for imitation in a number of non-human primates (chimpanzees, Whiten, 1998; Whiten et al., 1996; gorillas, Stoinski et al., 2001; capuchin monkeys *Cebus apella*, Custance et al., 1999).

Birds have also been demonstrated to process complex covered foods (such as shells and tough skins; Gibson, 1986). Huber et al. (2001) examined the ability of keas (*Nestor notabilis*), a New Zealand parrot, to imitate the actions of a demonstrator to gain food in an "artificial fruit". Entry to the fruit required the performance of three successive manipulations of three locking devices (a screw, split pin, and a bolt). Unlike the primate studies, conspecific demonstrators were trained to perform the appropriate actions used to gain entry to the fruit. One group of observers saw the demonstrators opening the fruit, whereas an additional control group was presented with the fruit without any experience of observing a demonstrator. The observers spent a longer time exploring the fruit, and the latency to first contact of the box was shorter in the observers. The observers also appeared to understand the goal of the task as they displayed greater perseverance in manipulating the locking devices on the fruit. Unfortunately, no bird succeeded in opening the fruit, but they displayed differences in success in opening the individual locking devices. All the observers ($n = 5$) succeeded in opening one device, and 2 out of 5 observers opened all devices. The observers did not differ from the controls in matching the particular demonstrated actions with the actions they performed themselves. This suggests that the keas were drawn to the fruit by localized stimulus enhancement, but they also appeared to learn about the goal and some of the actions required to gain access to the goal.

Primates Understand Others' Mental States

The term "theory of mind" (ToM) first appeared in a paper by Premack and Woodruff (1978) called *Do chimpanzees have a theory of mind?* In this paper, they described a study examining whether a language-trained chimpanzee, Sarah, could appreciate the correct solution to a problem presented to a human demonstrator. Sarah was presented with a video sequence of either an actor locked in a cage, shivering next to an unlighted heater, unable to clean a dirty floor, and unable to listen to music on an unplugged stereo. After each sequence, she was then presented with a number of alternative answers (as photographs), such as a key, a lighted paper wick, a connected hose, and a plugged in cord and she had to match the correct image with the appropriate video. A significant performance was interpreted by Premack and Woodruff (1978) as evidence for an understanding of the actor's intentions ("he wanted to get out of the cage," "he intended to listen to music", etc.). Although this interpretation has been criticized as demonstrating an understanding of the relationship between previously associated objects (i.e., lock and key) that the chimpanzee would have encountered during her life in captivity (Savage-Rumbaugh et al., 1978), it cannot explain all the abilities displayed by this particular chimpanzee (Premack and Premack, 1983). For the last 26 years, the question has remained open and has been tested in many species of primates (for a recent review see Tomasello and Call, 1997), but there has not yet been a persuasive demonstration of ToM in a non-human primate (Heyes, 1998). The precursors to ToM, such as gaze following have also begun to be tested in non-primate mammals, such as dolphins (Tschudin et al., 2001), goats (*Capra hircus*; Kaminski, 2002), dogs (*Canis familiaris*; Hare and Tomasello, 1999; Call, this volume), and horses (*Equus caballus*; McKinley and Sambrook, 2000). Birds have not been tested for gaze following as although birds are visually based animals, their visual system is very different from mammals, with a larger peripheral field of vision compared to some mammals, and depending on the avian species. It would therefore be extremely difficult, if not impossible to measure where a bird may be looking (however, see Dawkins, 2002). Hampton (1994), however, examined the propensity for sparrows (*Passer domesticus*) to intensify fear behavior when a human experimenter stared at them with two eyes, compared to one eye or turned backwards. This suggests at least a rudimentary appreciation of eye configuration representing forward attention (Emery, 2000).

ToM has also been investigated in pigs (*Sus scrofa*; Held et al., 2001) and dolphins (Tschudin, 2001), but conclusive answers to whether non-human mammals have a ToM remain wanting. This situation may be due to a prominent primatocentric view of complex cognition (discussed further), and the fact that the initial species chosen by Premack and Woodruff was a language-trained, enculturated chimpanzee.

Although birds display many of the forms of social organization seen in primates, ornithologists have tended to neglect research on complex social cognition in birds (Marler, 1997). There are a few studies, however, that do suggest that corvids and parrots, at least, have some of the hallmarks of ToM. Pepperberg and McLaughlin (1996), for example, discussed the important role of a precursor to ToM; shared attention, in the acquisition of language in African gray parrots, but they have yet to test whether parrots can understand another's mental states.

Tactical deception, or the intentional manipulation of another's beliefs leading to deception, has been proposed as another important indication of ToM (Whiten and Byrne, 1988). Very little experimental work has been performed on this cognitive capacity, with the predominant source being data compiled from a large number of anecdotes produced by primatologists (Whiten and Byrne, 1988). The use of anecdotes as a source of data has been criticized by Heyes (1998) and others, and very few experiments on tactical deception have been performed in any species (discussed in Tomasello and Call, 1997).

Ristau (1991) has argued that the broken-wing display in piping plovers (*Charadius melodus*), and Wilson's plovers (*Charadius Wilsonia*) on sand dunes in the eastern United States may provide some evidence of tactical deception in birds. The plovers utilize a so-called broken-wing display (BWD), where a plover feigns injury through a suite of behaviors that are concordant with injury, such as fanning of the tail, an increase in awkwardness of walking, arching of both wings, fluttering and dragging of the wings, and loud raucous squawks. It is argued that this collection of behaviors appears to cause a pretence that the bird is injured, and thus a potential predator follows the plover (as injured birds are easy prey) away from the plover's nest. Once the predator has been drawn hundreds of meters away from the nest, the plover flies away uninjured. Ristau argued that the plovers were intentionally deceiving predators (in this case humans) because in 98 percent of cases, the plovers moved in a direction away from the nest, never close to the nest, and in 98 percent of cases the plovers produced the BWD in front of the predator. In all cases, the plovers moved to an appropriate position before displaying, therefore suggesting that the display is not a reflex. Finally, the plovers appear to have monitored the predator's behavior by looking over their shoulder back towards them, and the plovers also modified their displays in response to the changing behavior of the predators. In 55 percent of cases, when the predator did not follow the plover, they stopped their display and re-approached the predator. In 29 percent of cases, the plovers increased the intensity of their display if the predators did not follow the plover.

The plovers' behavior has been interpreted using a number of alternative non-intentional explanations. Some have argued that the plovers perform a reflexive or a fixed action pattern response, but this seems unlikely given the plasticity of the plovers response that changes depending on the actions of the predator. What is not known is the role of learning in the development of such displays, in the absence of any information about the previous reinforcement history of these birds. And there is no evidence that plovers transfer this ability to other situations, such as social interactions with conspecifics, where tactical deception is more traditionally employed and more likely to be dependent on mental attribution. Ristau's behavioral studies were limited in scope, restricted to a single geographical population of plovers, and based only on a few trials (but this is likely to have been controlled for habituation effects).

Such field studies provide us with important insights into the behavior of animals in their natural environment, but tightly controlled experiments in the laboratory on a wide variety of species (not just the traditional subjects of comparative psychology experiments; rats, pigeons, and rhesus monkeys) are essential, especially when examining the question of complex cognitive abilities, and the presence of mental states in other individuals.

Recent laboratory studies of mental attribution have been conducted on some species of corvids. As discussed earlier, many corvids hide caches of food for later consumption. During harsh winter conditions efficient cache recovery can result in the difference between death and survival. Protection of caches from potential pilferers therefore presents an example where understanding the intentional behavior of conspecifics might be a useful attribute to possess.

Ravens, for example, are very cautious when hiding their caches when in a flock. This behavior has been studied in detail by Heinrich and colleagues (Heinrich, 1999; Heinrich and Pepper, 1998), and more recently by Bugnyar and Kotrschal (2002). It has been shown that storers will delay caching if other ravens are in the vicinity and wait until would-be thieves are distracted or have moved away before they resume caching. Heinrich has also observed ravens making false caches in the presence of observers, and repeatedly moving them around to distract the observer.

Bugnyar and Kotrschal (2002) have suggested that caching and raiding ravens may present an example of tactical deception through an attempt to manipulate another's attention, either to prevent opponents from gaining opportunities for stealing (see earlier section on social learning), or to gain opportunities for learning socially from the opponents. They examined two forms in which another's attention may be manipulated; withholding information and directing another's attention away from the caches (object of interest). During caching, storers tended to withdraw from conspecifics, and were at a greater distance from conspecifics during caching than during other activities, such as feeding and resting. The storers also cached close to large objects, with 80 percent of the caches between the storers and the observers, outside the view of the observers. Cachers moved their caches if an observer moved towards them, and also protected their cache sites. Cache raiders also employed a number of strategies to increase the potential for learning about cache sites, and for stealing caches. In 33 percent of cases, the observers changed their position relative to the structures that blocked their view of the cache sites and in 32 percent of cases the observers changed their orientation and distance away from the cachers. If a storer was close to a cache site, the observers delayed pilfering until the cacher was away from the caches, usually within 1-min of the cacher leaving the cache site. These behaviors suggest that caching and raiding ravens appreciate the visual perspective of one another, and produce strategies to counter the behavior of the other.

As with ravens and other corvids, western scrub-jays pilfer each other's food when presented with the opportunity, and pilferers have excellent recall of the location of another's caches seen being made during an earlier occasion (Griffiths, Duarte, and Clayton, unpublished observations). One of us (NSC) routinely observed western scrub-jays on the campus of the University of California at Davis, USA, compete over scraps of food left by humans. The birds would cache the food, but many would return later when other birds were distant from their cache sites and re-hide the caches in new locations.

With this in mind, we examined whether this re-caching behavior would be observed in a laboratory colony of hand-raised western scrub-jays (Emery and Clayton, 2001). We initially gave a group of seven birds the opportunity to cache worms either in private with another bird's view obscured or when another bird was present, and they could observe the location of the caches being made. After a 3-hr retention interval, the storing birds were allowed to recover their caches in private, independent of what they did during caching. During caching, the birds were provided with an ice-cube tray that was filled with sand so that the birds could make caches in it. Each tray was made trial unique by attaching novel configurations of Lego® bricks to the caching tray.

A bowl containing wax worms was also placed in the storer's cage. We recorded the number of caches the storer made and ate, counting the number of worms remaining in the bowl at the end of the 15-min trial. After the 3-hr retention interval, the birds were presented with the original tray containing their caches and an additional unique tray in which they could re-cache worms. We recorded the number of worms that were recovered, the number of looks made to cache sites (as a measure of recovery accuracy) and the number re-cached; either in old cache sites (those used to make the previous caches) or new cache sites (in the new tray or elsewhere in the storer's cage). We found that the storers recovered more caches when previously observed during caching than when they had previously cached in private. The storers also re-cached significantly more worms when observed during the previous caching episode than when they had cached in private, and almost all these re-caches were made in new sites unknown to the observer (Figure 4).

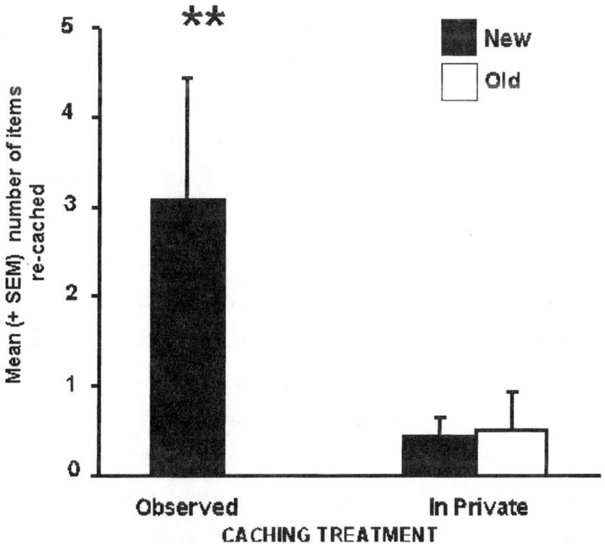

Figure 4. Mean (+SEM) number of caches that were re-cached in new or old sites, either when previously observed caching or when they had previously cached in private 3 hr earlier by O+P group birds in Davis, USA. $^{**}p<0.01$. Same statistical results when proportions were analyzed.

One interpretation of these results is that the storers predict the observer's intentions (i.e., to pilfer caches when given the opportunity), and use counter strategies to prevent this, by re-caching the worms in new sites. But being observed recently during caching could lead to an automatic tendency to re-cache when presented with the opportunity to do so at a later time, irrespective of whether the bird remembers anything specific about the caching event such as who was watching and where it had cached. An alternative explanation is that the bird remembers whether or not it had been watched during caching in a specific tray. We discriminated between these two possibilities by testing whether these jays could keep track of the social context of previous caching episodes that occurred in close temporal proximity.

We, therefore, tested this by presenting the same storers with a series of two interleaved trials, such that they were first observed caching in one unique tray, and then, 10-min later in a second unique tray, they cached in private. The order of the trials was counterbalanced. After a 3-hr retention interval, the two trays were returned to the storers, with an additional new tray for potential re-caching. Again, we found that the storers recovered more worms from the tray in which they had been observed caching, and they also re-cached more worms in new sites, specifically from the observed tray (Figure 5). This result shows that the birds remembered the specific social context during caching and did not automatically re-cache items if observed recently.

The jays described in the two previous experiments had had prior experience of watching another bird cache and were then given the opportunity to steal those caches (see earlier section on social learning). This Observer + Pilferer (O + P) group were almost as accurate at recovering caches they had observed being made as the birds that had made them. The storer birds in the observational learning experiment acted as observers in the experiments previously described (Observer; O group). An additional group in the observational learning study were given the opportunity to listen to another bird cache and then steal those caches, without ever observing the caches being made (Pilferer; P group). The O + P group were the birds in the previous two social context experiments. When the three groups were tested, either in private or when observed, the O + P group again demonstrated re-caching, with significantly more in new sites compared to old sites (Figure 6A). Surprisingly, the P group also displayed significant levels of re-caching, especially in new sites (Figure 6B). The O group, by contrast, with no pilfering

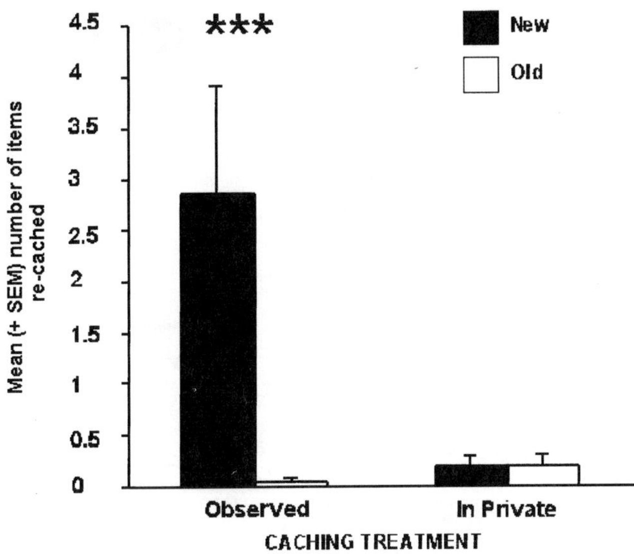

Figure 5. Mean (+SEM) number of caches that were re-cached in new or old sites when either observed caching in a trial-unique tray, or in a second trial-unique tray that was not observed during caching, by O + P group birds in Cambridge, UK. $***p<0.001$. Same statistical results when proportions were analyzed.

experience, did not demonstrate any re-caching, and the little re-caching they did was equally distributed in new and old sites (Figure 6C). This result suggests that the small amount of pilfering experience that the birds in the P and O + P groups had received was sufficient and necessary to trigger re-caching, whereas only observational experience failed to have this effect.

The O + P and P group birds appear to have transferred their pilfering experience to the current situation, and put themselves in the perspective of the observers, which may have the opportunity to pilfer the storers' caches in the future. Birds in the O group that had received no experience of pilfering another bird's caches did not do this. These experiments suggest the presence of a sophisticated level of social cognition in western scrub-jays, and one that depends on prior experience as a pilferer, but is it ToM? The storage and pilfering of food presents a case in which individuals have to play two different and competing roles; storer and pilferer, roles which require different behavioral strategies. So, in corvids, elements of ToM may be a byproduct of the

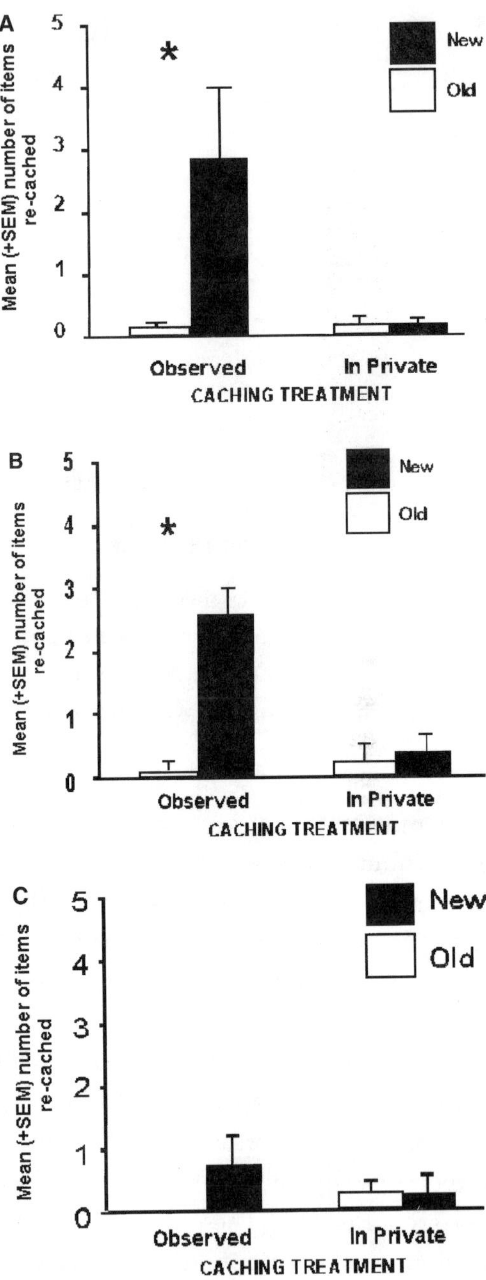

Figure 6. (A.) Mean (+SEM) number of caches that were re-cached in new or old sites when either observed during caching or when cached in private, by (A) O+P group birds. (B) P group birds and (C) O group birds in Cambridge, UK. * $p<0.05$. Same statistical results when proportions were analyzed.

arms race of strategies and counter-strategies in relation to the hiding and stealing of food (such as experience projection and perspective taking). Presumably other species in which individuals play competing roles in forms of behavior, be it storer and pilferer, or dominant and subordinate, may also possess rudimentary forms of ToM.

Primates Display Insight, Innovation, and they Construct and Use Tools

Insight and Innovation: Insight has been defined as "the sudden production of a new adaptive response not arrived at by trial behavior or as the solution of a problem by the sudden adaptive reorganization of experience" (Thorpe, 1963, p. 110). The classic studies on animal insight were performed during World War I on chimpanzees by Wolfang Köhler on the island of Tenerife (described in detail in Köhler, 1925). These studies have been described and discussed many times previously, but briefly, chimpanzees were presented with many different problems (usually associated with the procurement of inaccessible foods) and a number of potential objects to solve those problems. A famous example was a banana hanging out of reach from a string and a collection of boxes located underneath. The chimpanzees were described as attempting to reach the bananas first by standing on one box, then attempting to balance one box on top of another, and finally the chimpanzee placed one box on top of the other climbed up and hit the banana with a stick, thereby knocking the banana from its string and allowing it to fall to the ground to be retrieved. What remains to be determined is how long the chimpanzees took to produce this "insight," and whether they had previous experience with the boxes. In essence, the question is whether they could have learned through a series of instrumental responses, and thus what this "insight" actually represents. Indeed, it has been suggested that the chimpanzees could not have demonstrated insight in its classical terms because the solution to the problem would have been achieved by previous experience with sticks and boxes (Birch, 1945) and was not solved immediately after the problem was presented (i.e., arose as a consequence of many trial-and-error responses; Povinelli, 2000).

Is there any evidence of insight in birds, and if so, can supposed cases of "insight" be explained more simply in terms of associative learning or is there any evidence of something more complex? Although this problem has hardly been studied in many non-primates, there is some evidence that corvids (ravens and rooks) display some form of insight, at least similar to that displayed by primates.

Reid (1982) reported that a young rook used a plug to trap water in their aviary (plugging a man-made hole) for use in drinking and bathing on hot days. This has been presented as an example of tool-use, but may also provide an example of insight. As the author of the study did not look specifically at insight, and the plug and holes were present in the aviary for almost 1 year before the young rook put the two together, we do not know how much experience it had with the plug and the hole, whether the bird was just lucky the day the author noticed the occurrence or whether the rook actually demonstrated insightful behavior.

A more controlled series of experiments in ravens may present a clearer indication of insight in corvids. Heinrich (1996, 2000) provided hand-raised ravens with a novel opportunity to gain food, by attaching pieces of meat to string suspended from a branch. This constrained the ravens in their method of food acquisition, as they would be unable to hover under the food like hummingbirds. The ravens would have to use their grasping feet (present in all corvids and parrots) to pull the string up, and trap each piece of pulled-up string with their feet. This would have to occur at least five times before the bird could grab the meat in its bill. At first, birds attempted to grab at the food from underneath, or pecked at the string on the branch. Three birds, at their first attempt, pulled up the string and grabbed the meat (before being shooed off). The ravens did not attempt to fly off with the food when shooed off (i.e., had some concept that the meat was attached to the string). When presented with two strings; one attached to the meat and a second to a similar sized rock, the ravens tended to pull on the string attached to the meat (or immediately moved onto the correct string if they attempted to pull up the string with the rock attached). If the strings were crossed, most of the birds chose randomly, but eventually one bird pulled more strings attached to food than rocks. When the birds were presented with a novel string (dark green shoelaces rather than twine string), the birds almost exclusively pulled on the laces attached to food. This suggests that the birds have not just formed an association between a particular string and food, but have generalized to all string-like substrates with food. Finally, the ravens were presented with two strings, one attached to a small piece of meat, the other to a large sheep's head (that they would be unable to pull up). In all cases, the birds avoided pulling up the sheep's head (not due to fear of the head), and always pulled up the smaller piece of meat.

This series of experiments suggests that this group of hand-raised ravens do demonstrate some form of insight, given that the solution was largely achieved at the first attempt. Random chance may also be discounted as an explanation, because the probability of producing the five or six actions leading to a series of 30+ complex steps would be too high for actions performed on the first trial. These arguments cannot be said to be the case for Köhler's studies with chimpanzees described earlier (Köhler, 1925).

Manufacture and Use of Tools The propensity to manufacture, transport, and use tools was previously thought to be the exclusive realm of *Homo* sp. However, in the 1960s, Jane Goodall reported that chimpanzees living at Gombe in Africa, also used tools. Their tools were tree stems that were stripped of their leaves and poked into termite mounds. The termites would grab hold of the stem, the chimpanzee would pull the stem out and eat the termites (van Lawick-Goodall, 1968). The last 30 years have seen examples of tool-use in primates (either in the wild or the laboratory) increase to amazing levels (Tomasello and Call, 1997), with different populations of chimpanzees using different tools for different uses, such as using an anvil and hammer to crack nuts (Boesch and Boesch, 1983), or chewing leaves into a sponge for collecting liquids (Goodall, 1986). These variations in tool-use have been proposed as a potential cultural phenomenon (Whiten et al., 1999).

Some birds have also been described as creating and using tools. Some examples of animal tool-use, however, do not fulfill the strict criteria of tool-use demonstrated for non-human primates. Tool-use has been described as "the use of physical objects other than the animal's own body or appendages as a means to extend the physical influence realized by the animal" (Jones and Kamil, 1973, p. 1076). Vultures (*Neophron percnopterus*), for example, crack open eggs by dropping them onto rocks (van Lawick-Goodall and Lawick, 1966). This is not a demonstration of tool-use as the rock is not an extension of the vulture's body. However, vultures that throw stones at ostrich eggs are demonstrating tool-use (Thouless et al., 1987). Similarly, thrushes that open snail shells by smashing them onto stones (Gibson, 1986), or crows in Japan and California that open hard shelled walnuts by dropping them from great heights onto hard-surfaced roads (Cristol and Switzer, 1999; Nikei, 1995) are not demonstrating tool-use. These may be innate responses, and they may not require the mental manipulations required to transform an object with one distinct function into different functions.

A number of birds do manufacture and use tools in similar ways to primates. In the laboratory, Northern blue jays (*Cyanocitta cristata*) were found manipulating the shape of newspaper strips provisioned at the bottom of their cage, and using them to pull in inaccessible food pellets (Jones and Kamil, 1973). The jays did not use the paper tool when pellets were not present, and tended to use the tools more when the length of their food deprivation was greatest. The jays also were able to use a feather, thistle, straw grass, paper clip, and plastic bag tie in similar ways when presented with these objects. Finally, the jays also wet the strips of paper, placed the strips in their empty food bowl, and used them to collect food dust. Similar behavior was observed by Clayton and Joliffe (1996) in another food-storing species, the marsh tit (*Parus palustris*).

Tool-use and manufacture has also been demonstrated in wild birds. A species of Galapagos finch; the woodpecker finch (*Camarhynchus pallidus*) was reported to use a stick to probe for insects in the holes of trees (Millikan and Bowman, 1967). The finches would break off a twig, leaf stem, or cactus spine and then use it to dig into an inaccessible hole. The birds also transport the best tools with them when foraging and change the length of the tools when they are an inappropriate length for the next hole. Tebbich et al. (2001), using aviary housed finches, examined whether this tool-using was learned socially or through individual trial-and-error learning. Some captive adult finches learned to gain access to a beetle larva hidden in a hole in an artificial tree trunk using a twig. However, the non-tool users (when exposed to many weeks in the presence of tool-users) did not learn to use this technique. Hand-raised finches exposed to tool-users, however, did not learn to use tools any better than young exposed to non-tool-users, thereby suggesting that tool use in finches is independent of social learning, and may represent an example of learning during a critical period. Tebbich (2000) also found that the best finch tool-users were found in dry habitats (where prey is located under dry bark that is difficult to access), and virtually none in humid habitats (where prey is located under wet moss).

Perhaps the most spectacular use and manufacture of tools in an avian species is by the New Caledonian crows (*Corvus moneduloides*). Hunt (1996) described how he observed 4 crows manufacture two types of tools and 68 crows carry or use tools in three forests in New Caledonia. The tools were used for catching prey (insects) either in trees or under detritus leaves. Hunt collected all the examples of tools that were made by the crows (although some carried useful tools on foraging expeditions and secured them when resting). The tools could be categorized into two types; hooked-twig and stepped-cut tools. The hooked-twig tools were made from living secondary twigs that were stripped of their leaves and bark, and had a hook at their wider end. The stepped-cut tools, by contrast were fashioned from *Pandanus* leaves by tapering the ends into points. Different techniques were employed in using the tools depending on the location of the prey. If the prey was located under detritus, the tool was used with rapid back and forth movements, whereas if the prey was located at the base of holes and leaves, slow deliberate movements were used. Hunt (2000) has recently shown that tool manufacture in New Caledonian crows is lateralized at the population-level, that is most tool users tend to use the left-hand edge of the leaf to create the tool (see Chapter 10 by Hook).

New Caledonian crows have also recently been studied in the laboratory and appear able to choose the correct tool (a twig of certain length) from a "tool box" (collection of different length twigs) that is appropriate for a specific task, such as reaching food placed in the middle of a transparent tube (Chappell and Kacelnik, 2002). Most intriguingly, a New Caledonian crow has demonstrated the ability to modify an unnatural material, bending a straight piece of metal wire to form a hook at one end that was subsequently used to pull up a cup containing food (Weir, Chappell, and Kacelnik, 2002). This has not been seen in any other animal to date, including the great apes (Povinelli, 2000).

Primates Utilize Symbolic and Referential Communication

Perhaps the most famous demonstration of symbolic and referential communication in the wild is the alarm calling of vervet monkeys (*Cercopithecus aethiops*). Vervet monkeys use at least three different alarm calls to discriminate between different predator types: leopards, eagles, and snakes (Struhsaker, 1967). It has also been demonstrated that the signals are not just emotional reactions to the appearance of a predator. Through the use of alarm call playbacks, it was demonstrated that the vervets responded to each different alarm with different avoidance behaviors appropriate for the predator by which the call was elicited (Cheney and Seyfarth, 1990; Seyfarth, Cheney, and Marler, 1980). A leopard call caused the monkeys to run into the trees, a snake call caused a mobbing response towards the ground and an eagle call caused the monkeys to run into the bushes. This may suggest that the information contained within the different alarm calls represents an image of the associated predator type (symbolic representation). A second line of evidence suggesting that the calls were not reactive emotional exclamations was the fact that there were dramatic audience effects on calling rate (i.e., effects of social context). Alarm calls were not given if an individual experienced a predator when alone (Cheney and Seyfarth, 1990), and calling rate was also dependent on whether other monkeys in the vicinity were kin or non-kin (Cheney and Seyfarth, 1985).

Interestingly, a similar pattern of response has been demonstrated for domestic chickens (*Gallus gallus*). Using artificially constructed predators (such as video images of cardboard silhouettes of birds of prey), Evans et al. (1993) found that chickens also responded to the sight of the artificial predators with different alarm calls and responded to the playbacks of the different alarm calls in different appropriate ways, crouching after hearing an alarm call representing an aerial predator. Similar to vervets, Marler and colleagues also found significant audience effects in chickens (Gyger et al., 1986; Karakashian et al., 1988).

Tomasello and Call (1997) have suggested that when discussing the work on vervet monkey alarm calls "one must be careful in interpreting the cognitive underpinnings of these phenomena ... because several nonprimate species also give alarm calls differentially depending on the presence of others and differentiate kin from other conspecifics" (p. 252). We find this position difficult to reconcile based on the available evidence. Why should the presence of a complex cognitive ability in chickens be less plausible and less valid than the same ability in primates?

Although we do not have the space here to discuss this in great detail, a second form of symbolic communication found in some primates is the acquisition of human language by animals, particularly in enculturated apes. Earlier studies concentrated on the use of American Sign Language (ASL) that was acquired from observing human trainers sign, but never speak. Washoe, a chimpanzee, for example, learned over 100 signs and could combine signs into new combinations, such as "water bird" to represent swan (Gardner and Gardner, 1969). Terrace et al. (1979) however, using the same method of teaching ASL found that Nim, another chimpanzee, learned 125 signs and produced 4 sign combinations, but had no understanding of syntactical structure, as signs were often repeated, as were the most used personal pronouns, such as "me." Other studies in chimpanzees have examined directly the understanding of symbols as representations of objects, such as Sarah (described earlier) who learned to associate plastic tokens of various colors and shapes with the objects they represented (Premack, 1971) and Lana who learned that geometric designs on a keyboard represented different words ("Yerkish," Rumbaugh, 1977).

Most impressively, a bonobo named Kanzi, learned to use Yerkish after observing his foster mother being taught to use the lexigrams, and eventually learned to respond to human speech (equivalent to that of a 2-year-old human child; Savage-Rumbaugh and Lewin, 1994).

It is not our place in this chapter to criticize or champion these studies. Whether you agree or disagree that the previously described studies have demonstrated human language abilities in apes, for the purpose of this chapter the important point is that similar abilities have been reported in great detail for an African gray parrot, named Alex (Pepperberg, 1990). Pepperberg was successful in teaching human language to Alex and other gray parrots because parrots are excellent vocal mimics (Baylis, 1982). Also, the parrots were provided with live, interactive trainers who supplied referential and contextual use of each label associated with each object it represented, and the reward objects were inherently interesting (but not food).

Furthermore, Pepperberg used the model/rival technique (Todt, 1975), in which the parrots were trained in the presence of two humans; a model trainer who shows an object to the second human (model and rival). The trainer asks the model/rival questions about the object (such as "What's here?"), and gives praise and the object as a reward. The trainer also displays disapproval if the answer is incorrect. Therefore, the parrot can learn which is the correct label for each object. Alex, to date, has demonstrated object identification by requesting specific objects, categorizing objects based on their color and shape, forming abstract concepts, such as same/different or absent, and appears to have a concept of number of objects (reviewed in Pepperberg, 1999). All these abilities have largely been demonstrated only in non-human primates (reviewed in Shettleworth, 1998; Tomasello and Call, 1997).

Primates Demonstrate Elements of Mental Time Travel

One of the remaining bastions of human uniqueness may be the ability to travel backwards and forwards mentally in time; so-called "mental time travel"

(Suddendorf and Corballis, 1997). The mental time travel hypothesis posits that, unlike humans, animals cannot travel backwards in time to re-experience and recollect specific past episodes (episodic memory) or travel forwards in time in order to anticipate future states of affairs (future planning). The idea that animals do not share with us a sense of the past and of the future is something that has been suggested several times within comparative psychology, but the idea that mental time travel might be exclusive to humans was initially proposed by Köhler. Although he believed his chimpanzees were capable of many complex cognitive feats such as insight (see earlier section), he suggested that there was one important limitation "The time in which the chimpanzee lives is limited in past and future" (Köhler 1927/1917, p. 272). The same idea was later expressed by Bischof (1978) and Bischof-Köhler (1985).

Mental time travel has two components: retrospective and prospective.

1. The retrospective component: "Episodic memory receives and stores information about temporally dated phases or events, and temporal-spatial relations among those events" (Tulving, 1972).
2. The prospective component: "... animals other than humans cannot anticipate future need or drive states, and are therefore bound to a present that is defined by their current motivational state. We shall refer to this as the Bischof-Köhler hypothesis ..." (Suddendorf and Corballis, 1997).

***The Retrospective Component—Episodic Memory*:** Both humans and animals are capable of learning from past experience and using that information at a later date. But there is one form of memory that is thought to differentiate us from other animals, and that is the ability to episodically recall and re-experience unique past events. Cognitive psychologists such as Tulving make the distinction between semantic factual knowledge and episodic recall (Tulving, 1972, 1983). Thus, knowing that Paris is the capital of France, or the date and time we were born, are examples of semantic memory. But remembering what we did when we went to Paris, or what we did on our wedding day, are examples of episodic memory. Aside from the distinction between knowing (semantic memory) and remembering (episodic memory), the two types of memory differ in whether or not they involve chronesthesia (Tulving, 2002). Whereas semantic memories transcend space and time, episodic memories are concerned with specific events in one's personal past that involve information about when as well as what-and-where. Furthermore, language-based reports of episodic recall suggest that the retrieved experiences are not only explicitly located in the past but are also accompanied by the conscious experience of one's recollections, so-called autonoetic consciousness (e.g., Wheeler, 2000). But this definition makes it impossible to demonstrate episodic memory in animals because there are no agreed nonlinguistic behavioral markers of conscious experience. The dilemma can be resolved to some degree by using Tulving's original definition that episodic memory "receives and stores information about temporally dated episodes or events, and temporal–spatial relations among these events" (Tulving, 1972). Episodic memory provides information about the what and when of events (i.e., temporally-dated experiences) and where they happened (i.e., temporal–spatial relations). An animal's ability to fulfill the behavioral criteria regardless of autonoetic consciousness is termed episodic-like memory (Griffiths et al., 1999).

Do Animals have Episodic-like Memory? Most studies of animal memory have not distinguished between episodic recall of events from semantic knowledge for facts, but it was generally assumed that in animals events are remembered with no specific past-time reference. Typically, the tasks require the animal to retrieve information about only a single feature of the episode as opposed to testing its ability to form an integrated memory of temporal–spatial relations. It is also common for the animal to be given multiple training experiences thereby removing the trial-uniqueness of the task (see Griffiths et al., 1999). Furthermore, in many of the tasks that have been used, the animal does not need to recall the what, where, and when of an event. Instead, the task may be solved by discriminating on the basis of relative familiarity, a process that is dissociable both psychologically and neurobiologically from episodic memory recall. For example, a face can appear highly familiar without any recall of where and when one has previously met its owner.

In many studies that appear to demonstrate episodic-like recall in animals, the results observed can be explained more in terms of simple familiarity (Griffiths et al., 1999). Consider the case of monkeys that have been trained to choose between two complex objects on the basis of whether they are same as (delayed matching-to-sample—DMS) or different from (delayed non-matching-to-sample or oddity—DNMS) an object they were shown some time previously at the start of the trial. The monkey may have recalled episodically the events at the start of the trial. A simpler explanation, however, is that the monkey learned to choose- or avoid- the most familiar object. By only requiring the monkeys to recognize a stimulus, but not recollect where and when it was previously seen, these matching tasks are most readily solved by familiarity rather than episodic recall. Indeed, the fact that monkeys learned more rapidly when novel objects are used supports this account. The point is that experiments such as this do not provide convincing evidence for episodic recall because they can be explained in simpler terms.

Episodic-like Memory in Scrub-Jays: An alternative strategy is to consider cases in nature in which an animal might need to retrieve and integrate simultaneously information about what happened, where, and when during a specific previous experience. When considering the evolutionary history of episodic-like memory in animals, it might seem intuitively logical to focus on the abilities of non-human primates. The study of non-human primate coalition or alliance formation might provide a useful starting point given that the social partners need to keep track of cognitively complex social relationships. Males are distinguished from each other based on size, strength, fighting ability, etc., and the greater these attributes, the higher the individual's social status, and the greater their access to resources, such as food and mating partners (Tomasello and Call, 1997). The need to keep track of trial unique events concerning who did what to whom and where might, therefore, provide a candidate for episodic memory, though it is less clear whether a when component is critical. It is also difficult to think of a good experimental paradigm for testing episodic memory in such complex social interactions.

The food-caching paradigm, however, provides an opportunity to combine ethological validity with rigorous laboratory control because food-caching animals readily hide food caches in the laboratory and rely on memory to recover their caches at a later date. Experiments in the laboratory and field show that birds can remember accurately the location of their caches based upon a single caching experience, but there is also good reason to believe that at least some species form richer representations of caching events than others. Some species cache different types of food and can remember the contents of their caches as well as the location; and some species also cache perishable foods. It therefore may be adaptive for these species to encode and recall information about when a particular food item was cached, as well as what was cached and where. As discussed earlier, experiments on food-caching memory in scrub-jays do provide evidence of episodic-like memory. On the basis of a single caching episode, scrub-jays remember when and where they cached a variety of foods that differ in the rate at which they degrade, in a way that is inexplicable by relative familiarity (Clayton and Dickinson, 1998, 1999a; Clayton et al., 2001a).

Griffiths et al. (1999) have argued that previous demonstrations of trial-unique memory in animals could have been mediated by relative familiarity rather than the temporal encoding of the episode and do not require an integrated memory of the features of the episode. Recently, however, Menzel (1999) demonstrated that a chimpanzee can remember both the what and where of food sources, and experiments by Schwarz and Evans (2001) suggest that a gorilla can also remember who, but it remains to be seen whether this memory has an integrated what-where-when or who-where-when structure. ("Who" information is not likely to be processed differently from other kinds of what information.)

The Prospective Component—Future Planning: As with episodic memory, the ability to anticipate future states of affairs (future planning) has been thought to be unique to humans, in part because of the reliance on language for assessing such abilities. Is there any evidence that animals show any signs of forethought?

In order to address this question it is important to distinguish mental time travel into the future from simple prospective behavior. Any species-specific behavior that appears to involve the anticipation of future states of affairs may not actually involve any planning (learning), but may simply result from natural selection, in much the same way as some fixed action patterns such as nest provisioning by hunting wasps (Fabre, 1916). And in some cases a behavior that is orientated towards the future is inherited. Consider the case of migratory behavior in black capped warblers attempting to migrate. Although migration might appear to have some of the features of prospective behavior, because the birds migrate in order to acquire a new, more profitable territory for a given time of year, the "decision" of whether to migrate, and in which direction, does not involve future planning (Berthold et al., 1992).

So in order to test whether animals are capable of mental time travel into the future one needs to rule out simple explanations of prospective behavior that need not involve learning and, as Suddendorf and Corballis (1997) state, "many behaviors involve anticipation of future events in some way, but need not involve the actual simulation or imagining of future events." Thus experiments using delayed reinforcement, where the animal is trained to discriminate between two choices, and different groups of animals receive varying delays between the choice response and the rewarded or non-rewarded outcome, do not provide evidence of future planning. The fact that a reinforcer can be delayed by a few seconds or even minutes and yet still be viewed as relevant to the preceding behavior is not evidence for future planning because it need not involve anything more complex than simply associating the reward and the response. With short time intervals into the future, the learning can be promoted by the contiguity between the reinforcement and the memory of the correct response. Furthermore, implicit in the Bischof-Köhler hypothesis that mental time travel is unique to humans is the idea that non-human animals may be unable to dissociate a previous or future mental state from their current one, and therefore unable to take actions to ensure that a future need will be satisfied. Thus a sated animal may be unable to understand that it will be hungry again later and thus "a full-bellied lion is no threat to nearby zebras, but a full-bellied human may be" (Suddendorf and Corballis, 1997).

A number of anthropologists and primatologists have suggested that some great apes might share with humans the ability to plan for the future. Unfortunately most of these claims rest on anecdotal evidence. Byrne (1995), for example, recounts the following anecdote as an example of future planning in chimpanzees: A group of Mahale chimpanzees surrounded a cave in which a leopard mother and infant had hidden. After several attempts, an old male lunged into the cave and stole the cub. Afterwards it pummeled the cub, bit it, and eventually killed it. But none of the chimpanzees attempted to eat the cub so this behavior cannot be explained in terms of simple foraging. Byrne suggests that this is evidence of future planning and that the

> Humans might well carry out this action, with the ultimate ends of reducing the population of dangerous carnivores and deterring the mother from continuing to inhabit their particular range. It is not easy to account for the chimpanzees' actions without attributing similar goals to them, which implies long-term anticipatory planning (Byrne, 1995, pp. 157–158).

But a number of other explanations are possible, none of which assume anything about the possible thought processes underlying the behavior. Perhaps the chimpanzee was showing re-directed aggression following a fight with other more dominant males, for example.

The chimpanzee's response could have been controlled by currently observable stimuli. A better candidate for future planning is the ability of some animals, notably chimpanzees and New Caledonian crows, to manufacture and use tools (discussed earlier). Chimpanzees, for example, have been described carrying stones over long distances to be used as tools for cracking nuts found in an area where no stones were available. Although this example might be taken to suggest that (some) animals can anticipate a future event, their ability to do so may be limited. Critics have argued that the preparation of a tool to crack nuts may be largely bound up in the act of consuming the food item only a short time into the future. Importantly, these signs of anticipatory thought concern items that are relevant to the individual's current motivational state: in this example, the hungry chimpanzee's anticipatory use of tools is bound by the context of the animal's current hunger state. The same argument applies to the acquisition of mental maps for future use in foraging. Suppose an animal were capable of taking the shortest route to an invisible goal, and across previously untraversed terrain, as a result of having previously formed a mental map of the environment. This need not involve future planning for there is no reason to believe that it requires an explicit reference to a future-need state. For future planning, then, the critical component is evidence that the animals can anticipate future needs and desires, "independent" of their current needs.

Inevitably with field observations, the problem is that one has no idea whether a particular action is learned or not. And in the absence of any information about the animals' previous reinforcement history, and other important control procedures, it is not possible to assess whether or not these instances of chimpanzee behavior involve future planning. No field study can establish whether or not animals are capable of anticipatory thought. To do so one needs carefully controlled experiments to establish that the behavior in question is learned, in the absence of any previous reinforcement, and that it genuinely involves the anticipation of a future need state not merely the current one.

Is there any Evidence of Future Planning in Animals? Laboratory studies have seldom addressed the issue of whether animals can travel mentally in time. There are some studies that claim to have tested future planning with emotive titles such as "Medial prefrontal lesions in the rat and spatial navigation: Evidence for impaired planning" (Granon and Poucet, 1995) and "Anticipation of incentive gain" (Flaherty and Checke, 1982). But none of these studies test the animal's ability to anticipate a future-need state, and most of these studies involve short retention intervals that only appeal to the animal's immediate future.

As with episodic memory, an alternative approach is to consider natural behaviors that might be likely to require the ability to travel forwards in time as well as backwards. And food-caching might also be a good candidate for testing the Bischof-Köhler hypothesis. At first sight, food caching would seem to contradict this claim—functionally at least, it is a behavior that is oriented towards future needs because food-caching animals hide food for future consumption, and over long retention intervals. A Clark's nutcracker for example, may cache food in October and recover it up to 9 months later. Of course food-caching would not be an example of future planning if the individuals instinctively cached all food items irrespective of the consequences of their actions, or if they were insensitive to the difference between their current-need state at caching and their future one at recovery. But a couple of recent experiments on the caching behavior of scrub-jays suggest that this might be a productive paradigm for testing mental time travel in animals.

Food Caching by Scrub-Jays: A Candidate for Future Planning in Animals? If caching is controlled by prospective cognition, we should expect the birds to anticipate the conditions at future recovery opportunities on the basis of past recoveries and to use these anticipated conditions to control present caching. Preliminary experiments by Clayton, Dally, Gilbert, and Dickinson (in prep.) have tested whether jays are sensitive to the state of the food caches at recovery. Jays were given the opportunity to cache two foods, worms and peanuts, simultaneously in the morning and then recover their caches later that afternoon. All birds received fresh food items to cache, but the groups differed in whether the worms were still fresh at recovery or whether they had degraded by the time of recovery. If jays can anticipate the recovery conditions at the time of caching then birds that are only able to recover degraded worms should reduce the amount of worms cached. The results were consistent with this hypothesis in that birds whose worm caches were degraded at recovery substantially reduced the number of worms cached and increased the number of peanuts cached. Note that the birds in all groups continued to eat worms which rules out any explanation in terms of a conditioned taste aversion. And interestingly, all birds continued to cache worms in their home cage, suggesting that a reduction in worm caching was restricted to those sites in which the worms appeared to degrade. The important next step is to test whether jays are sensitive to a future motivational state as opposed to the current one. Can animals learn that they should cache now even when they are not hungry in order to fulfill a future-need state? And is there any evidence of future planning in other behaviors, in primates or in birds?

THE PERILS OF PRIMATOCENTRISM AND "SCALA NATURAE"

Why have primates achieved this special status when the evidence suggests that species of birds have cognitive abilities that are equal to or more sophisticated than have been demonstrated for primates? We believe that this harks back to the beginnings of primatology as a discipline. Early studies of primates in zoos or the laboratory were performed to determine whether the cognitive abilities of primates could be compared to the cognitive abilities of humans (stemming from the evolutionary arguments of Darwin and others). This approach has continued unabated since, culminating in the work of David Premack and Duane Rumbaugh that was concerned with whether chimpanzees have the capacities for human language, relational learning, and ToM. The tradition of comparing primates with humans on tests of cognition continues to this day.

Early field researchers studying primates in their natural habitat may have had a similar aim, with almost all being anthropologists investigating the behavior patterns of monkeys and apes as an indicator of how our extinct ancestors may have behaved (see papers in DeVore, 1965 and papers in a more recent volume edited by de Waal, 2001, that continues this tradition in other disciplines in addition to anthropology). Of particular note were Jane Goodall, Dian Fossey, and Birute Galdikas, who were hired and directed by Louis Leakey to study the natural behavior of three great apes (chimpanzees, gorillas, and orangutans) and use the data to form models of early human behavior. Although all field workers are not anthropologists, and ethologists tend to study their selected animal as a means to its own end, the tendency to study primates as models of human behavior still pervades the animal behavioral sciences.

The social or Machiavellian Intelligence hypothesis revolution, for example, started in the 1960s (Jolly, 1966) and 1970s (Humphrey, 1976), and gaining pace in the late 1980s (Byrne and Whiten, 1988) continued this line of research, using anecdotes recorded in the field to establish testable hypotheses relating primate social behavior to the evolution of social behavior in humans. (We would like to note that these researchers have never claimed that anecdotes can wholly replace experiments and the collection of "proper" data.)

Although our aim is not to criticize these fields of endeavor, we would like to make the comment that this way of thinking about the cognitive abilities of primates may have had detrimental effects on constructive thoughts about non-primate animal cognition. The argument follows thus. Primates are our closest relatives (discussed previously) and therefore are likely to have similar cognitive abilities to humans. Anything that does not resemble human cognition must be viewed as less "intelligent." As primate cognition is structurally similar to human cognition (Tomasello and Call, 1997), primates must be cognitively more advanced than non-primate species. Although space does not permit us to discuss the abuses of the concept of degrees of animal intelligence (i.e., the incorrect assumption that one animal is more intelligent than another based on experimental tests; Macintosh, 1988) and the concept that this idea was based upon, that is, the flawed idea that animals can be ranked on a continuous scale (the scala naturae; Hodos and Campbell, 1969), we would like to suggest that the assumption that primates are a special case arises as a direct consequence of the propagation of these two concepts.

USES AND ABUSES OF THE ECOLOGICAL/ETHOLOGICAL APPROACH TO COGNITION

One concept that the majority of primate researchers have so far failed to incorporate into their research programs is the ecological/ethological approach (EEA) to cognition (Balda et al., 1997; Kamil, 1988; Shettleworth, 1998). This approach utilizes conspecifics (or images/vocalizations of conspecifics) as stimuli, and also utilizes an appreciation of the natural behavior and ecology of the test species in the design of experiments. The EEA is used primarily to reduce any potential confounds from using synthetic objects or heterospecifics (such as humans) as stimuli.

Many cognitive experiments in non-human animals are drawn from studies in humans, particularly those in developmental psychology, such as the preferential looking time and expectancy violation paradigms. This has been especially true for non-human primates (Hauser, 2000). This methodology, however, may not be compatible for use with other non-human species, which do not have the same visual system as primates. For example, bird vision is as well developed as primate vision, but their field of vision is >180° (much larger than primates), and their eyes are located at the sides of their head, rather than forward facing (although this is not always the case, such as in owls). This presents a problem for measuring where the birds are looking. The ethological approach may be particularly appropriate for species that are of a greater evolutionary distance from humans, such as birds.

And clearly, an understanding of the species' natural history allows the experimenter to design an ethologically relevant task by predicting what sort of motor responses may occur, what kind of stimuli the experimental subjects are more sensitive to, and what kind of cognitive constraints characterize their learning abilities (see Gerlai and Clayton, 1999). That said, adopting an ethological approach poses some problems of its own. The ethological view is that species differ because they have adapted to different environments that have different sets of problems to solve. But in order to test this hypothesis we need to make certain assumptions about what the different demands imposed by the different environments in which the animals live might be. Which ecological variables are important, and to what extent are they good predictors of how species might differ in cognitive abilities, if at all?

Species Differences in Ecology and Cognition

Western scrub-jays provide an example of the first case in point. In the comparative studies of spatial and social cognition in four species of North American corvids by Balda and Kamil's laboratories; Clark's nutcrackers and Pinyon jays are thought to have the greatest reliance on stored food, Mexican jays are intermediate and western scrub-jays are the least reliant on stored food. And, whereas the Pinyon jays and Mexican jays are social, western scrub-jays and Clark's nutcrackers are territorial. According to the adaptive specialization hypothesis then, one would predict that the nutcrackers and Pinyon jays should perform more accurately on spatial tests of memory because they are more dependent on accurately relocating their caches than Mexican jays and western scrub-jays. And for social tasks, then one would predict that the two social species, Pinyon jays and Mexican jays, would perform more accurately than territorial Clark's nutcrackers and western scrub-jays.

Is there any evidence for this? Western scrub-jays do perform least accurately of the four species at a number of the tasks requiring spatial memory, including an adapted radial arm maze and delayed non-matching to sample tasks (Balda and Kamil, 1989; Balda et al., 1997; Kamil et al., 1994; but see Gould-Beierle, 2000, for a counter-example). As we described earlier, these results seem surprising when contrasted with their seemingly sophisticated ability to remember the "what, where, and when" of caching events (see earlier section on mental time travel; Clayton and Dickinson, 1998, 1999; Clayton et al., 2001b). The mechanism for cache recovery in these scrub-jays, however, does not appear to be based on a sophisticated appreciation of spatial location, rather the binding of "what," "where," and "when" a unique event occurred (see earlier section, and Clayton et al., 2001b). Also, the relative size of the hippocampal formation in western scrub-jays is larger than would be predicted from body size or size of the telencephalon (Basil et al., 1996).

Western scrub-jays do not appear to have a complex social system, when compared with Pinyon jays, for example. Individual western scrub-jays come together in pairs largely only during the breeding season and during territory disputes. The adults mate during the breeding season, form short-term selective bonds, and the father contributes to the raising of his offspring. Group size is therefore small compared to related corvid species and is even different from the closely related Florida scrub-jay, that demonstrates cooperative breeding (helpers at the nest), and possess a sentinel system (Woolfenden and Fitzpatrick, 1984). Balda et al. (1997), therefore, predicted from the standpoint of adaptive specialization in the social domain, as in the spatial domain, that the social cognition of *Aphelcoma californica* should be relatively unsophisticated compared to the more sociable Mexican and Pinyon jays. However, as we described earlier, western scrub-jays form complex representations of individual social events, remember the context in which they occurred and use the information to influence their future behavior (Emery and Clayton, 2001). It remains to be seen whether social cognition is an adaptive specialization. But clearly, any hypothesis about the adaptive specialization of social cognition based on group size or traditional measures of social complexity may not be the most productive avenue for constructing testable predictions based on selective adaptation.

Ethologically Relevant Stimuli are Difficult to Control

One productive use of the ethological approach to cognition may be the use of naturalistic stimuli in experiments. A problem with this is how these stimuli are accumulated and presented. Conspecifics as demonstrators, for example in social learning and social cognition experiments provide a higher ethological relevance than use of human demonstrators. Those studies in primates, for example, that have either used videos of conspecifics (Emery et al., 1997) or live conspecifics (Hare et al., 2000, 2001; Tomasello et al., 1998) have produced less ambiguous data than those that relied on humans as stimuli (Povinelli and Eddy, 1996; Povinelli et al., 1990). Conspecifics have been used in studies of bird song learning for many years (Catchpole and Slater, 1995, for review), as have playbacks of primate vocalizations in free-ranging primates (e.g., Cheney and Seyfarth, 1990).

There are some problems with using conspecifics as stimuli, specifically that they are very difficult to control when a particular behavior will occur. This can be overcome through the use of video, although video playback has its own problems, such as a lack of realism, no potential for social interaction, and the question of whether some species perceive video stimuli as representatives of conspecifics (as some monitors have a high flicker-fusion rate that disrupts perception; D'Eath, 1998).

How Far can the Natural Behavior of an Animal be Translated to the Laboratory?

Some experimental psychologists who utilize primates (primarily the great apes) as subjects have taken an extreme position with negligible external validity in their cognitive experiments, such as using enculturated chimpanzees and household objects (see Povinelli, 2000, or Povinelli and Eddy, 1996, for examples). This situation, however, is not exclusive to primate studies, as it has also been adopted in long-term studies using African gray parrots (Moore, 1992; Pepperberg, 1990). There may also be the question that hand-raised birds represent an unnatural case of avian cognition, but the birds are not raised as humans. Although an EEA should clearly increase the external validity of experimental designs in cognitive studies (for a primate example, see Hare, 2001), studies in the laboratory do not equate to the behavior of an animal in its natural environment. We therefore have to appreciate this fact when we interpret the data accumulated from cognitive experiments performed in the laboratory, and we have to maintain a balance between ecological validity and appropriate control in our experiments.

THE GREAT DIVIDE: AWARENESS OF "SELF"

So what may separate the corvids and parrots from primates in mental ability? As yet, the only evidence that an animal may have an awareness of the "self" versus awareness of other individuals has been demonstrated in chimpanzees (and possibly orangutans and dolphins; Reiss and Marino, 2001; Suarez and Gallup, 1981). Although the mark test developed by Gallup (1970) to demonstrate mirror guided self-recognition (MSR) has been criticized on many counts (Heyes, 1998), these have been convincingly countered (Gallup et al., 1995). The use of mirrors still provides the only convincing demonstration of self-directed behavior in relation to specific body locations, which has been suggested to be an indication of a self-concept (Gallup, 1970, 1982; also see papers in Parker et al., 1994).

MSR has been tested in corvids (jungle crows; *Corvus macrorhynchos*), where the crows were exposed to mirrors in different orientations (horizontal and vertical). The crows aggressively attacked their reflection in the mirror as if a novel, same sex conspecific, therefore not demonstrating any aspect of self-awareness (Kusayama et al., 2000). MSR has also been tested in African gray parrots but, although they were shown to use a mirror to locate hidden objects, they did not demonstrate any self-exploratory behavior (Pepperberg et al., 1995).

An earlier study (Gallup and Capper, 1970) compared the orientation behavior of parakeets (*Melopsittacus undulates*) to either a mirror, a conspecific, food, or a blank piece of cardboard. The parakeets spent a significantly greater amount of time sitting on the perch in front of the mirror compared to the other conditions; however the birds were found to display and vocalize toward their reflection, thereby demonstrating a failure to understand that the reflection represented themselves. A study that claimed to train pigeons to demonstrate self-awareness through associative learning (Epstein et al., 1981) had little to do with self-awareness per se, and has yet failed to be replicated (Thompson and Contie, 1994).

Gallup (1982) has suggested that an awareness of self is an important prerequisite for understanding others' mental states. Understanding one's own actions and mental states may provide the basis for attributing those same states to others. In fact, Gallup (1994) stated that "in addition to being able to recognize themselves in mirrors, organisms that can conceive of themselves ought to be able to use their own experience and knowledge to infer comparable experiences and knowledge in other organisms" (Gallup, 1994, p. 48).

As discussed above, our recent experiments on social reasoning in scrub-jays (Emery and Clayton, 2001) may provide convincing evidence of mental attribution. But these results suggest that if Galllup's hypothesis is correct, western scrub-jays should also demonstrate mirror self-recognition. We decided that an aesthetic mark test for mirror-guided self-examination (similar to previous studies) was not ethologically relevant to the jays. So in a preliminary experiment by Emery, Gilbert, and Clayton we examined whether the presence of a mirror during caching would alter the jays' re-caching behavior later at recovery, compared to when caching in private or when observed by another bird. We predicted that if scrub-jays had a concept of self, they would interpret the image in the mirror at caching as themselves and forgo re-caching at the time of recovery (as the potential for pilfering would have been removed). If the jays interpreted the mirror image as a conspecific, they would re-cache the earlier hidden food, especially in new locations, as demonstrated in the previous study by Emery and Clayton (2001).

Scrub-jays have also been observed to cache non-food items, such as stones, so we provided the jays with a bowl containing both hazelnuts (a preferred food) and ceramic baking beans (a standardized equivalent to a stone). This was added to the design to determine whether the jays would attempt to "deceive" the observer (or mirror reflection) by caching more baking beads than hazelnuts, and to determine whether the non-food items have similar motivational value to the food items. Heinrich and Smolker (1998) described caching of inedible items by ravens as a form of play, although adults cached such items out of visual contact of conspecifics and were found to defend the caches. This may suggest that the inedible items hold a significant motivation value to the ravens. Clayton and colleagues (1994) found that when provided with stones, Eurasian jays cached them only when there was no food available or all had been eaten or cached. They also tended to cache those stones that resembled the food items, suggesting that they were tuned to the properties of the stones that were similar to the properties of the food (such as color, size, and shape).

We predicted that the stones should retain the same motivational value independent of the social context, and therefore they should either not be re-cached, or re-cached at the same rate across conditions (Observed, In Private, and Mirror). The stones were re-cached at the same rate in each of the three conditions (Figure 7A). If the hazelnuts had a significant motivational value attached to them (as would be predicted for a preferred food, compared to an inedible stone), and the reflection in the mirror was treated as a conspecific rather than a reflection of the caching scrub-jay, then the jay should have re-cached hazelnuts at the same rate as the Observed condition. If the caching scrub-jay regarded the image in the mirror as themselves (self-concept) then they should have re-cached hazelnuts at the same rate as the In Private condition. We found that the jays re-cached hazelnuts at the same rate in the Mirror condition as the Observed condition, suggesting that they regarded the reflected image in the mirror as a conspecific rather than themselves (similar to other birds and monkeys; Figure 7B).

These results and the earlier results from jungle crows, parakeets, and African gray parrots tend to suggest that self-awareness may be out of the capacity of the avian brain. This does not mean that Gallup's hypothesis is incorrect, as we have not demonstrated human ToM in the scrub-jays only some of the hallmarks of ToM. We need to examine the scrub-jays' reactions

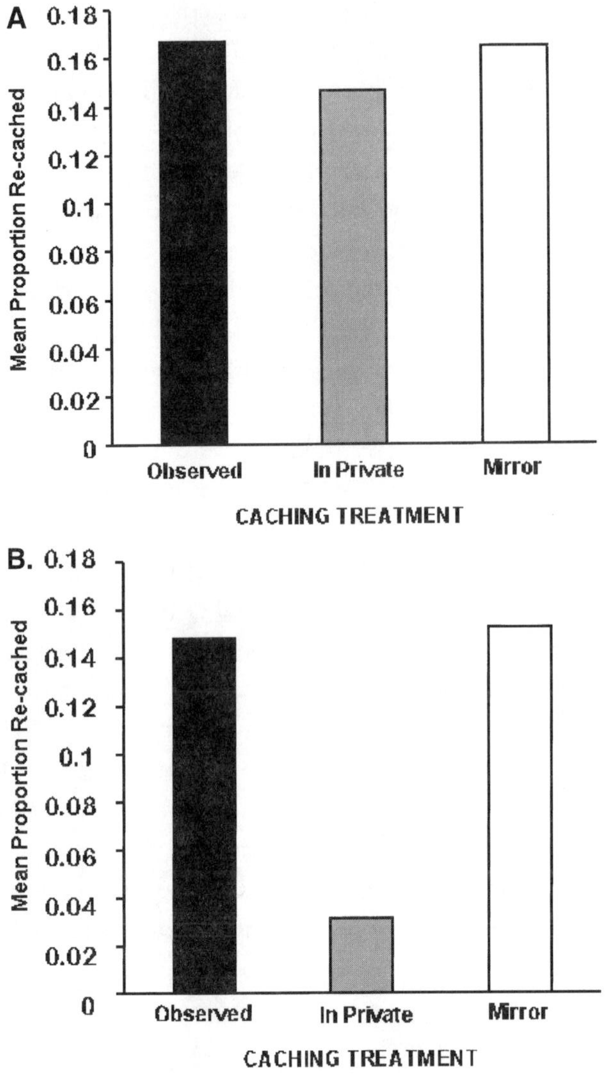

Figure 7. Mean proportion of (A) stones and (B) hazelnuts that were re-cached when the scrub-jays were either observed during caching, when they had cached in private or when they had cached in front of a large mirror. Unpublished data by Emery, Gilbert, and Clayton.

to the presence of the mirror more closely. The jays received very little experience with mirrors, in comparison to the studies with non-human primates, and this may have been insufficient. But a human-like ToM may still require the concept of self-recognition.

IS THERE A CASE FOR CONVERGENT COGNITIVE EVOLUTION AND DIVERGENT NEUROLOGICAL EVOLUTION?

In this chapter, we have described how some of the supposed "uniqueness" of primates in the structure of their brains and their advanced cognitive abilities are also present in a number of bird species, primarily the corvids and parrots. Initial thoughts on the distinct evolutionary histories of birds and mammals led us to presume that these neural and cognitive similarities do not exist between corvids, parrots, and primates. A number of significant factors draw us to the conclusion that although the common ancestor of mammals and birds lived over 280 million years ago, the recent evolution of these three taxonomic groups would facilitate the enhancement of cognition not demonstrated in other species (with the exception of Cetaceans). We have shown that some species of birds have forebrains larger than would be predicted from their body size, and that the neostriatum of the forebrain displays the greatest expansion, possibly representing an equivalent structure to the primate prefrontal cortex. We have also demonstrated that some of the bastions of primate intelligence; tool use, insight, symbolic communication, social learning, mental attribution, and mental time travel are also present in similar degrees to the primates, and in some cases surpass the available evidence for the primates. We suggest that this may largely stem from a lack of research in avian cognition, a primatocentric view of animal cognition inherited from field anthropology, and a paucity of studies in primate cognition that utilize EEA, the ecological/ethological approach.

The brains of birds and mammals are very different, and the neural substrates of cognition are located within such diverse brain structures, however the behavioral outcome of cognitive processing is very similar and likely based on similar principles. Surely then, one can argue a case for divergent neurological evolution yet convergent mental evolution (particularly within the realms of mental attribution and mental time travel).

We have a long way to go before birds and primates can be compared properly to a full extent on identical cognitive skills. New experimental studies in the field and in the laboratory need to be initiated, ones based on a high degree of ecological validity. One group therefore that would benefit from collaborating with comparative psychologists interested in animal cognition would be behavioral ecologists (Yoerg, 1991), especially those working with birds (Marler, 1997).

ACKNOWLEDGMENTS

The research was supported by grants from the National Institutes of Health (NS35465 & MH62602), the Grindley Trust, the Biotechnology & Biological Sciences Research Council (S16565), and the University of Cambridge. NJE was supported by a Royal Society University Research Fellowship and a Medical Research Council grant to Barry Keverne. We thank Dan Griffiths, James Gilbert, Sam Hettige, Dan Jennings, and Victoria Metcalf for help in running some of the scrub-jay experiments.

REFERENCES

Akins, C. K., and Zentall, T. R., 1996, Imitative learning in male Japanese quail (*Coturnix japonica*) using the two-action method, *J. Comp. Psychol.* **110**:316–320.

Balda, R. P., and Kamil, A. C., 1989, A comparative study of cache recovery by three corvid species, *Anim. Behav.* **38**:486–495.

Balda, R. P., Kamil, A. C., and Bednekoff, P. A., 1997, Predicting cognitive capacity from natural history: Examples from species of corvids, in: *Current Ornithology*, Volume 13, V. Nolan and E. D. Ketterson, eds., Plenum Press, New York, pp. 33–66.

Basil, J. A., Kamil, A. C., Balda, R. A., and Fite, K. V., 1996, Differences in hippocampal volume in food-storing corvids, *Brain Behav. Evol.* **47**:156–164.

Baylis, J. R., 1982, Avian vocal mimicry: Its function and evolution, in: *Acoustic Communication in Birds*, D. E. Kroodsma, E. H. Miller, and H. Ouellet, eds., Academic Press, London, pp. 51–83.

Bednekoff, P. A., and Balda, R. A., 1996a, Observational spatial memory in Clark's nutcrackers and Mexican jays, *Anim. Behav.* **52**:833–839.

Bednekoff, P. A., and Balda, R. A., 1996b, Social caching and observational spatial memory in Pinyon jays, *Behaviour* **133**:807–826.

Berthold, P., Heilbig, A. J., Mohr, G., and Querner, U., 1992, Rapid microevolution of migratory behaviour in a wild bird species. *Nature* **360**:668–670.

Birch, H. G., 1945, The relation of previous experience to insightful problem solving, *J. Comp. Psych.* **38**:367–383.

Bischof, N., 1978, On the phylogeny of human morality, in: *Morality as a Biological Phenomenon*, G. Stent, ed., Berlin, Abakon, pp. 53–74.

Bischof-Koehler, D., 1985, Zur Phylogenese menschlicher motivation [On the phylogeny of human motivation], in: *Emotion und Reflexivitaet*, L. H. Eckensberger and E. D. Lantermann, eds., Vienna, Urban & Schwarzenberg, pp. 3–47.

Boesch, C., and Boesch, H., 1983, Optimization of nut-cracking with natural hammers by wild chimpanzees, *Behaviour* **83**:265–286.

Brockmann, H. J., and Barnard, C. J., 1979, Kleptoparasitism in birds, *Anim. Behav.* **27**:487–514.

Bugnyar, T., and Huber, L., 1997, Push or pull: An experimental study on imitation in marmosets, *Anim. Behav.* **54**:817–831.

Bugnyar, T., and Kotrschal, K., 2001, Do ravens manipulate others' attention in order to prevent or achieve social learning opportunities? *Adv. Ethol.* **36**:106.

Bugnyar, T., and Kotrschal, K., 2002, Observational learning and the raiding of food caches in ravens, *Corvus corax*: Is it "tactical" deception? *Anim. Behav.* **64**:185–195.

Byrne, R. W., 1995, *The Thinking Ape: Evolutionary Origins of Intelligence*, Oxford University Press, Oxford.

Byrne, R. W., 1996, Relating brain size to intelligence in primates, in: *Modelling the Early Human Mind*, P. A. Mellars and K. R. Gibson, eds., MacDonald Institute for Archaeological Research, Cambridge, pp. 49–56.

Byrne, R. W., and Whiten, A., eds., 1998, *Machiavellian Intelligence: Social Expertise and the Evolution of Intellect in Monkeys, Apes and Humans*, Clarendon Press, Oxford.

Campbell, F. M., Heyes, C. M., and Goldsmith, A. R., 1999, Stimulus learning and response learning by observation in the European starling, in a two-object/two action test, *Anim. Behav.* **58**:151–158.

Catchpole, C. K., and Slater, P. J. B., 1995, *Bird Song: Themes and Variations*, Cambridge University Press, Cambridge.

Chappell, J., and Kacelnik, A., 2002, Tool selectivity in a non-primate, the New Caledonian crow (*Corvus moneduloides*), *Animal Cognition* **5**:71–78.

Cheney, D. L., and Seyfarth, R. M., 1985, Vervet monkey alarm calls: Manipulation through shared information? *Behaviour* **93**:150–166.

Cheney, D. L., and Seyfarth, R. M., 1990, *How Monkeys See the World: Inside the Mind of Another Species*, University of Chicago Press, Chicago.

Clayton, D. A., 1978, Socially facilitated behavior, *Q. Rev. Biol.* **53**:373–391.

Clayton, N. S., 1998, Memory and the hippocampus in food-storing birds: A comparative approach, *Neuropharmacol.* **37**:441–452.

Clayton, N. S., and Dickinson, A., 1998, Episodic-like memory during cache recovery by scrub jays, *Nature* **395**:272–278.

Clayton, N. S., and Dickinson, A., 1999a, Scrub jays (*Aphelocoma coerulescens*) remember the relative time of caching as well as the location and content of their caches, *J. Comp. Psychol.* **113**:403–416.

Clayton, N. S., and Dickinson, A., 1999b, Memory for the contents of caches by Scrub jays, *J. Exp. Psychol.: Anim. Behav. Proc.* **25**:82–91.

Clayton, N. S., and Jolliffe, A., 1996, Marsh tits (*Parus palustris*) use tools to store food., *Ibis* **138**:554.

Clayton, N. S., Griffiths, D. P., and Bennett, A. T. D., 1994, Storage of stones by jays *Garrulus glandarius, Ibis* **136**:331–334.

Clayton, N. S., Yu, K., and Dickinson, A., 2001a, Scrub jays (*Aphelocoma coerulescens*) can form an integrated memory for multiple features of caching episodes, *J. Exp. Psychol.: Anim. Behav. Proc.* **27**:17–29.

Clayton, N. S., Griffiths, D. P., Emery, N. J., and Dickinson, A., 2001b, Elements of episodic-like memory in animals, *Phil. Trans. Roy. Soc. Lond.: B.* **356**:1483–1491.

Cristol, D., and Switzer, P. V., 1999, Avian prey-dropping behavior. II. American crows and walnuts, *Behav. Ecol.* **10**:220–226.

Custance, D., Whiten, A., and Freedman, T., 1999, Social learning of an artificial fruit task in capuchin monkeys (*Cebus apella*), *J. Comp. Psychol.* **113**:13–23.

Dawkins, M. S., 2002, What are birds looking at? Head movements and eye use in chickens, *Anim. Behav.* **63**:991–998.

Dawson, B. V., and Foss, B. M., 1965, Observational learning in budgerigars, *Anim. Behav.* **13**:470–474.

Deacon, T., 1990, Rethinking mammalian brain evolution, *Am. Zool.* **30**:629–705.

D'Eath, R. B., 1998, Can video images imitate real stimuli in animal behaviour experiments? *Biol. Rev.* **73**:267–292.

De Vore, I., ed., 1965, *Primate Behavior: Field Studies of Monkeys and Apes*, Holt Reinhart Winston, New York.

de Waal, F. B. M., ed., 2001, *Tree of Origin: What Primate Behavior can tell us about Human Social Evolution*, Harvard University Press, Cambridge.

Durstewitz, D., Kroner, S., and Gunturkun, O., 1999, Dopamine innervation of the avian telencephalon, *Prog. Neurobiol.* **59**:161–195.

Emery, N. J., 2000, The eye have it: The neuroethology, evolution and function of social gaze, *Neurosci. Biobehav. Rev.* **24**:581–604.

Emery, N. J., and Clayton, N. S., 2001, Effects of experience and social context on prospective caching strategies by scrub jays, *Nature* **414**:443–446.

Emery, N. J., Lorincz, E. N., Perrett, D. I., Oram, M. W., and Baker, C. I., 1997, Gaze following and joint attention in rhesus monkeys (*Macaca mulatta*), *J. Comp. Psychol.* **111**:286–293.

Epstein, R., Lanza, R. P., and Skinner, B. F., 1981, "Self-awareness" in the pigeon, *Science* **212**:695–696.

Evans, C. S., Evans, L., and Marler, P., 1993, On the meaning of alarm calls: Functional reference in an avian vocal system, *Anim. Behav.* **46**:23–38.

Fabre, J. H., 1916, *The Hunting Wasps*. Hodder and Stoughton.

Flaherty, C. F., and Checke, S., 1982, Anticipation of incentive gain, *Anim. Learn. Behav.* **10**:177–182.

Fritz, J., and Kotrschal, K., 1999, Social learning in common ravens, *Corvus corax, Anim. Behav.* **57**:785–793.

Gallup, G. G., Jr., 1970, Chimpanzees: Self-recognition, *Science* **167**:341–343.

Gallup, G. G., Jr., 1982, Self-awareness and the emergence of mind in primates, *Am. J. Primatol.* **2**:237–248.

Gallup, G. G., Jr., 1994, Self-recognition: Research strategies and experimental design, in: *Self-Awareness in Animals and Humans: Developmental Perspectives*, S. T. Parker, R. W. Mitchell, and M. L. Boccia, eds., Cambridge University Press, Cambridge, pp. 35–50.

Gallup, G. G., Jr., and Capper, S. A., 1970, Preference for mirror-image stimulation in finches (*Passer domesticus domesticus*) and parakeets (*Melopsittacus undulates*), *Anim. Behav.* **18**:621–624.

Gallup, G. G., Jr., Povinelli, D. J., Suarez, S. D., Anderson, J. R., Lethmate, J., and Menzel, E. W., Jr., 1995, Further reflections on self-recognition in primates, *Anim. Behav.* **50**:1525–1532.

Gardner, R. A., and Gardner, B. T., 1969, Teaching sign language to a chimpanzee, *Science* **165**:664–672.

Gerlai, R., and Clayton, N. S., 1999, Analyzing hippocampal function in transgenic mice: An ethological perspective, *TINS* **22**:47–51.

Gibson, K. R., 1986, Cognition, brain size, and the extraction of embedded food resources, in: *Primate Ontogeny, Cognition and Social Behavior*, J. G. Else and P. C. Lee, eds., Cambridge University Press, Cambridge, pp. 93–105.

Goldman-Rakic, P. S., Lidow, M. S., Smiley, J. F., and Williams, M. S., 1992, The anatomy of dopamine in monkey and human prefrontal cortex, *J. Neural Trans. Suppl.* **36**:163–177.

Goodall, J., 1986, *The Chimpanzees of Gombe. Patterns of Behavior*. Harvard University Press, Cambridge.

Gould-Beierle, K., 2000, A comparison of four corvid species in a working and reference memory task using a radial arm maze, *J. Comp. Psychol.* **114**:347–356.

Granon, S., and Poucet, B., 1995, Medial prefrontal lesions in the rat and spatial navigation: Evidence for impaired planning, *Behav. Neurosci.* **109**:474–484.

Griffiths, D., Dickinson, A., and Clayton, N. S., 1999, Episodic and declarative memory: What can animals remember about their past? *Trends Cog. Sci.* **3**:74–80.

Güntürkün, O., 1997, Cognitive impairments after lesions of the neostriatum caudolaterale and its thalamic afferent in pigeons: Functional similarities to the mammalian prefrontal system? *J. Hirnforsch* **38**:133–143.

Gyger, M., Karakashian, S., and Marler, P., 1986, Avian alarm calling: Is there an audience effect? *Anim. Behav.* **34**:1570–1572.

Hampton, R. R., 1994, Sensitivity to information specifying the line of gaze of humans in sparrows (*Passer domesticus*). *Behaviour* **130**:41–51.

Hare, B., 2001, Can competitive paradigms increase the validity of experiments on primate social cognition? *Anim. Cog.* **4**:269–280.

Hare, B., and Tomasello, M., 1999, Domestic dogs (*Canis familiaris*) use human and conspecific social cues to locate hidden food, *J. Comp. Psychol.* **113**:173–177.

Hare, B., Call, J., and Tomasello, M., 2001, Do chimpanzees know what conspecifics know? *Anim. Behav.* **61**:139–151.

Hare, B., Call, J., Agnetta, B., and Tomasello, M., 2000, Chimpanzees know what conspecifics do and do not see, *Anim. Behav.* **59**:771–785.

Hartmann, B., and Güntürkün, O., 1998, Selective deficits in reversal learning after neostriatum caudolaterale lesions in pigeons: Possible behavioral equivalencies to the mammalian prefrontal system, *Behav. Brain Res.* **96**:125–133.

Hauser, M. D., 2000, *Wild Minds: What Animals Really Think.* Henry Holt and Company, New York.

Heinrich, B., 1996, An experimental investigation of insight in common ravens, *Corvus corax, Auk* **112**:994–1003.

Heinrich, B., 1999, *The Mind of the Raven: Investigations and Adventures with Wolf-Birds.*

Heinrich, B., 2000, Testing insight in ravens, in: *The Evolution of Cognition*, C. M. Heyes and L. Huber, eds., MIT Press, Cambridge, pp. 289–309.

Heinrich, B., and Pepper, J. W., 1998, Influence of competitors on caching behavior in the common raven, *Anim. Behav.* **56**:1083–1090.

Heinrich, B., and Smolker, R., 1998, Play in common ravens (*Corvus corax*), in: *Animal Play: Evolutionary, Comparative and Ecological Perspectives*, M. Bekoff and J. A. Byers, eds., Cambridge University Press, Cambridge, pp. 27–44.

Held, S., Mendl, M., Devereux, C., and Byrne, R. W., 2001, Behaviour of domestic pigs in a visual perspective taking task, *Behaviour* **138**:1337–1354.

Heyes, C. M., 1996, Genuine imitation? in: *Social Learning in Animals: The Roots of Culture*, C. M. Heyes, and B. G. Jr. Galef, (eds.) Academic Press, San Diego, pp. 371–404.

Heyes, C. M., 1998, Theory of mind in nonhuman primates, *Behav. Brain Sci.* **21**:101–148.

Heyes, C. M., and Dawson, G. R., 1990, A demonstration of observational learning using a bi-directional control, *Q. J. Exp. Psychol.* **42B**:59–71.

Heyes, C. M., and Galef, B. G. Jr., eds., 1996, *Social Learning in Animals: The Roots of Culture*, Academic Press, Inc., San Diego.

Hodos, W., and Campbell, C. B. G., 1969, Scala naturae: Why there is no theory in comparative psychology, *Psych. Rev.* **76**:337–350.

Horn, G., 1985, *Memory, Imprinting and the Brain: An Inquiry into Mechanisms.* Oxford University Press, Oxford.

Huber, L., Rechberger, S., and Taborsky, M., 2001, Social learning affects object exploration and manipulation in keas, *Nestor notabilis, Anim. Behav.* **62**:945–954.

Humphrey, N. K., 1976, The social function of intellect, in: *Growing Points in Ethology*, P. P. G. Bateson and R. A. Hinde, eds., Cambridge University Press, Cambridge, pp. 303–317.

Hunt, G. R., 1996, Manufacture and use of hook-tools by New Caledonian crows, *Nature* **379**:249–251.

Hunt, G. R., 2000, Human-like, population-level specialization in the manufacture of pandanus tools by the New Caledonian crows (*Corvus moneduloides*), *Proc. Roy. Soc. Lond: B.* **267**:403–413.

Jerison, H. J., 1971, *Evolution of Brain and Intelligence*, Academic Press, Inc., New York.

Jolly, A., 1966, Lemur social behavior and primate intelligence, *Science* **153**:501–507.

Jones, T. B., and Kamil, A. C., 1973, Tool-making and tool-using in the northern blue jay, *Science* **180**:1076–1078.

Kamil, A. C., 1988, A synthetic approach to the study of animal intelligence, in: *Nebraska Symposium on Motivation*, Volume 35, D. W. Leger, ed., University of Nebraska Press, Lincoln, pp. 257–308.

Kamil, A. C., Balda, R. P., and Olson, D. J., 1994, Performance of four seed-caching corvid species in the radial-arm maze analog, *J. Comp. Psychol.* **108**:385–393.

Kaminski, J., 2002, Do animals know what others can and cannot see. Paper presented at the 2nd International Symposium on Comparative Cognitive Science, Inuyama, Japan.

Karakashian, S. J., Gyger, M., and Marler, P., 1988, Audience effects on alarm calling in chickens (*Gallus gallus*), *J Comp. Psychol.* **102**:129–135.

Karten, H. J., 1969, The organization of the avian telencephalon and some speculations on the phylogeny of the amniote telencephalon, in: *Comparative and Evolutionary Aspects of the Vertebrate Central Nervous System*, J. N. Petras and C. Noback, eds., New York Academy of Sciences, New York, pp. 164–179.

Karten, H. J., 1991, Homology and the evolutionary origins of the "neocortex". *Brain, Behav. Evol.* **38**:264–272.

Köhler, W., 1927, *The Mentality of Apes*. Routledge & Kegan Paul (original work published in 1917, English translation by E. Winter, 1927).

Kusayama, T., Bischoff, H.-J., and Watanabe, S., 2000, Responses to mirror-image stimulation in jungle crows (*Corvus macrorhynchos*), *Anim. Cog.* **3**:61–64.

Lefebvre, L., Nicolakakis, N., and Boire, D., 2002, Tools and brains in birds. *Behaviour* **139**:939–973.

Lefebvre, L., Whittle, P., Lascaris, E., and Finkelstein, A., 1997b, Feeding innovations and forebrain size in birds, *Anim. Behav.* **53**:549–560.

Macintosh, N. J., 1988, Approaches to the study of animal intelligence, *Br. J. Psychol.* **79**:509–525.

Madge, S., and Burn, H., 1999, *Crows and Jays: A Guide to the Crows, Jays and Magpies of the World*, Houghton Mifflin Co., Boston.

Marler, P., 1997, Social cognition: Are primates smarter than birds? in: *Current Ornithology*, Volume 13, V. Nolan and E. D. Ketterson, eds., Plenum Press, New York, pp. 1–32.

McKinley, J., and Sambrook, T. D., 2000, Use of human-given cues by domestic dogs (*Canis familiaris*) and horses (*Equus caballus*), *Anim. Cog.* **3**:13–22.

Menzel, C. R., 1999, Unprompted recall and reporting of hidden objects by a chimpanzee (*Pan troglodytes*) after extended delays, *J. Comp. Psychol.* **113**:1–9.

Millikan, G. C., and Bowman, R. I., 1967, Observations on Galapogos tool-using finches in captivity, *Living Bird* **6**:23–41.

Moore, B. R., 1992, Avian movement imitation and a new form of mimicry: Tracing the evolution of a complex form of learning, *Behaviour* **122**:231–263.

Nicolakakis, N., and Lefebvre, L., 2000, Forebrain size and innovation rate in European birds: Feeding, nesting and confounding variables, *Behaviour* **137**:1415–1429.

Nikei, Y., 1995, Variations of behavior of carrion crows (*Corvus corone*) using automobiles as nutcrackers, *Jap. J. Ornithol.* **44**:21–35.

Panksepp, J., 1998, *Affective Neuroscience*, Oxford University Press, New York.

Parker, S. T., Mitchell, R. W., and Boccia, M. L., eds., 1994, *Self-Awareness in Animals and Humans*, Cambridge University Press, Cambridge.

Passingham, R. E., 1982, *The Human Primate*, WH Freeman, Oxford.

Pepperberg, I. M., 1990, Some cognitive capacities of an African Grey Parrot (*Psittacus erithacus*), *Adv. Study Behav.*, **19**:357–409.

Pepperberg, I. M., 1999, *The Alex Studies: Communication and Cognitive Capacities of an African Grey Parrot.* Harvard University Press, Cambridge.

Pepperberg, I. M., and McLaughlin, M. A., 1996, Effects of avian-human joint attention in allospecific vocal learning by grey parrots (*Psittacus erithacus*), *J. Comp. Psych.* **110**:286–297.

Pepperberg, I. M., Garcia, S. E., Jackson, E. C., and Marconi, S., 1995, Mirror use by African grey parrots (*Psittacus erithacus*), *J. Comp. Psychol.* **109**:182–195.

Portmann, A., 1946, Etude sur la cérébralisation chez les oiseaux I, *Alauda* **14**:2–20.

Portmann, A., 1947, Etude sur la cérébralisation chez les oiseaux II. Les indices intra-cérébraux, *Alauda* **15**:1–15.

Povinelli, D. J., 2000, *Folk Physics for Apes*, Oxford University Press, New York.

Povinelli, D. J., and Eddy, T. J., 1996, What young chimpanzees know about seeing, *Mono. Soc. Res. Child Dev.* **61**(3) (Serial No. 247).

Povinelli, D. J., Nelson, K. E., and Boysen, S. T., 1990, Inferences about guessing and knowing by chimpanzees (*Pan troglodytes*), *J. Comp. Psychol.* **104**:203–210.

Premack, D., 1971, Language in chimpanzee? *Science* **172**:808–822.

Premack, D., and Premack, A. J., 1983, *The Mind of an Ape*, W. W. Norton & Co., New York.

Premack, D., and Woodruff, G., 1978, Does the chimpanzee have a theory of mind? *Behav. Brain Sci.* **1**:515–526.

Raleigh, M. J., and Steklis, H. D., 1981, Effects of orbitofrontal and temporal neocortical lesions on the affiliative behavior of vervet monkeys (*Cercopthecus aethiops sabaeus*), *Exp. Neurol.* **73**:378–389.

Rehkamper, G., Frahm, H. D., and Zilles, K., 1991, Quantitative development of brain and brain structures in birds (*Galliformes* and *Passeriformes*) compared to that in mammals (Insectivores and Primates), *Brain, Behav. Evol.* **37**:125–143.

Reid, J. B., 1982, Tool-use by a rook (*Corvus frugilegus*) and its causation, *Anim. Behav.* **30**:1212–1216.

Reiss, D., and Marino, L., 2001, Mirror self-recognition in the bottlenose dolphin: A case of cognitive convergence, *Proc. Nat. Acad. Sci.* **98**:5937–5942.

Ristau, C. A., 1991, Aspects of the cognitive ethology of an injury feigning plover, in: *Cognitive Ethology: The Minds of Other Animals*, C. A. Ristau, ed., Erlbaum, Hillsdale, pp. 91–126.

Rumbaugh, D. M., ed., 1977, *Language Learning by a Chimpanzee*, Academic Press, New York.

Savage-Rumbaugh, E. S., and Lewin, R., 1994, *Kanzi, the Ape at the Brink of the Human Mind.* John Wiley & Sons, New York.

Savage-Rumbaugh, E. S., Rumbaugh, D. M., and Boysen, S. T., 1978, Sarah's problems of comprehension. Commentary to Premack and Woodruff (1978), *Behav. Brain Sci.* **4**:555–557.

Schwartz, B. L., and Evans, S., 2001, Episodic memory in primates, *Am. J. Primatol.* **55**:71–85.

Semendeferi, K., Lu, A., Schenker, N., and Damasio, H., 2002, Humans and great apes share a large frontal cortex, *Nature Neurosci.* **5**:272–276.

Seyfarth, R. M., Cheney, D. L., and Marler, P., 1980, Vervet monkey alarm calls: Semantic communication in a free-ranging primate, *Science* **210**:801–803.

Shettleworth, S. J., 1998, *Cognition, Evolution, and Behavior.* Oxford University Press, New York.

Stephan, H. D., Frahm, H., and Baron, G., 1981, New and revised data on volumes of brain structures in insectivores and primates, *Folia primatol.* **35**:1–29.

Stoinski, T. S., Wrate, J. L., Ure, N., and Whiten, A., 2001, Imitative learning by captive Western lowland gorillas (*Gorilla gorilla gorilla*) in a simulated food-processing task, *J. Comp. Psychol.* **115**:272–281.

Struhsaker, T. T., 1967, Auditory communication among ververt monkeys (*Cercopithecus aethiops*), in: *Social Communication among Primates*, S. A. Altmann, ed., University of Chicago Press, Chicago.

Stuss, D. T., and Alexander, M. P., 2000, Executive functions and the frontal lobes: A conceptual view, *Psychol. Res.* **63**:289–298.

Suarez, S. D., and Gallup, G. G., Jr., 1981, Self-recognition in chimpanzees and orangutans but not gorillas, *J. Comp. Psychol.* **343**:35–56.

Suddendorf, T., and Corballis, M. C., 1997, Mental time travel and the evolution of the human mind, *Genetic Soc. Gen. Psychol. Monographs* **123**:133–167.

Tebbich, S., 2000, *Tool Use in the Woodpecker Finch Cactospiza Pallida: Ontogeny and Ecological Relevance.* Unpublished Doctoral thesis, University of Vienna, Austria.

Tebbich, S., Taborsky, M., Fessl, B., and Blomqvist, D., 2001, Do woodpecker finches acquire tool-use by social learning? *Proc. Roy. Soc. Lond.: B.* **268**:2189–2193.

Templeton, J. J., Kamil, A. C., and Balda, R. P., 1999, Sociality and social learning in two species of corvids: the Pinyon jay (*Gymnorhinus cyanocephalus*) and the Clark's nutcracker (*Nucifraga columbiana*), *J. Comp. Psych.* **113**:450–455.

Terrace, H. S., Pettito, L. A., Sanders, R. J., and Bever, T. G., 1979, Can an ape create a sentence? *Science* **206**:891–902.

Thompson, R. K. R., and Contie, C. L., 1994, Further reflections on mirror usage by pigeons: Lessons from Winnie-the-Pooh and Pinocchio too, in: *Self-Awareness in Animals and Humans: Developmental Perspectives*, S. T. Parker, R. W. Mitchell, and M. L. Boccia, eds., Cambridge University Press, Cambridge, pp. 392–409.

Thorpe, W. H., 1963, *Learning and Instinct in Animals*, Second edition, Methuen, London.

Thouless, C. R., Fanshawe, J. H., and Bertram, C. R., 1987, Egyptian vultures *Neophron percnopterus* and Ostrich *Struthio camelus* eggs: The origins of stone-throwing behaviour, *Ibis* **131**:9–15.

Timmermans, S., Lefebvre, L., Boire, D., and Basu, P., 2000, Relative size of the hyperstriatum ventrale is the best predictor of feeding innovation rate in birds, *Brain, Behav Evol.* **56**:196–203.

Todt, D., 1975, Social learning of vocal patterns and modes of their application in grey parrots (*Psittacus erithacus*), *Z. Tierpsychol.* **39**:178–188.

Tomasello, M., and Call, J., 1997, *Primate Cognition*, Oxford University Press, New York.

Tomasello, M., Call, J., and Hare, B., 1998, Five primate species follow the visual gaze of conspecifics, *Anim. Behav.* **55**:1063–1069.

Tschudin, A., 1999, Relative neocortex size and its correlates in dolphins: comparisons with humans and implications for mental evolution. Unpublished Doctoral thesis, University of Natal, Pietermaritzburg, South Africa.

Tschudin, A., Call, J., Dunbar, R. I. M., Harris, G., and van der Elst, C., 2001, Comprehension of signs by dolphins (*Tursiops truncates*), *J. Comp. Psych.* **115**:100–115.

Tulving, E., 1972, Episodic and semantic memory, in: *Organisation of Memory*, E. Tulving and W. Donaldson, eds., New York, Academic Press, pp. 381–403.

Tulving, E., 1983, *Elements of Episodic Memory.* Clarendon Press, Oxford.

Tulving, E., 2002, Chronesthesia: Conscious awareness of subjective time, in: *The Age of the Frontal Lobes*, D. T. Stuss and R. C. Knight, eds.

van Lawick-Goodall, J., 1968, The behaviour of free-living chimpanzees in the Gombe Stream reserve, *Anim. Behav. Mono.* **1**:161–311.

van Lawick-Goodall, J., and van Lawick, H., 1966, Use of tools by the Egyptian vulture, *Neophron percnopterus, Nature* **212**:1468–1469.

Voronov, L. N., Bogoslovskaya, L. G., and Markova, E. G., 1994, A comparative study of the morphology of forebrain in corvidae in view of their trophic specialization, *Zoo. Z.* **73**:82–96.

Weir, A. A. S., Chappell, J., and Kacelnik, A., 2002, Shaping of hooks in New Caledonian crows, *Science* **297**:981.

Wheeler, M. A., 2000, Episodic memory and autonoetic awareness, in: *The Oxford Handbook of Memory*, E. Tulving and F. I. M. Craik, eds., Oxford, Oxford University Press, pp.597–625.

Whiten, A., 1998, Imitation of the sequential structure of actions by chimpanzees (*Pan troglodytes*), *J. Comp. Psychol.* **112**:270–281.

Whiten, A., and Byrne, R. W., 1988, Tactical deception in primates, *Behav. Brain Sci.* **11**:233–244.

Whiten, A., Custance, D. M., Gomez, J. C., Texidor, P., and Bard, K. A., 1996, Imitative learning of artificial fruit processing in children (*Homo sapiens*) and chimpanzees (*Pan troglodytes*), *J. Comp. Psychol.* **110**:3–14.

Whiten, A., Goodall, J., McGrew, W. C., Nishida, T., Reynolds, V., Sugiyama, Y., Tutin, C. E. G., Wrangham, R. W., and Boesch, C., 1999, Cultures in chimpanzees, *Nature* **399**:682–685.

Whiten, A., and Ham, R., 1992, On the nature and evolution of imitation in the animal kingdom: Reappraisal of a century of research, in: *Advances in the Study of Behavior*, Volume 21, P. J. B. Slater, J. S. Rosenblatt, C. Beer, and M. Milinski, eds., Academic Press, Inc., New York, pp. 239–283.

Woolfenden, G. E., and Fitzpatrick, J. W., 1984, *The Florida Scrub Jay: Demography of a Cooperative-breeding Bird*, Princeton University Press, Princeton.

Yoerg, S. I., 1991, Ecological frames of mind: The role of cognition in behavioral ecology, *Q. Rev. Biol.* **66**:287–301.

Zilles, K., and Rehkamper, G., 1988, The brain, with special reference to the telencephalon, in: *Orang-utan Biology*, J. H. Schwartz, ed., Oxford University Press, New York, pp. 157–176.

CHAPTER TWO

Visual Cognition and Representation in Birds and Primates

Giorgio Vallortigara

INTRODUCTION

Birds and primates depend greatly on vision for their survival. Their visual systems present suggestive similarities, such as the subdivision into a thalamofugal and a tectofugal pathways (i.e., geniculo-striate and collicular systems in mammals; cf. Husband and Shimizu, 2001), but also striking differences, such as the uniformly laminated structure of the telencephalon of primates as contrasted with the (mostly) non-laminated, nuclear organization of the telencephalon of birds (Karten and Shimizu, 1989). Moving from anatomy to the structure of visual experiences, our visual world appears to consist of stable, meaningful, and unified objects that move in spatially and temporally predictable ways. "Objects," therefore, constitute the functional units of perception and action to the human primate. Is this true for the other primate species? Is it true even for birds? Primates and birds appear to be as visual and mobile animals as we are. Do they experience an object-filled world like our own?

Giorgio Vallortigara • Department of Psychology and B.R.A.I.N Centre for Neuroscience, University of Trieste, Trieste, Italy.

Comparative Vertebrate Cognition, edited by Lesley J. Rogers and Gisela Kaplan. Kluwer Academic/Plenum Publishers, 2004.

Visual cognition of objects in birds and primates has been investigated extensively. This chapter, however, is not aimed at providing a comprehensive review, but rather at discussing a few selected topics (a selection based mainly on the author's own research interests). Basically, I shall follow the vicissitudes of the birds' and primates' visual cognitive systems facing the problem of recognizing, representing, and localizing objects in the environment when these objects are partly occluded by other objects (in the next section), or auto-occluded because of motion in three-dimensional space (in the third section), or completely unavailable to direct sensory experience because they are entirely hidden behind (or within) other objects (in the fourth and fifth sections).

INTEGRATION AND INTERPOLATION OF VISUAL INFORMATION IN THE SPATIAL DOMAIN

Animals that use vision face a fundamental computational problem, namely that information falling on the retina is unavoidably incomplete because of occlusion phenomena. Given that most objects are opaque, depending on an observer's vantage point, some objects may partly hide other objects. Nevertheless, our perception suffers little when objects are partly occluded because the parts that are directly visible seem to continue (or to be completed) behind the occluders (see Figure 1). The phenomenon has been dubbed "amodal completion" by perceptual psychologists and has been extensively investigated (Grossberg and Mingolla, 1985; Kanizsa, 1979; Michotte, 1963; Michotte et al., 1964). In non-human species, evidence is more contradictory both in primates and birds.

The problem with non-human species is that we cannot ask them directly whether they perceive completion of partly occluded objects; some sort of

Figure 1. An example of amodal completion. The Belgian psychologist Albert Michotte (Michotte et al., 1964) reported that even after extensive experience of the covering and uncovering of the middle of the display shown on the left, people report that it looks like a triangle when partly occluded by the finger (on the right). Non-human primates and at least some species of birds have been shown to perceive continuity of partly occluded objects as well (see text).

nonverbal trick is needed to reveal this information. Most studies with birds have used conditioning procedures and the pigeon as a model. Cerella (1980), for instance, after training pigeons to respond to a triangle, found that responses to an amputated triangle (i.e., lacking a piece) exceeded those to a partially occluded triangle. He also reported that after learning to discriminate figures of Charlie Brown, pigeons responded to pictures representing only parts of Charlie Brown's figure, but also emitted many responses to random mixtures of these parts. These results seem to suggest that pigeons perceive complex stimuli as an assembly of local features, and that responses to partly occluded objects depend only on the visual information remaining after fragmentation of the stimulus.

It is interesting to observe that mammals like mice behaved quite differently from pigeons in similar tests. Kanizsa et al. (1993) trained mice to discriminate between complete and amputated disks. After fulfilling the criterion, the mice performed test trials in which outlined rectangles were either exactly juxtaposed or only placed close to the missing sectors of the disks in order to produce or not produce the impression (to a human observer) of an occlusion of the missing sectors by the rectangles. Mice responded in these tests as if they were experiencing completion of the partly occluded disks; pigeons, in contrast, responded on the basis of local, visible features and failed to complete (Sekuler et al., 1996) or even to perceive continuation of the figure behind the occluder (Fujita, 2001).

One theoretical problem that arises from these observations concerns the way in which pigeons manage to negotiate objects in their visual world if they do not complete partly occluded stimuli. When a mobile organism, such as a pigeon, moves in the environment, different parts of the objects that are partly covered in the visual scene will be occluded and disoccluded as a result of the changes in the pigeon's vantage point. Only objects, which are in front view, not occluded by any other object, would retain their shape and size, the rest of the visual world would be continuously changing (actually, even retinal projections of objects in front view undergo modifications with changes in the vantage point). Thus, completion phenomena, in humans at least, seem to be strictly related to the perception of a world of stable, unchanging visual objects. However, what about the visual world of an animal that responds only to the fragments, the single unstructured pieces of the actual retina stimulation? That would appear as a very weird visual world to live in.

However, it has been shown that using natural stimuli rather than figures of Charlie Brown, pigeons do not respond only to local features. Watanabe and

Ito (1991) trained pigeons to discriminate color slides of different individuals, and then tested them with the full face, separate parts, and randomly connected parts of the original stimuli. In this case, pigeons emitted very few responses to scrambled figures. Evidence that the global relations among component parts can be critical in discriminative control has been obtained also by Wasserman et al. (1993) who found that scrambling the component parts of complex objects reduced their discrimination, indicative of partial control by the spatial configuration of the component parts (see also Kirkpatrick [2001], Cook [2001], and Towe [1954] for some old evidence). These somewhat contrasting results suggest that pigeons can perceive and discriminate complex stimuli based on either the local parts or the global configuration, much like humans—a point to which we shall return soon.

Evidence in primates seems to favor the view that they perceive object unity in partly occluded displays, though negative results have sometimes been reported. For instance, Sato et al. (1997) trained a chimpanzee on a matching-to-sample of one rod and two rods. In probe test trials, the chimpanzee matched two rods moving in concert with the central portion occluded with one rod, whereas it matched two rods moving out of concert with two rods. Thus, the chimpanzee's performance resembles that of humans in similar tasks. On the other hand, Sugita (1999) showed in Japanese macaques that some orientation-selective cells of the primary visual cortex fired when a patch had cross disparities, so that it appeared in front of an occluded bar, but stopped firing when the patch had uncrossed disparities, so that it appeared to be behind a bar. Intriguingly, there was no cell response when the patch appeared at the same depth as the stimulus bar (zero disparity), thus questioning the existence of amodal completion under normal viewing conditions. Could it be that amodal completion occurs only in great apes because of their superior cognitive abilities, but not in monkeys? Behavioral results seem to tell another story. Deruelle et al. (2000) conducted five experiments to verify whether baboons perceive partly occluded objects as complete. The first three experiments used a go/no-go procedure and a video monitor for stimulus presentation. These experiments failed to reveal any amodal completion. In contrast, completion was demonstrated in a fourth experiment with cardboard stimuli in a two-alternative forced-choice discrimination task. Interestingly, completion was again absent when, in a fifth experiment, stimuli were shown with a computer graphic system. These results suggest that baboons share with humans the ability for amodal completion, but also underline some procedural factors that might affect the elicitation of this capacity. Amodal completion in other sensory modalities (i.e., acoustic signals) has been reported in cotton-top tamarins by Miller et al. (2001); I do not know of any study that has investigated the same issue in birds.

Studies on visual discrimination learning have demonstrated that animals, including non-human primates (Fagot and Deruelle, 1997), frequently attend to the featural aspects of the stimuli rather than to global properties of the visual scene. In an attempt to circumvent this problem, Regolin and Vallortigara (1995) used the naturalistic setting made available by filial imprinting, the process whereby young birds (usually of precocial species) form attachment to their mothers or some artificial substitute. Newborn chicks were reared singly with a red cardboard triangle, to which they rapidly imprinted and therefore treated it as a social partner. On day 3 of life, the chicks were presented with pairs of stimuli composed of either isolated fragments or occluded parts of the imprinting stimulus. Chicks consistently chose to associate with complete or with partly occluded versions of the imprinting stimulus rather than with separate fragments of it. Moreover, chicks reared with a partly occluded triangle chose to associate with a complete triangle rather than with a fragmented one, whereas chicks reared with a fragmented triangle chose to associate with a fragmented triangle and not with a complete one. Newly hatched chicks thus appear to behave as if they could experience amodal completion.

The results of Regolin and Vallortigara (1995) have been duplicated by Lea et al. (1996). These authors were interested in comparing the abilities of chicks with those of newborn human infants. Human newborns provide scientists with the same type of challenge offered by non-human species: we cannot simply ask them what they are perceiving, we must use some tricks to obtain such an answer. The trick used by developmental scientists is a procedure called habituation/dishabituation (e.g., see Kellman and Spelke, 1983; Spelke, 1998a, 1998b). Infants were habituated to a rod, which moved back and forth behind a central occluder, so that only the top and bottom ends of the rod were visible. After habituation had occurred, babies were shown either one of two stimuli without the occluder, one being a complete rod, the other being the top and bottom parts of the rod, with a gap where the occluder had been. Dishabituation (as measured by looking times) when viewing the complete rod would indicate that this display was novel and that the infants had not seen a complete rod during the habituation trials, whereas dishabituation (surprise) for the rod pieces was taken to indicate that the infants had been perceiving object unity.

It took about 4–7 months, depending on details of procedure (i.e., use of moving or stationary stimuli), for the human infants to show evidence of completion of partly occluded objects (Kellman and Arterberry, 1998). Chicks, in contrast, did that soon after hatching. In the Lea et al. (1996) experiments, newly hatched chicks imprinted on a complete rod preferred at test the complete rod to the fragmented rod, and chicks imprinted on the fragmented rod preferred the fragmented to the complete one; in the crucial condition, however, chicks imprinted on the partly occluded rod preferred the complete rod over the fragmented rod.

The difference between the two species in the developmental time-course is probably not surprising, however. Recognition of a partly occluded mother would be useful when an organism can move by itself to rejoin the mother and thus reinstate social contact; this is the case for the highly precocial young chick, but not for an highly altricial species such as the human newborn. Development of recognition of partly occluded objects in our species can probably be postponed, allowing the nervous system extra time for neural development.

Research with other species of birds has been conducted within the framework of Piagetian studies of object permanence (discussed in the fourth section). Standardized tests of object permanence included, for the very early stages of development of this concept, tasks in which the subject had to respond to partly occluded objects. Psittacine birds, such as parrots and parakeets (Funk, 1996; Pepperberg and Funk, 1990), mynahs (Plowright et al., 1998) and magpies (Pollok et al., 2000) pass these tests easily (as well as much more advanced stages of object permanence; discussed later). Interestingly, pigeons in contrast lose interest in food when it becomes invisible behind a screen (Plowright et al., 1998).

Could it be that the difference between chicks and pigeons resides in the use of a more ecological procedure (filial imprinting) and/or to age differences? The latter is doubtful, for evidence has been obtained for completion in adult hens using conditioning procedures. Forkman (1998) trained hens to peck at a touchscreen. On the touchscreen there was a grid providing pictorial depth cues and two stimuli, a square, and a circle. The hens were rewarded for pecking at the stimulus that was highest up on the grid irrespective of its shape, that is, at the stimulus that to a human observer appeared to be as the furthest. During probe trials, the hens were presented with either the circle overlapping the square or vice versa at the same height on the touchscreen; the hens pecked more at the stimulus that was occluded; that is, in the absence of any other cues they used occlusion to determine which of the two stimuli was the furthest.

Deruelle et al. (2000) on the basis of the pattern of failures/successes of baboons in performing amodal completion, suggested that the different outcomes could depend on the mode of stimulus presentation. With baboons, use of cardboard forms has been crucial for revealing amodal completion (discussed earlier). Cardboard stimuli were used in the experiment carried out by Regolin and Vallortigara (1995). However, in the experiments of Lea et al. (1996) and Forkman (1998), presentation using a video monitor was used, as in the case of most of the experiments with pigeons (see Fujita, 2001). Screen flicker may be perceptible by pigeons (flicker fusion: 140 Hz; for a review see D'Eath, 1998). Recent work with chickens suggests that at five luminance levels (10, 100, 200, 500, and 1000 cd/m^2), the overall chicken flicker sensitivity was considerably lower than for humans, except at high frequencies; critical frequency fusion values were however either similar or slightly higher for chickens compared to humans (40.8, 50.4, 53.3, 58.2 versus 39.2, 54.0, 57.4, and 71.5 Hz, respectively for humans and chickens) for increasing stimulus luminance level (see Jarvis et al., 2002). Deruelle et al. (2000) however, suggested that amodal completion might be experienced even with digitized images provided extra depth cues are added to the display. Indeed, in the Forkman (1998) experiments, there was a background grid depicting a linear perspective and in the Lea et al.'s (1996) experiments moving stimuli were used; both can provide the necessary depth cues for processing the images as three-dimensional (3-D) objects.

It could be that work using real objects will prove eventually that pigeons do perceive amodal completion, but I am quite inclined to believe that these experiments reveal some real difference between species, though it is probably not an all-or-one difference, but a more fuzzy one. My point will become clearer when considering a phenomenon that is strictly related to amodal completion, that is, perception of subjective contours.

Contours represent an essential source of information in visual perception. Most objects can in fact be recognized by their silhouettes alone. However, in natural scenes, in which objects frequently occlude one another, contours may vanish. Clearly, the visual system would fail miserably if it reflected only information directly present in the retinal images. Contour perception requires active organization of visual information to achieve a perceptually coherent representation of shape. The ability to perceive subjective (or illusory) contours (see Figure 2) that lack a physical counterpart is one of the most striking illustrations of the ability of the human visual system to interpolate visual information, to a level well beyond the available sensory information.

It has been suggested that a single unit-formation process underlies "modal" (i.e., "real" and "subjective" contours) and "amodal" completion (Shipley and Kellman, 1992). In Figure 2, you can appreciate the perception of the subjective triangle associated with the impression that the interrupted circles continue, and are completed behind the illusory triangle. If amodal completion and subjective contours are manifestations of the same basic process, then we can predict that those species that perceive subjective contours should also manifest completion of partly occluded objects and vice versa.

Evidence indicates that primates (macaque monkeys), as well as at least one other mammalian species (i.e., the cat; see Bravo et al., 1988), do perceive subjective contours (Peterhans and von der Heydt, 1989; von der Heydt and Peterhans, 1989; von der Heydt et al., 1984). Neuron cells in the striate cortex of macaque monkeys that selectively respond to subjective contours have been found (Peterhans and von der Heydt, 1989; von der Heydt and Peterhans, 1989; von der Heydt et al., 1984). Turning to birds, young chickens (about 2 weeks old) have been proved to be able to perceive subjective contours (Zanforlin, 1981), which would agree with evidence for completion of partly occluded objects in this species. Barn owls have also been shown to perceive subjective contours defined by both grating gaps and phase-shifted abutting gratins (Nieder and Wagner, 1999).

Figure 2. An example of subjective contours: the Kanizsa triangle (see Kanizsa, 1979). Several species of birds and non-human primates have been shown to perceive subjective contours (see text).

The work with the barn owl is particularly interesting because it allows a comparison with the electrophysiological data obtained in primates. The primary visual forebrain structures of owls and primates, the visual Wulst and the striate cortex, respectively, though evolved independently, have been considered equivalent in that they share important physiological properties, such as neurons with small and retinotopically organized receptive fields, binocular interaction, and selectivity for orientation and movement direction (Pettigrew and Konishi, 1976). Nieder and Wagner (1999) reported that single neuron recordings from the visual Wulst of awake, behaving owls revealed a high proportion of neurons signaling subjective contours, independent of local stimulus attributes. There was an average proportion of 59 percent of neurons sensitive to subjective contours in the hyperstriatum accessorium of the owl's visual Wulst, quite consistent with the 60 percent found in area V2 of macaque monkeys. Interestingly, evidence of perception of subjective contours and neural cells performing interpolation operations have been found even in insects (Horridge et al., 1992, van Hateren et al., 1990), suggesting that the ability to perceive subjective contours may have evolved primarily to detect partly occluded or somewhat degraded images of objects in a variety of animal species with basically similar (though independently evolved) mechanisms.

What about pigeons, however? In a study on subjective contours with Kanizsa's triangles and squares, Prior and Güntürkün (1999) were able to demonstrate that some of the pigeons that they tested reacted to the test stimuli as if they were seeing subjective contours. Four out of fourteen animals succeeded. It is interesting that only a minority of pigeons responded to subjective contours. As indicated by control tests, pigeons responding to subjective contours were attending to the "global" pattern of the stimuli, whereas pigeons not responding to subjective contours were attending to extracted elements of the stimuli. Perception of subjective contours is closely linked to amodal completion. In natural situations, in which objects occlude one another, boundaries may vanish and interpolation mechanisms to reconstruct contours absent from retinal images are sometimes needed. The fact that only pigeons attending to the more "global" aspects of the stimulation respond to subjective contours suggests that such individual variability in attending "globally" or "locally" to visual scenes can explain why pigeons fail in amodal completion tests which are effective in other species (see Sekuler et al., 1996). It is as if for pigeons a "featural" style of analysis would be more natural than a "global" one, although pigeons apparently can, with some effort, switch to such a global style of analysis. Recent research in visual categorization learning is in accordance with this quite focused attending to and remembering of details in visual scenes (Huber, 2001). So, I would guess that probably pigeons can respond to amodal completion, but only if strongly forced to do so. It is important to stress that the possibility for such a switch is inherent even to our own visual perception. We can, with some effort, turn to a featural, mosaic-like perception of visual scene, in which we look at fragments of partly occluded objects without completing them (visual artists, because of training and perhaps natural inclination, do this routinely). There is also evidence from perceptual psychology that such a mosaic stage normally occurs during very early stages of visual processing (see Sekuler and Palmer, 1992) and in human newborns before 4–7 months of age (above). The crucial point, however, is what functions, if any, could be subserved by the two strategies of perceptual analysis and what mechanisms could underlie the major emphasis on one or another strategy.

From the functional point of view, Fujita (2001) stressed that pigeons are grain eaters; grain is a type of food, which is usually abundant and does not require the animal search behind obstacles. Fowls, in contrast, appreciate eating worms and insects that often hide under leaves or soil, becoming only partly visible. Thus, there could be ecological differences favoring perception based on response to parts or on reconstruction of the whole objects on the basis of their parts. We are not aware of similar differences among primates, but the issue is worth testing.

As to mechanisms, it is interesting to observe that in experiments using conditioning procedures, such as those performed with pigeons, the stimuli fall into the frontal binocular visual field of the animals, a portion of the visual field which is mainly represented within the tectofugal pathway in pigeons (Güntürkün and Hahmann, 1999; Hellmann and Güntürkün, 1999). Unlike the condition in pigeons (Hodos et al., 1984), lesions to the thalamofugal visual system affect tasks that rely on frontal viewing in chicks to a marked degree (Deng and Rogers, 1997, 1998a, 1998b). This suggests that, different from pigeons, the frontal field is represented within the thalamofugal system in chicks (as also in the barn owl and other birds that are, dissimilar from chicks, frontally-eyed). These differences could be also associated with brain asymmetry (see the Chapter 9 by Rogers, this volume). Research using temporary occlusion of one eye, which takes advantage of complete decussation of optic nerve fibers and of large segregation of function between the two hemispheres in the avian brain (see Andrew, 1991; Rogers, 1995; Vallortigara et al., 1999, 2001) has revealed that the right eye (largely sending input to the left hemisphere) is dominant in pigeons' visual discrimination learning (Güntürkün, 1997) and presumably favors a featural strategy of analysis of visual scene. Chicks, in contrast, have shown a more balanced and complementary use of the two eyes, with the left eye (and right hemisphere) being dominant when more global strategies of analysis are needed (such as in spatial analyses, see Vallortigara [2000] for a review). Although the pattern emerging in monocular left and right stimulation in the study by Prior and Güntürkün on perception of subjective contours was similar, suggesting that in pigeons that responded to subjective contours both the left and the right hemispheres were capable of "filling in" processes, it could be that a basic difference between the hemispheres still exists, but is revealed by the difference between pigeons that respond and pigeons that do not respond to subjective contours. For instance, it could be that dominance by one or other hemisphere favors a "global" or "local" strategy of analysis of visual stimuli. We are currently testing these ideas in chicks, using the imprinting paradigm applied to recognition of partly occluded objects in monocularly tested birds. Preliminary results (Regolin et al., 2001) suggest that behavioral responses based on amodal completion seem to be more frequent in chicks using their left rather than their right eye (binocular chicks tend to be more similar to left-eyed chicks, suggesting that in this species, the right hemisphere is dominant in this task). Interestingly, even in primates (humans), the right hemisphere seems to play a more important part in amodal completion (Corballis et al., 1999).

Before moving on, I would like to mention another basic computational problem in perceiving occlusion, which deals with establishing the direction of depth stratification, that is, which surface is in front and which is behind. Usually, when two objects differ in color, brightness, or texture, occlusion indeterminacy can be solved using T-junctions (see Cavanagh, 1987), that is, by determining, on the basis of contour collinearity, what boundaries belong to each other and thereby allowing the formation of modal (occluding) and amodal (occluded) contours (see Kanizsa, 1979; Michotte, 1963). However, humans can perceive unconnected and depth-stratificated surfaces even in chromatically homogenous patterns, with only L-junctions and no T-junctions at all. Although it would be possible, in principle, to perceive a unitary object, in Figure 3, the hen appears as being behind the fence when the region of the legs is inspected (because of the differences in color that specify the direction of occlusion), but it appears to be in front of the fence when the region of the upper part of the body is inspected!

Figure 3. The hen appears as if it is standing in front of the fence in the region of the upper body but, if one inspects the region of the legs, it appears to be standing behind the fence. Domestic hens experience this sort of illusions (see text and see Figure 4 for an illustration of the principle underlying the visual illusion).

Why is it that the upper body of the hen appears to most observers (and/or for most of the time) to be in front of the fence rather than the other way round? The reason why larger surfaces (such as this) tend to be seen modally as being in front, rather than behind, might depend on the geometrical property that overlapping objects, in which larger surfaces are closer, present shorter occluding boundaries than when smaller surfaces are closer (see Figure 4). Shorter modal (occluding) contours are needed to account for the occlusive effect of the hen on the fence, whereas larger modal contours would be needed to account for the occlusive effect of the fence on the upper body of the hen. This "rule", first described by Petter (1956), according to which the visual system tends to minimize the formation of interpolated modal contours, has been largely confirmed in studies of human visual perception (Shipley and Kellman, 1992; Singh et al., 1999; Tommasi et al., 1995). It has been shown that the rule is independent of the empirical depth cue of relative size (Tommasi et al., 1995) and can be made to play against information based on other depth cues thus generating intriguing visual paradoxes such as the hen/fence illusion (see Kanizsa [1979] for further examples).

Is Petter's rule a geometric regularity that has been incorporated in the design of all vertebrate brains or is it limited only to the human visual machinery? Apart from humans I am not aware of any study involving other primates species. The use of the hen in Figure 4 is not incidental, however, for domestic hens have proved to behave in visual discrimination tests as if they would experience the same phenomenon. Forkman and Vallortigara (1999) presented domestic hens with two chromatically identical patterns, a diamond

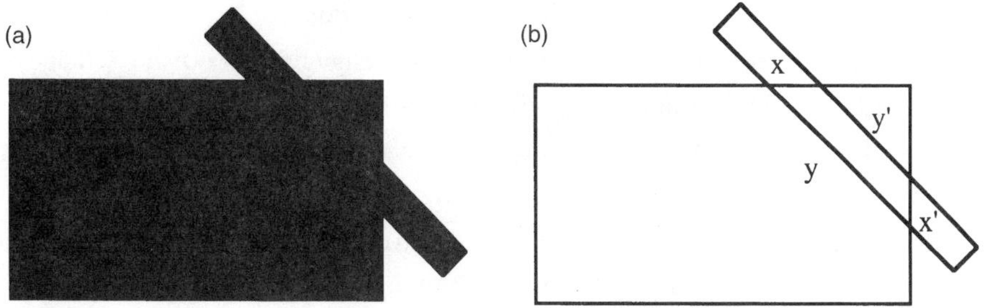

Figure 4. An illustration of the Petter's principle. Two overlapping, chromatically homogeneous figures. (a) The oblique bar is seen as continuing behind the rectangle. (b) This outcome can be predicted on the basis of a minimization of occluding subjective contours: perceiving the bar as in front would require longer occluding contours (y, y′) than perceiving the rectangle as in front (x, x′).

and a ladder, shown on a computer touchscreen. Hens were reinforced for pecking at the pattern that was highest up on a grid that provided pictorial depth information, that is, on the pattern that to a human appears as being the furthest away. Every tenth trial was a non-rewarded probe trial with the two patterns partially overlapping. In the absence of other cues, depth stratification can occur on the basis of a minimization of interpolated occluding contours. In humans, the diamond is usually perceived to be in front of the ladder because shorter interpolated contours are needed to account for the occlusive effect of the diamond on the ladder. The hens pecked more often at the ladder during the probe trial. These findings suggest that there are important constraints which are related to the geometrical and physical properties of the world that must be incorporated in the design of any efficient biological visual system, either in the primate (human, at least) or avian (hen) brain (see also Vallortigara and Tommasi, 2002).

INTEGRATION AND INTERPOLATION OF VISUAL INFORMATION IN THE TEMPORAL DOMAIN

Many adaptive responses to objects in the natural environment depend on the way these objects are moving, and this is particularly true for the movement of biological creatures. When a vertebrate moves about its environment, in fact, its limbs and torso undergo a pattern of movement in typical synchrony. Johansson (1973) has shown that in humans, an animation sequence consisting of just a few points of light strategically placed on the articulations is enough to create an immediate, automatic perception of someone engaged in coordinated activity such as walking. Discrimination in these stimuli must depend on motion cues alone, and in humans they sustain a range of discriminations including the gender of the person (Kozlowski and Cutting, 1977), the identity of a friend (Cutting and Kozlowski, 1977), the nature of the action (Johansson, 1973), and the weight of an object being lifted (Runeson and Frykholm, 1981).

Little is known about perception of complex moving patterns in other species. Among mammals, Blake (1993) has proved that cats can be trained to discriminate a point-light animation sequence depicting biological motion (a cat walking) from an equivalent animation in which local motion vectors were variously "scrambled." The ability to discriminate the so-called "structure-from-motion"—of which biological motion would represent an instance (discussed later)—has been documented in macaque monkeys (Siegel and Andersen, 1988). Cells in area TE, located in the monkeys' superior temporal sulcus respond to biological motion, for example, to a walking human figure covered with point lights (Perrett et al., 1990). Even more interesting, it has been reported that some cells in the upper bank of the superior temporal sulcus (areas TPO and PGa) in monkeys are selectively responsive to the perspective view of the human form (facing left or right) as well as its direction of movement (Perrett et al., 1990). This selectivity is quite impressive because the two perspective views of the human form are hardly distinguishable in terms of their parts: they differ only at a global level.

Among birds, Emmerton (1986) has demonstrated that pigeons can discriminate complex motion patterns (Lissajous figures), being able to differentiate two successively presented cyclic trajectories of a single moving dot. This suggests that pigeons are capable of temporally integrating a path of motion as a discriminative cue per se. Discrimination of point-light animation sequences can be obtained in newborn chicks using an imprinting procedure (Regolin et al., 2000). Day-old chicks were exposed to point-light animation sequences depicting either a walking hen or a rotating cylinder. On a subsequent free-choice test, the chicks approached the novel stimulus, irrespective of this being the hen or the cylinder (choice for slight novelty is not infrequent in imprinting experiments, also being dependent on the sex of the animals, see Vallortigara [1992], Vallortigara and Andrew [1991], and see below). In order to obtain equivalent local motion vectors, in another experiment, newly hatched chicks were exposed either to a point-light animation sequence depicting a walking hen, or to a positionally scrambled walking hen (i.e., an animation in which set of dots moved in the way employed by a walking hen, but with spatially randomized starting positions; see Figure 5). Chicks proved to be able to discriminate the two animation sequences, males preferentially approaching the novel stimulus and females the familiar one. These results indicate that discrimination was not based on local motion vectors, but rather on the temporally integrated motion sequence. However, the data do not tell us what the birds perceived in these patterns, but simply that they could discriminate between them.

There is however some evidence that pigeons perceive structures similar to those conveyed by real (or videotaped) stimuli in animated point-light displays. Omori and Watanabe (1996) trained pigeons to discriminate between the motions of three dots pasted on a pigeon and those pasted on a toy dog: pigeons could learn the discrimination of Johansson stimuli and some of them could also transfer from the dot motion to the real movement. Similarly, Dittrich et al. (1998) found that pigeons trained to discriminate between categories of moving stimuli such as walking and pecking presented on videotape could show

Figure 5. Schematic representation of the point-light animation sequences used for assessing chicks' discrimination of biological motion. The dots were strategically placed on to eight specific positions of a hen's body and limbs (a sample of the sequence is shown at the top of the figure). Below are shown three frames of point-light sequences of, respectively, the moving hen and (at the bottom) the positionally scrambled hen. (Redrawn from Regolin et al., 2000.)

some transfer to new stimuli in which the same movements were represented by a small number of point lights; the reverse, however, transfer from a point-light's display to fully detailed displays, was not observed.

Perception of biological motion in point-light displays is one example of the more general problem of "structure-from-motion" perception, that is, of the ability of the primate visual system to extract 3-D shape information from two-dimensional (2-D) transformations in the image (Ullman, 1979; Wallach and O'Connell, 1953). When considering this ability, one is faced with the inherent mathematical ambiguity of a changing 2-D image: indeed, there are an infinite number of object transformations in 3-D space that are projectively equivalent to any given moving image on a 2-D surface. However, in most circumstances, the primate's visual system is little affected by this potential stimulus ambiguity. What about birds?

Cook and Katz (1999) trained pigeons to discriminate computer-generated projections of cube and pyramid objects. The authors found that pigeons showed evidence of recovering the structure of these objects from just the pattern of their motion and were sometimes better at discriminating when all contour and surface information was removed (i.e., using 2-D monochromatic colored blobs moving consistently with the rigid projective geometry of either a cube or pyramid).

Unfortunately, we currently lack information about perception of meaningful motions in birds, such as actions of conspecifics, of the sort that has been investigated in primates by Perrett and colleagues (discussed earlier). One step in this direction, however, is provided by the recent work of Frost and colleagues (e.g., Frost et al. [1998]; see also Shimizu [1998]), who showed that pigeons do show robust courtship responses not only to real partners but also to video images. This encouraged them to take this a step further and construct a virtual pigeon using high-level computer graphics tools that give complete control on the actions and appearance of the model bird. The authors reported that miniature telemetry units are constructed to transmit electrophysiological data from pigeons as they observe and respond to these well-controlled video and virtual conspecifics. No published data are available on this project yet.

REPRESENTING OBJECTS

In the previous sections we discussed the case in which visibility of objects is somewhat degraded by partial occlusion or auto-occlusion associated with motion in 3-D space. However, obviously objects frequently are entirely occluded by other objects, and thus completely unavailable to direct sensory experience. In this case, some sort of visual representation of objects is needed to guide purposeful behavior.

This problem has been traditionally investigated within the Piagetian framework of "object permanence." In humans, object permanence develops in stages according to Piaget (1953). During Stage 1, children do not search for an object they have seen disappear; then they track the object's movement (Stage 2); recover a partly occluded object (Stage 3); and finally, a fully occluded object (Stage 4). Later, children can retrieve an object that has been hidden successively in several locations (i.e., hidden, exposed, and rehidden several times, Stage 5) and, at the end, children master invisible displacements (i.e., an object is hidden in a container, the container is moved behind another occluding device, the object is transferred to this second device, the children are shown that the first container is empty, and the children successfully infer where the object now resides; Stage 6). Comparative studies have shown that, although species differences surely exist, there is no overall superiority of primates over birds in mastering the object permanence concept. For instance, great apes (Natale et al., 1986) and parrots (psittacids; Pepperberg and Kozak, 1986; Pepperberg et al., 1997) both reached Stage 6 object permanence, magpies provided ambiguous evidence (Pollok et al., 2000), whereas monkeys (Natale et al., 1986), ring doves (*Streptopelia risoria*; Dumas and Wilkie, 1995), and domestic chickens (Etienne, 1973) apparently did not master Stage 6. I believe, however, that extreme caution in interpreting these data is necessary. Beautiful evidence has been provided in recent years showing that human subjects ostensibly in Stage 1 can demonstrate some understanding that objects exist when occluded if the test paradigm does not require an active search for the hidden object (Baillargeon et al., 1985). Investigations using these techniques, based on the so-called "expectance-violation procedure," have been just started in primates (see Hauser, 2000) and in our lab we are currently trying to develop avian versions of these tasks. Moreover, research carried out on the earlier stages of object permanence (e.g., Stage 4) has shown that simple modifications of the behavioral tasks can produce dramatic changes in animals' performance. Domestic chicks can provide an interesting case in point. Regolin et al. (1995b) presented chicks of 2 and 6 days of age with a goal-object that was made to disappear behind one of two screens opposite each other. Chicks proved able to choose the correct screen when the goal-object was a social partner (i.e., a red ball on which they had been imprinted), whereas they searched at random behind either screen when the goal-object was a palatable prey (i.e., a mealworm). Chicks, however, appeared capable of making use of the directional cue provided by the movement of the mealworm when tested in the presence of a cagemate. These results suggest that previous failure to obtain detour behavior in the double screen test in the chick (e.g., Etienne, 1973) was not due to a cognitive limitation, but rather to the evocation of fear responses to the novel environment that interfered with the correct execution of the spatial task. Further work using detour behavior has shown that 2-day-old chicks master certain aspects of Stage 4 of object permanence (see also Regolin et al., 1994, 1995a; Vallortigara and Regolin, 2002), but not all aspects. For instance, although chicks do have an object concept that maintains a representation of the object in the absence of direct sensory cues, it seems that they are not able to predict the resting position of an imprinted ball from its direction of movement prior to occlusion (Freire and Nicol, 1997, 1999). It is not clear whether this reflects a basic cognitive limitation or an adaptation to ecological demands (for instance, when prey or other interesting objects hide themselves behind an occluder, it is more likely that they would reappear, after some time, in the same location where they were seen to disappear rather than at the other side of the occluder; e.g., see Haskell and Forkman, 1997).

One problem with classical Piagetian or other object-permanence tests is that they provide evidence that animals represent and maintain in memory "something," but little can be deduced as to the precise nature of the representation. Consider the case of food-storing birds (Clayton and Krebs, 1995). The ability to represent completely hidden objects (Stage 4) would appear rather obvious for these birds because successful food storing does not depend so much on the ability of the birds to remember places where they have been, but on the capacity to remember hidden food items at particular places (Clayton and Krebs, 1994). However, how do we know that these birds remember a specific food item (a "what") rather than simply a location in space (a "where")? One interesting procedure exploits the observation, made on primates and other mammals, that prior feeding on one type of food selectively reduces the value of that food (see Hetherington and Rolls [1996] and Rolls [1990] for reviews). For instance, Clayton and Dickinson (1999a) reported an experiment in which scrub jays (*Aphelocoma coerulescens*) cached peanuts and kibbles in two distinct caching trays. The relative incentive value of food was manipulated by prior feeding one of the foods immediately before cache recovery. It was found that scrub jays preferentially searched for food caches on which they had received no prior feeding (see also Clayton [1998], Clayton and Dickinson [1998, 1999b, 1999c]). This demonstrates that these birds can indeed remember the content of food caches apart from their position.

Recently, evidence that even non-"specialized" bird species can remember the content of food caches has been reported. Individual hens (Forkman, 2000) and 5-day-old chicks (Cozzutti and Vallortigara, 2001) were fed in an enclosure with two food plates, each with a new type of food. The food was devalued by pre-feeding with one of the food types. Both hens and young chicks were found to move to the location previously occupied by the non-devalued food. There were obviously methodological differences with respect to the experiments with the scrub jays, in that hens do not cache food (but were trained to recover previously encountered food) and they were given several training trials in order to learn which food plate contained which type of food. Scrub jays, in contrast, were tested on one-trial tests and they actually cached food by themselves (see Clayton and Dickinson, 1999a). Nonetheless, hens and young chicks appeared to be able to maintain a memory of the content of the food plates, a sort of "declarative-like" representation.

The notion that objects are separate entities that continue to exist when out of sight of the observer is relevant when considering that these "represented" objects serve mainly to guide the course of action of animals. When a prey has disappeared from sight, the predator can maintain for some time a representation of its continuing presence and thus actively search for it. But how long can the representation be maintained? The issue has been investigated early in this century, mainly in primates, using the so-called "delayed response problem," developed by Hunter (1913). In a typical test, the subject is watching the experimenter while the latter places a preferred food incentive under one of two identical, or different, objects. During a subsequent delay period, the animal has no physical access to the objects (in some studies they have visual access to the objects, whereas in others the objects are hidden from view). After the delay period, the animal is allowed to choose between the two objects; obtaining the food more frequently than expected by chance is assumed to indicate that the animal remembers where it was placed initially and chooses on the basis of that information (Wu et al., 1986). Using food as incentive, even primates (e.g., macaques) perform at chance after delays longer than 30–40 s (Fletcher [1965], Rumbaugh and Gill [1975], but see also

Chapter 1 for results with other primate species). Similar results have been reported for several other species of mammals, although results varied considerably depending on procedural variables (Fletcher, 1965; Tinklepaugh, 1928). Using social stimuli as the incentive, infant pigtailed macaque monkeys reliably chose the correct stimulus with delays up to 60 s (Wu et al., 1986).

Very little is known about delayed responding in avian species. Studies on object permanence have been performed in birds (discussed earlier), but without investigating the delayed response problem. Obviously, the so-called matching-to-sample task is derived from the delayed response problem and has been largely used in pigeon and other avian species but, typically, the delays used are very short (of the order of seconds or milliseconds). Using a 1.5-s delay interval, domestic hens seem to behave well and similar to pigeons in delayed matching-to-sample tasks (Foster et al., 1995). However, no data are available for longer delay periods.

Vallortigara et al. (1998) trained 5-day-old chicks to follow an imprinted object (a small red ball with which they had been reared) that was moving slowly in a large arena, until it disappeared behind an opaque screen. At test, each chick was initially confined in a transparent cage, from where it could see and track the ball while moving towards, and then behind, one of two screens. The screens could be either identical or differ in color and pattern. Immediately after the disappearance of the ball (or with a certain delay), the chick was released and allowed to search for its imprinted object behind either screen. Results showed that chicks could take into account the directional cue provided by the ball movement and its concealment, up to a delay period of about 180 s, independently from the perceptual characteristics of the two screens. If an opaque partition was positioned in front of the transparent cage immediately after the ball had disappeared so that, throughout the delay, neither the goal-object nor the two screens were visible, chicks were still capable of remembering and choosing the correct screen, though over a much shorter period of about 60 s. Similar effects of the visibility of the test tray during the delay phase have been documented in primates, and have been usually attributed to visual orientation, or to attentional factors associated with visual orientation (Fletcher, 1965).

Even when employing the opaque screen during the delay, the performance of the chicks remained nonetheless impressive. One minute is a very long period of delay, fully comparable in length with the retention intervals observed in primate species under similar testing conditions (Fletcher, 1965; Wu et al., 1986). Obviously, it can be claimed that chicks might rely very little on cognitive and representative skills in this task. They could have simply learnt to associate proximity of the ball to a screen as a cue to direct approach responses toward that screen. Nevertheless, particularly in the condition in which the screens are not visible, in order to solve the problem, the chicks needed to maintain some form of representation of the position of the correct screen, and to continuously update the content of the representation from trial to trial on the basis of the directional cues provided by the movement of the ball. In mammals, the maintenance of information "on-line" during short temporal intervals is usually described as "working memory," and is believed to be implemented into some neural circuitry within the prefrontal cortex (Fuster, 1989; Goldman-Rakic, 1987). Within the primate prefrontal cortex there are neurons that display sustained elevated firing while holding active an internal representation of the relevant stimulus during its physical absence (Funahashi et al., 1989). An involvement of the prefrontal cortex in Piagetian object-permanence tasks has been suggested for both humans and monkeys (Diamond and Goldman-Rakic [1989]; see also Johnson [1997] for a review). There is evidence that a region resembling the mammalian prefrontal cortex does exist in the avian telencephalon—a semilunar area in the caudalmost part of the forebrain, called "neostriatum caudolaterale" (NCL) (see Mogensen and Divac, 1982). In pigeons, it has been shown that temporary blocking of D1 receptors (which are the prevailing dopamine receptor subtype in the mammalian prefrontal cortex) in the NCL strongly affects working memory while leaving reference memory unimpaired (Güntürkün and Durstewitz, 1998). Moreover, neurons in the NCL have been found that respond selectively during the delay period of a working memory task and show activity patterns identical to those described for the delay cells of the primate prefrontal cortex (Kalt et al., 1999). Thus, although the anatomical morphological structure of the avian NCL is very different from the primate neocortical architecture of the prefrontal cortex, the neuronal mechanisms that have been evolved to master analogous cognitive demands seem to be very similar.

OBJECTS IN SPACE: USE OF GEOMETRIC AND NONGEOMETRIC INFORMATION

Animals can make use of stored information about their environment to go back to places already visited for the most various purposes, such as mating, feeding, nesting, and so on. Research has pointed out that these abilities are much more developed in those species (e.g., food-storing birds) that need to deal with large collection and intentional hiding of large amounts of food items (see Sherry et al., 1992 for a review). The remarkable spatial learning abilities of food-storing birds will be covered in another chapter of this book (see Chapter 1 by Emery and Clayton). However, spatial learning seems to be a rather general ability, possessed by most species of animals, and related to the obvious necessity of an organism to orient itself in its own environment. In fact, spatial memory was demonstrated in primates and birds that do not show the amazing abilities at retrieving huge amounts of hidden items such as food-storing birds, but that nonetheless have to deal with their environment in a way that requires at least some basic form of spatial cognition (Thinus-Blanc, 1996 for a review). Direct comparisons in similar tasks, however, have rarely been performed. One interesting exception is the so-called test on the "geometric sense of space" (Cheng, 1986; Gallistel, 1990).

When disoriented in an environment with a distinctive geometry—such as a rectangular-shaped arena (Figure 6a)—animals can (partly) reorient themselves even in the absence of any extra-arena cues by simply using the geometry of the environment. Suppose the target (say food) is located at corner A and then made to disappear. Apparently, following passive disorientation, animals should choose at random between the four corners. But as a matter of fact, a partial disambiguation of the problem is possible: corner A (the food location) stands in the same geometric relations to the shape of the environment as corner C. Geometric information alone, therefore, cannot specify unambiguously between corners A and C, which are geometrically equivalent, but is sufficient to distinguish between corners AC and corners BD.

Several species of primates and birds have been shown to be able to reorient using this "purely geometric" information (e.g., Gouteux et al., 2001; Kelly et al., 1998; Vallortigara et al., 1990). Interestingly, however, it has been reported that human infants (Hermer and Spelke, 1994) and adult rats (Cheng, 1986) failed to reorient by nongeometric information, such as a distinctive differently colored wall in the rectangular cage, in spite of the fact that this featural information would have allowed fully successful reorientation (Figure 6b). Given that rats are able to use nongeometric information for solving spatial tasks that do not involve spatial disorientation (e.g., Morris, 1981; Suzuki et al., 1980), these findings have been interpreted to suggest that spatial reorientation depends on an encapsulated, task-specific mechanism, a "geometric module" (Cheng and Gallistel, 1984; Cheng, 1986; see also Fodor, 1983). The module

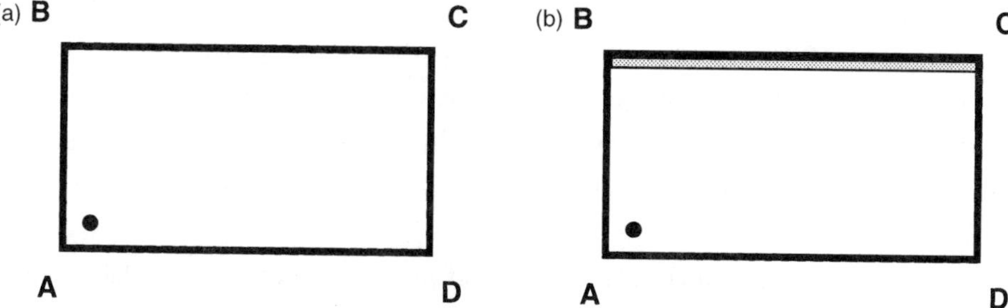

Figure 6. Schematic representation of the test apparatus used to investigate geometric representations in primates and birds. The animal could find food (or other desired targets) in the rectangular enclosure at, say, corner A. (a) In one version of the test, all walls were white and the task for the animal is to distinguish between corners A, C and corners B, D using purely geometric information (corner A and its rotational equivalent, corner C, are in fact indistinguishable on the basis of purely geometric information, but can be distinguished from corners B and D, which, in turn, are geometrically equivalent and cannot be distinguished from each other). (b) In another version of the task, one wall (indicated by the fine line) was made of a different color (with some species, panels with different features and colors were positioned in the four corners instead of using a colored wall as nongeometric information). The animal in this case could disambiguate between the two geometrically equivalent corners A and C using the nongeometric information provided by the colored wall (or by the panels). Chicks, pigeons, and rhesus monkeys proved capable of conjoining geometric and nongeometric information to reorient themselves (see text).

would encode only the geometric properties in the arrangement of surfaces as surfaces: in the case of the spatial reorientation task in the rectangular environment, for instance, the geometric module would use only "metric properties" (i.e., distinction between a long and a short wall) and what is known in geometry as "sense" (i.e., distinction between right and left). Use of geometric information for spatial reorientation makes sense ecologically. The large-scale shape of the landscape does not change across seasons, whereas there are important seasonal changes in the nongeometric properties of the landscape (e.g., appearance of grass and vegetation, snowfall and melting, and so on; see also Cheng and Gallistel, 1984).

Human adults, in contrast to young children and rats, readily solved the blue-wall version of the reorientation task in the rectangular environment (i.e., when both geometric and nongeometric information are available; see Hermer and Spelke, 1994), suggesting that the most striking limitations of the geometric module are overcome during human development. Hermer and Spelke (1994, 1996) also went on with a more specific and strong hypothesis, namely that the performance of human adults, when compared with that of rats and human infants, would suggest that some representational systems become more accessible and flexible over development and evolution.

It has been suggested that language, and more specifically spatial language, may provide the medium for representing conjunctions of geometric and nongeometric properties of the environment (Hermer-Vasquez et al., 1999). Indeed, the ability to correctly orient in the blue-wall task (Hermer and Spelke, 1994) correlated with the ability of children to produce and use phrases involving "left" and "right" when describing the locations of hidden objects (MacWhinney, 1991). The developmental time course of the ability to conjoin geometric and nongeometric information thus suggests that language acquired by children (starting at 2–3 years of age) would allow them to perform as well as adults (at about 5–7 years of age).

Gouteux et al. (2001) recently demonstrated, however, that rhesus monkeys also combine geometric and nongeometric information; these authors thus proposed a less stronger version of the original Hermer and Spelke's claim according to which joint use of geometric and nongeometric information, though not strictly dependent on language, nonetheless would become accessible only to advanced mammalian species. However, even this hypothesis does not appear to be correct. It has been proved that young chickens (Vallortigara et al., 1990) and pigeons (Kelly et al., 1998), which are able to reorient themselves using "purely geometric" information (discussed earlier), can also easily combine geometric and nongeometric information. The performance of birds in these tasks is identical to that of rhesus monkeys (and human adults) and clearly surpasses that of rats or human infants. (More recent work has demonstrated that even fish reorient themselves by conjoining geometric and nongeometric information in the rectangular arena-task; see Sovrano et al., 2002, 2003.)

Work carried out in my laboratory has aimed at exploring further the geometric properties of avian spatial cognition, using the domestic chicks. We found (Tommasi et al., 1997) that chicks can learn to localize the central position of a closed environment in the absence of any external cues. After some days of training, during which food-deprived chicks were allowed to eat food that was progressively buried under sawdust in the center of the floor of an arena, they developed a ground-scratching strategy to uncover the food in order to eat it. With training, chicks became more and more accurate in finding food so that, when they were eventually tested in the absence of any food, their pattern of ground-scratching concentrated in a very limited central area. We also showed that chicks were able to generalize among arenas of different shapes. For instance, when trained to find the center in a square-shaped arena and then tested in a triangular- or circular-shaped one of nearly the same size, chicks searched in the central region of the novel arena (Figure 7).

Previously, similar studies on spatial localization in animals provided varying results using expansion tests. Cartwright and Collett (1982) successfully trained bees to find a goal at a particular direction and distance from an array of three identical landmarks. When the landmarks were spread farther apart ("array expansion" test), the bees searched farther from the landmark array, whereas when the landmarks were moved closer ("array contraction" test), the bees searched closer to the landmark array. Gerbils trained to locate a virtual point relative to two tall, narrow cylinders in an otherwise featureless environment behaved quite differently (Collett et al., 1986): when tested with the landmarks moved further apart, the gerbils showed two peaks of searching, one at each location defined by the training vector from an individual landmark. Using a

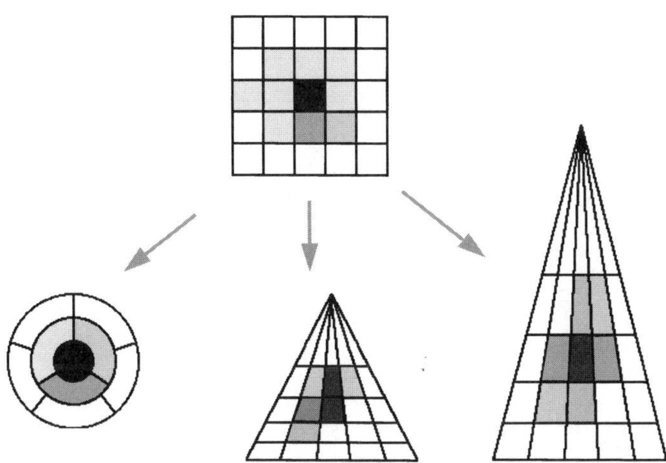

Figure 7. Chicks were trained within the square-shaped arena and then tested in the other arenas. The darker a cell, the more searching behavior the chicks showed in that location. Chicks showed quite precise searching behavior during training and good transfer to all testing arenas of similar size and different shapes. (Redrawn from Tommasi et al., 1997.)

touch-screen procedure, Spetch et al. (1996) compared pigeons and humans' search for an unmarked goal located relative to an array of landmarks. Humans searched in the middle of expanded arrays, whereas pigeons preserved the distance and direction relatively to a single landmark (see also Spetch et al., 1997).

When the environmental change involved a substantial modification in the size of the arena, as was the case for the transition from a square-shaped arena to an arena of the same shape but of a larger size, the scratching bouts of chicks in the test (larger) arena were localized in two regions: in the actual center of the test arena and (in part) at a distance from the walls that was equal to the distance from the walls to the center in the training (smaller) arena (Figure 8). Similarly, when the transfer was from a square-shaped arena to a rectangular-shaped one (the latter obtained by doubling one side of the square-shaped arena), their searching behavior was concentrated both in the center of the rectangular arena

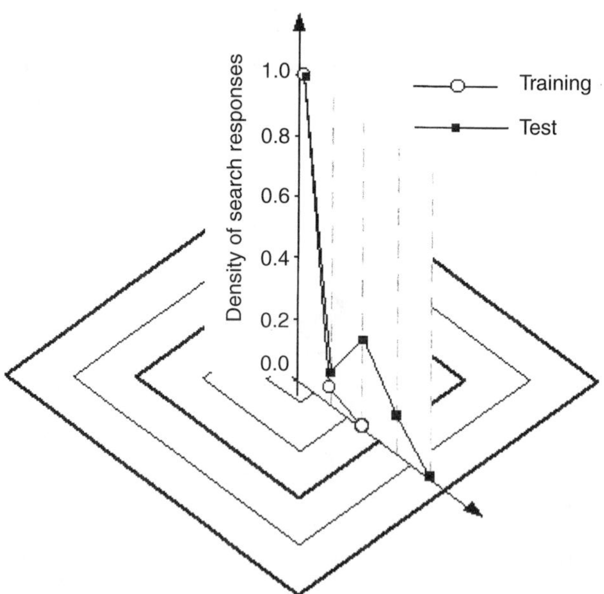

Figure 8. Density of searching behavior as a function of the distance from the center when chicks were first trained to find the center in a small square-shaped arena (inner square in bold) and then tested with a larger arena of the same shape (outer square in bold). As can be seen, during testing, chicks showed one peak located in the center of the larger arena and one peak located at a distance from the walls corresponding to the previously learned distance from the center in the training (smaller) arena. (Redrawn from Tommasi and Vallortigara, 2000.)

and in the two centers of each "component" square-shaped portions. Apparently, the two behavioral strategies which were, in principle, available to the chicks were simultaneously displayed, that is, encoding a goal location in terms of absolute distance and direction to the walls, and encoding a goal location in terms of ratios of distances (whatever their absolute values) from the walls. Interestingly, tests carried out under monocular viewing (after binocular training) revealed striking asymmetries of brain function: encoding of absolute distance being predominantly attended to by the left hemisphere and encoding of relative distance being predominantly attended to by the right hemisphere (Tommasi and Vallortigara, 2001; see Chapter 9 of Rogers, this volume, for more evidence concerning cerebral lateralization in birds and primates).

When training was performed in the presence of a conspicuous landmark (a red cylinder) located at the center of the arena, animals searched at the central location even after the removal of the landmark. Apparently, domestic chicks seem to be able to use the geometrical relationships between the walls of the arena as well, though they were not explicitly trained to do so. Marked changes in the height of the walls of the arena produced some displacement in the spatial location of searching behavior, suggesting that chicks used the angular size of the walls to estimate distances within the arena. These results provide evidence that chicks are able to encode information on the absolute and relative distance of the food from the walls of the arena, and that they encode this large-scale spatial information even when orientation by a single landmark alone would suffice for food localization.

The center-localization task studied in chicks is clearly an example of a "perceptual" concept of center, in the sense that it is based on use of visual information to estimate distances. Chicks can learn (quite spontaneously) that food is located in a point in space that minimizes (as much as possible) distances from surfaces of the enclosure, irrespective of the dimensions of the enclosure. Recently, however, evidence that birds can manage even a more abstract notion of "center" has been collected. It has been reported that Clark's nutcrackers can learn to find the point halfway between two landmarks that vary in the distance between them (Kamil and Jones, 1997, 1999). It seems that the birds learned a general rule in this task, in order to find the halfway point correctly when the landmarks were presented with new distances between them (see Biegler et al., [1998] for a discussion of possible mechanisms). Again, no obvious general differences between primates and birds could be observed in the use of relational rules, as against absolute information. When trained to find a hidden goal in the middle of an array of identical landmarks, adult humans learn an abstract "middle rule," as revealed by expansion tests in which the landmarks are spread farther apart (Spetch et al., 1996, 1997). In contrast, pigeons (Spetch et al., 1996, 1997) trained with similar landmark arrangements did not learn relational rules but instead responded to absolute spatial relationships between landmarks and goal. Interestingly, squirrel monkeys also failed to learn a relative middle rule in similar tasks (Sutton et al., 2000). On the other hand, we have seen (discussed earlier) that chicks and Clark's nutcrackers can learn abstract spatial relational rules. Task requirements associated with the use of local landmarks or global large-scale spatial information are likely to be more important than species differences, however. For instance, with the rectangular environment used to test the geometric sense of space (discussed earlier), pigeons appear to be able to encode relative geometry, as revealed by their good performance when tested in environments that preserved the relative geometry but altered the absolute geometry (Kelly and Spetch, 2001).

CONCLUSIONS

This rather brief survey of some crucial topics in object perception and cognition in birds and primates has probably revealed more similarities than differences. Differences, when they emerge, seem to be the result of the specialized structure and functions of the avian visual system. There are hints, for instance (discussed in the second section), that some species of birds, like the pigeon, could be more prone to use a local, rather than a global, style of analysis of visual scene. This could be expected from the fact that pigeons have two specialized areas, or foveas, in their eyes, which may serve different functions (cf. Güntürkün, 1996). The frontal visual field seems to be specialized for (myopic) foraging for food on the ground, whereas the lateral visual field seems to be specialized for predator detection and flight control. Near-sighted acuity would favor examination of fine stimulus details and may be responsible for the local advantages observed in most experiments that used frontal presentations of visual stimuli; the lateral visual fields, in contrast, may be more concerned with the larger scale integration of scene and flight control information (Martinoya et al., 1984), thus showing more sensitivity to global information (see also Cook, 2001). Quantitative differences in the temporal integration and interpolation of moving stimuli are also evident between birds and primates (discussed in the third section). For instance, Bishof et al. (1999) reported that the coherence and velocity threshold, as measured through discrimination of coherent from random motion in dynamic random dot displays, is substantially higher for pigeons than for humans. Nonetheless, such differences seem to be more a matter of degree than of kind and could vary considerably between different species within each class.

Other aspects of visual cognition, such as representation of objects that have disappeared from sight, delayed responses, working memories, and geometric spatial cognition, though showing differences within each of the two taxonomic groups, do not reveal any basic general difference between birds and primates. To a certain extent, independent evolution of basically similar computational neural mechanisms should be expected when organisms face basically similar problems (e.g., the parallel between the primate prefrontal cortex and the avian NCL discussed in the fourth section). Birds, in fact, have evolved independently from mammals a series of characteristics such as bipedalism, parental care, omeothermism, vocal communication, and so on.

Obviously, the absence of proof is not proof of absence. I am quite sure that the more we progress in the comparative analyses of cognition, the more we shall find differences in the relative efficiency of different cognitive mechanisms as a result of specific ecological adaptations (see also Rogers, 1997). However, I doubt that these differences will reveal an overall superiority (or inferiority) of birds with respect to primates. Rather I expect that, as a result of specific adaptations, certain species of birds and certain species of primates might reveal, in specific domains and for specific tasks, quite specific capacities, since every species is special in its own way.

ACKNOWLEDGMENTS

I thank Lucia Regolin and Lesley J. Rogers for reading and commenting on the manuscript.

REFERENCES

Andrew, R. J. 1991, The chick in experiment: Techniques and tests. General, in: *Neural and Behavioural Plasticity* R. J. Andrew, ed., Oxford University Press, Oxford, pp. 6–11.

Baillargeon, R., Spelke, E. S., and Wasserman, S. 1985, Object permanence in five-month-old infants, *Cognition* **20**:191–208.

Blake, R. 1993, Cats perceive biological motion, *Psychol. Sci.* **4**:54–57.

Biegler, R., McGregor, A., and Healy, S. D. 1998, How do animals "do" geometry? Animal Behaviour, F4-F8 (online forum: http://www.academicpress.com/anbehav/forum).

Bishof, W. F., Reid, S. L., Wylie, D. R. G., and Spetch, M. L. 1999, Perception of coherent motion in random dot displays by pigeons and humans, *Percept. Psychophy.* **61**:1089–1101.

Bravo, M., Blake, R., and Morrison, S. 1988, Cats see subjective contours, *Vision Res.* **28**:861–865.

Cartwright, B. A., and Collett, T. S. 1982, How honey bees use landmarks to guide their return to a food source, *Nature* **295**:560–564.

Cavanagh, P. 1987, Reconstructing the third dimension: Interactions between color, texture, motion, binocular disparity and shape, *Comp. Vision: Graphic Images Process* **37**:171–195.

Cerella, J. 1980, The pigeon's analysis of pictures, *Pattern Recognition* **12**:1–6.

Cheng, K. 1986, A purely geometric module in the rat's spatial representation, *Cognition* **23**:149–178.

Cheng, K., and Gallistel, C. R. 1984, Testing the geometric power of an animal's spatial representation, in: *Animal Cognition*, H. L. Roitblatt, T. G. Bever, and H. S. Terrace, eds., Erlbaum, Hillsdale, NJ, pp. 409–423.

Clayton, N. S. 1998, Memory and the hippocampus in food-storing birds: A comparative approach, *Neuropharmacology* **37**:441–452.

Clayton, N. S., and Dickinson, A. D. 1998, Episodic-like memory during cache recovery by scrub jays, *Nature* **395**:272–274.

Clayton, N. S., and Dickinson, A. D. 1999a, Memory for the content of caches by scrub jays (*Aphelocoma coerulescens*), *J. Exp. Psychol.: Anim. Behav. Process* **25**:82–91.

Clayton, N. S., and Dickinson, A. D. 1999b, Motivational control of caching behaviour in the scrub jay *Aphelocoma coerulescens, Anim. Behav.* **57**:435–444.

Clayton, N. S., and Dickinson, A. D. 1999c, Scrub jays (*Aphelocoma coerulescens*) remember the relative time of caching as well as the location and content of their caches, *J. Comp. Psychol.* **113**:403–416.

Clayton, N. S., and Krebs, J. R. 1994, Memory for spatial and object-specific cues in food storing and non-storing birds, *J. Comp. Physiol. A* **174**:371–379.

Clayton, N. S., and Krebs, J. R. 1995, Memory in food-storing birds: From behaviour to brain, *Curr. Opin. Neurobiol.* **5**:149–154.

Collett, T. S., Cartwright, B. A., and Smith, B. A. 1986, Landmark learning and visuospatial memories in gerbils, *J. Comp. Physiol. A* **154**:835–851.

Cook, R. G. 2001, Hierarchical stimulus processing by pigeons, in: *Avian Visual Cognition [On-line]*, R. G. Cook, ed., available: www.pigeon.psy.tufts.edu/avc/cook/

Cook, R. G., and Katz, J. S. 1999, Dynamic object perception in pigeons, *J. Exp. Psychol.: Anim. Behav. Process.* **25**:194–210.

Corballis, P. M., Fendrich, R., Shapley, R., and Gazzaniga, M. 1999, Illusory contour perception and amodal boundary completion: Evidence of a dissociation following callostomy, *J. Cogn. Neurosci.* **11**:459–466.

Cozzutti, C., and Vallortigara, G. 2001, Hemispheric memories for the content and position of food caches in the domestic chick, *Behav. Neurosci.* **115**:305–313.

Cutting, J. E., and Kozlowski, L. 1977, Recognizing friends by their walk: Gait perception without familiarity cues, *Bull. Psychon. Soc.* **9**:353–356.

D'Eath, R. B. 1998, Can video images imitate real stimuli in animal behaviour experiments? *Biol. Rev.* **73**:267–292.

Deruelle, C., Barbet, I., Dépy, D., and Fagot, J. 2000, Perception of partly occluded figures by baboons (*Papio papio*), *Perception* **29**:1483–1497.

Deng, C., and Rogers, L. J. 1997, Differential contributions of the two visual pathways to functional lateralization in chicks, *Behav. Brain Res.* **87**:173–182.

Deng, C., and Rogers, L. J. 1998a, Organisation of the tectorotundal and SP/IPS-rotundal projections in the chick, *J. Comp. Neurol.* **394**:171–185.

Deng, C., and Rogers, L. J. 1998b, Bilaterally projecting neurons in the two visual pathways of chicks, *Brain Res.* **794**:281–290.

Diamond, A., and Goldman-Rakic, P. S. 1989, Comparison of human infants and infant rhesus monkeys on Piaget's AB task: Evidence for dependence on dorsolateral prefrontal cortex, *Exp. Brain Res.* **74**:24–40.

Dittrich, W. H., Lea, S. E. G., Barrett, J., and Gurr, P. R. 1998, Categorization of natural movements by pigeons: Visual concept discrimination and biological motion, *J. Exp. Anal. Behav.* **70**:281–299.

Dumas, C., and Wilkie, D. M. 1995, Object permanence in Ring Doves, *J. Comp. Psychol.* **109**:142–150.

Emmerton, J. 1986, The pigeon's discrimination of movement patterns (Lissajous figures) and contour-dependent rotational invariance, *Perception* **15**:573–588.

Etienne, S. A. 1973, Searching behaviour towards a disappearing prey in the domestic chick as affected by preliminary experience, *Anim. Behav.* **21**:749–761.

Fagot, J., and Deruelle, C. 1997, Processing of global and local information and hemispheric spacialization in humans (*Homo sapiens*) and baboons (*Papio papio*), *J. Exp. Psychol.: Hum. Percept. Perform.* **23**:429–442.

Fletcher, H. J. 1965, The delayed response problem, in: *Behavior of Nonhuman Primates*, Vol. 1, A. M. Schrier, H. F. Harlow, and F. Stollnitz, eds., Academic Press, New York, pp. 129–165.

Fodor, J. A. 1983, *The Modularity of Mind. An Essay on Faculty Psychology*, MIT Press, Cambridge, MA.

Forkman, B. 1998, Hens use occlusion to judge depth in a two-dimensional picture, *Perception* **27**:861–867.

Forkman, B. 2000, Domestic hens have declarative representations, *Anim. Cogn.* **3**:135–137.

Forkman, B., and Vallortigara, G. 1999, Minimization of modal contours: An essential cross species strategy in disambiguating relative depth, *Anim Cogn.* **4**:181–185.

Foster, T. M., Temple, W., MacKenzie, C., Demello, L. R., and Poling, A. 1995, Delayed matching-to-sample performance of hens. Effects of sample duration and response requirements during the sample. *J. Exp. Anal. Behav.* **64**:19–31.

Freire, F., and Nicol, C. J. 1997, Object permanence in chicks: Predicting the position of an occluded imprinted object. Abstracts of the ASAB Meeting "Biological Aspects of Learning," St Andrews July 1–4, 1997, Scotland, p. 21.

Freire, F., and Nicol, C. J. 1999, Effect of experience of occlusion events on the domestic chick's strategy for locating a concealed imprinting object, *Anim. Behav.* **58**:593–599.

Frost, B. J., Troje, N. F., and David, S. 1998, Pigeon courtship behaviour in response to live birds and video presentations, *5th Int. Cong. Neuroethol.*

Fujita, K. 2001, Perceptual completion in rhesus monkeys (*Macaca mulatta*) and pigeons (*Columba livia*), *Percept. Psychophys.* **63**:115–125.

Funahashi, S., Bruce, C., and Goldman-Rakic, P. S. 1989, Mnemonic coding of space in the monkey's dorsolateral prefrontal cortex, *J. Neurophysiol.* **61**:331–349.

Funk, M. S. 1996, Development of object permanence in the New Zealand parakeet (*Cyanoramphus auriceps*), *Anim. Learn. Behav.* **24**:375–383.

Fuster, J. M. 1989, *The Prefrontal Cortex*, 2nd edn. Raven Press, New York.

Gallistel, C. R. 1990, *The Organization of Learning*, MIT Press, Cambridge, MA.

Goldman-Rakic, P. S. 1987, Development of cortical circuitry and cognitive function, *Child Dev.* **58**:601–622.

Gouteux, S., Thinus-Blanc, C., and Vauclair, J. 2001, Rhesus monkeys use geometric and nongeometric information during a reorientation task. *J. Exp. Psychol.: General* **130**:505–519.

Grossberg, S., and Mingolla, E. 1985, Neural dynamics of form perception: Boundary completion, illusory figures, and neon colour spreading, *Psychol. Rev.* **92**:173–211.

Güntürkün, O. 1996, Sensory physiology: Vision, in: *Sturkie's Avian Physiology*, G. C. Whittow, ed., Academic Press, Orlando, pp. 2–50.

Güntürkün, O. 1997, Avian visual lateralization: A review, *Neuroreport* **8**:3–11.

Güntürkün, O., and Durstewitz, D. 2000, Multimodal areas of the avian forebrain—Blueprints for cognition? in: *Evolution of the Brain and Cognition*, G. Roth and M. Wulliman, eds., Spektrum Akademischer Verlag, 2000, pp. 431–450.

Güntürkün, O., and Hahmann, U. 1999, Functional subdivisions of the ascending visual pathways in the pigeon, *Behav. Brain Res.* **98**:193–201.

Haskell, M., and Forkman, B. 1997, An investigation into object permanence in the domestic hen. Abstracts of the ASAB Meeting "Biological Aspects of Learning," St Andrews July 1–4, 1997, Scotland, p. 22.

van Hateren, J. H., Srinivasan, M. V., and Wait, P. B. 1990, Pattern recognition in bees: Orientation discrimination, *J. Comp. Physiol. A: Sensory, Neural Behav. Physiol.* **167**:649–654.

Hauser, M. D. 2000, *Wild Minds. What Animals Really Think*, Henry Holt and Co., New York.

Hellmann, B., and Güntürkün, O. 1999, Visual field specific heterogeneity within the tectofugal projection of the pigeon, *Eur. J. Neurosci.* **11**:1–18.

Hermer, L., and Spelke, E. S. 1994, A geometric process for spatial reorientation in young children, *Nature* **370**:57–59.

Hermer, L., and Spelke, E. S. 1996, Modularity and development: The case of spatial reorientation, *Cognition* **61**:195–232.

Hermer-Vasquez, L., Spelke, E. S., and Katsnelson, A. S. 1999, Sources of flexibility in human cognition: Dual-task studies of space and language, *Cogn. Psychol.* **39**:3–36.

Hetherington, M. M., and Rolls, B. J. 1996, Sensory-specific satiety: Theoretical issues and central characteristics, in: *Why We Eat What We Eat*, E. D. Capaldi, ed., American Psychological Association, Washington, DC, pp. 267–290.

von der Heydt, R., and Peterhans, E. 1989, Mechanisms of contour perception in monkey visual cortex: I. Lines of pattern discontinuity, *J. Neurosci.* **9**:1731–1748.

von der Heydt, R., Peterhans, E., and Baumgartner, G. 1984, Illusory contours and cortical neuron responses, *Science* **224**:1260–1262.

Hodos, W., Macko, K. A., and Bessette, B. B. 1984, Near-field acuity changes after visual system lesions in pigeons. II. Telencephalon, *Behav. Brain Res.* **13**:15–30.

Horridge, G. A., Zhang, S. W., and O'Carroll, D. 1992, Insect perception of illusory contours, *Phil. Trans. Royal Soc. London, Series B* **337**:59–64.

Huber, L. 2001, Visual categorization in pigeons, in: *Avian Visual Cognition [on-line]*, R. G. Cook, ed., available: www.pigeon.psy.tufts.edu/avc/huber/

Husband, S., and Shimizu, T. 2001, Evolution of the avian visual system, in: *Avian Visual Cognition [on-line]*, R. G. Cook, ed., available: www.pigeon.psy.tufts.edu/avc/husband/

Hunter, W. S. 1913, The delayed reaction in animals and children, *Behav. Monogr.* **II**:1.

Jarvis, J. R., Taylor, N. R., Prescott, N. B., Meeks, I., Meeks, I., and Wathes, C. M. 2002, Measuring and modelling the photopic flicker sensitivity of the chicken (*Gallus g. domesticus*), *Vision Res.* **42**:99–106.

Johansson, G. 1973, Visual perception of biological motion and a model for its analysis, *Percept. Psychophys.* **14**:201–211.

Johnson, M. H. 1997, *Developmental Cognitive Neuroscience*, Blackwell, Oxford.

Kalt, T., Diekamp, B., and Güntürkün, O. 1999, Single unit activity during a go/nogo task in the "prefrontal cortex" of the pigeon, *Brain Res.* **839**:263–278.

Kamil, A. C., and Jones, J. E. 1997, The seed-storing corvid Clark's nutcracker learns geometric relationships among landmarks, *Nature* **390**:276–279.

Kamil, A. C., and Jones, J. E. 1999, How do they, indeed? A reply to Biegler et al. *Animal Behaviour*, F9-F10 (online forum: http://www.academicpress.com/anbehav/forum).

Kanizsa, G. 1979, *Organization in Vision*, Praeger, New York.

Kanizsa, G., Renzi, P., Conte, S. Compostela, C., and Guerani, L. 1993, Amodal completion in mouse vision, *Perception* **22**:713–722.

Karten, H. J., and Shimizu, T. 1989, The origins of neocortex: Connections and laminations as distinct events in evolution, *J. Cogn. Neurosci.* **1**:291–301.

Kellman, P. J., and Arterberry, M. E. 1998, *The Cradle of Knowledge*, MIT Press, Cambridge, MA.

Kellman, P. J., and Spelke, E. S. 1983, Perception of partly occluded objects in infancy, *Cogn. Psychol.* **15**:483–524.

Kelly, D. M., and Spetch, M. L. 2001, Pigeons encode relative geometry, *J. Exp. Psychol.* **27**:417–422.

Kelly, D. M., Spetch, M. L., and Heth, C. D. 1998, Pigeons' (*Columba livia*) encoding of geometric and featural properties of a spatial environment, *J. Comp. Psychol.* **112**:259–269.

Kirkpatrick, K. 2001, Object recognition, in: *Avian Visual Cognition [On-line]*, R. G. Cook, ed., available: www.pigeon.psy.tufts.edu./avc/kirkpatrick/

Kozlowski, L. T., and Cutting, J. E. 1977, Recognising the sex of a walker from a dynamic point-light display, *Percept. Psychophys.* **21**:575–580.

Lea, S. E. G., Slater, A. M., and Ryan, C. M. E. 1996, Perception of object unity in chicks: A comparison with the human infant, *Infant Behav. Dev.* **19**:501–504.

MacWhinney, B. 1991, *The CHILDES Project: Tools for Analyzing Talk*, Erlbaum, Hillsdale, NJ.

Martinoya, C., Rivaud, S., and Bloch, S. 1984, Comparing frontal and lateral viewing in pigeons: II. Velocity thresholds for movement discrimination, *Behav. Brain Res.* **8**:375–385.

Michotte, A. 1963, *The Perception of Causality*, Basic, New York.

Michotte, A., Thines, G., and Crabbe, G. 1964, Les complements amodaux des structures perceptives (Amodal completion of perceptual structures), *Studia Psychologica*, Publications Universitaires de Louvain, Louvain.

Miller, C. T., Dibble, E., and Hauser, M. D. 2001, Amodal completion of acoustic signals by a nonhuman primate, *Nature Neurosci.* **4**:783–784.

Mogensen, J., and Divac, I. 1982, The prefrontal "cortex" in the pigeon. Behavioral evidence, *Brain Behav. Evolution* **21**:60–66.

Morris, R. G. M. 1981, Spatial localisation does not depend on the presence of local cues, *Learn. Motiv.* **12**:239–260.

Natale, F., Antinucci, F., Spinozzi, G., and Poti, P. 1986, Stage 6 object concept in nonhuman primates cognition: A comparison between gorilla (*Gorilla gorilla*) and Japanese macaque (*Macaca fuscata*), *J. Comp. Psychol.* **100**:335–339.

Nieder, A., and Wagner, H. 1999, Perception and neuronal coding of subjective contours in the owl, *Nature Neurosci.* **2**:660–663.

Omori, E., and Watanabe, S. 1996, Discrimination of Johansson's stimuli in pigeons, *Int. J. Comp. Psychol.* **9**:92.

Pepperberg, I. M., and Kozak, F. A. 1986, Object permanence in the African Grey parrot (*Psittacus erithacus*), *Anim. Learn. Behav.* **14**:322–330.

Pepperberg, I. M., and Funk, M. S. 1990, Object permanence in four species of psittacine birds: An African Grey parrot (*Psittacus erithacus*), an Illiger mini macaw (*Ara maracana*), a parakeet (*Melopsittacus undulatus*), and a cockatiel (*Nymphicus hollandicus*), *Anim. Learning Behav.* **18**:97–108.

Pepperberg, I. M., Willner, M. R., and Gravitz, L. B. 1997, Development of Piagetian object permanence in a Grey Parrot (*Psittacus erithacus*), *J. Comp. Psychol.* **111**:63–75.

Perrett, D. I., Harries, M. H., Benson, P. J., Chitty, A. J., and Mistlin, A. J. 1990, Retrieval of structure from rigid and biological motion: An analysis of the visual responses of neurones in the macaque temporal cortex, in: *AI and the Eye*, A. Blake and T. Troscianko, eds., John Wiley and Sons, Chichester, pp. 181–201.

Peterhans, E., and von der Heydt, R. 1989, Mechanisms of contour perception in monkey visual cortex: II. Contours bridging gaps, *J. Neurosci.* **9**:1749–1763.

Petter, G. 1956, Nuove ricerche sperimentali sulla totalizzazione percettiva, *Rivista di Psicologia* **50**:213–227.

Pettigrew, J. D., and Konishi, M. 1976, Neurons selective for orientation and binocular disparity in the visual Wulst of the barn owl (*Tyto alba*), *Science* **193**:675–678.

Piaget, J. 1953, *Origin of Intelligence in the Child*, Routledge and Kegan Paul, London.

Plowright, C. M. S., Reid, S., and Kilian, T. 1998, Finding hidden food: Behavior on visible displacement tasks by mynahs (*Gracula religiosa*) and pigeons (*Columba livia*), *J. Comp. Psychol.* **112**:13–25.

Pollok, B., Prior, H., and Güntürkün, O. 2000, Development of object permanence in food-storing magpies (*Pica pica*), *J. Comp. Psychol.* **114**:148–157.

Prior, H., and Güntürkün, O. 1999, Patterns of visual lateralization in pigeons: Seeing what is there and beyond. *Perception* (Supplement) **28**:22.

Regolin, L., and Vallortigara, G. 1995, Perception of partly occluded objects in young chicks, *Percept. Psychophys.* **57**:971–976.

Regolin, L., Vallortigara, G., and Zanforlin, M. 1994, Perceptual and motivational aspects of detour behaviour in young chicks, *Anim. Behav.* **47**:123–131.

Regolin, L., Vallortigara, G., and Zanforlin, M. (1995a), Object and spatial representations in detour problems by chicks, *Anim. Behav.* **49**:195–199.

Regolin, L., Vallortigara, G., and Zanforlin, M. 1995b, Detour behaviour in the domestic chick: Searching for a disappearing prey or a disappearing social partner, *Anim. Behav.* **50**:203–211.

Regolin, L., Tommasi, L., and Vallortigara, G. 2000, Visual perception of biological motion in newly hatched chicks as revealed by an imprinting procedure, *Anim. Cogn.* **3**:53–60.

Regolin, L., Marconato, F., Tommasi, L., and Vallortigara, G. 2001, Do chicks complete partly occluded objects only with their right hemisphere? *Behav. Pharmacol.* **12**(Suppl. 1):S82. (First Joint Meeting of the European Brain and Behaviour Society and European Behavioural Pharmacology Society, Marseille, France, September 8–12, 2001.)

Rogers, L. J. 1995, *The Development of Brain and Behaviour in the Chicken*, CAB International, Wallingford.

Rogers, L. J. 1997, *Minds of their Own*, Allen and Unwin, St. Leonards, Australia.

Rolls, B. J. 1990, The role of sensory-specific satiety in food intake and food selection, in: *Taste, Experience and Feeding*, E. D. Capaldi and T. L. Powley, eds., American Psychological Association, Washington, DC, pp. 197–209.

Rumbaugh, D. M., and Gill, T. V. 1975, The learning skill of the rhesus monkey, in: *The Rhesus Monkey*, G. H. Bourne, ed., Academic Press, New York, pp. 303–321.

Runeson, S., and Frykholm, G. 1981, Visual perception of lifted weight, *J. Exp. Psychol.: Hum. Percept. Perform.* 7:733–740.

Sato, A., Kanazawa, S., and Fujita, K. 1997, Perception of object unity in chimpanzee (*Pan troglodytes*), *Jpn. Psychol. Res.* **39**:191–199.

Sekuler, A. B., Lee, J. A. J., and Shettleworth, S. J. 1996, Pigeons do not complete partly occluded figures, *Perception* **25**:1109–1120.

Sekuler, A. B., and Palmer, S. E. 1992, Perception of partly occluded objects: A microgenetic analysis, *J. Exp. Psychol.: General* **121**:95–111.

Sherry, D. F., Jacobs, L. F., and Gaulin, S. J. C. 1992, Spatial memory and adaptive specialization of the hippocampus, *Trends Neurosci.* **15**:298–303.

Shimizu, T. 1998, Conspecific recognition in pigeons (*Columba livia*) using dynamic video images, *Behaviour* **135**:43–53.

Shipley, T. F., and Kellman, P. J. 1992, Strength of visual interpolation depends on the ratio of physically specified to total edge length, *Percept. Psychophys.* **52**:97–106.

Siegel, R. M., and Andersen, R. A. 1988, Perception of three-dimensional structure from visual motion in monkey and man, *Nature* **331**:259–261.

Singh, M., Hoffman, D., and Albert, M. 1999, Contour completion and relative depth: Petter's rule and support ratio, *Psychol. Sci.* **10**:423–428.

Sovrano, V. A., Bisazza, A., and Vallortigara, G. 2002, Modularity and spatial reorientation in a simple mind: Encoding of geometric and nongeometric properties of a spatial environment by fish, *Cognition* **85**:51–59.

Sovrano, V. A., Bisazza, A., and Vallortigara, G. 2003, Modularity as a fish views it: Conjoining geometric and nongeometric information for spatial reorientation, *J. Exp. Psychol.: Anim. Behav. Process.*, **29**:199–210.

Spelke, E. 1998a, Nativism, empiricism, and the origins of knowledge, *Infant Behav. Dev.* **21**:181–200.

Spelke, E. 1998b, Nature, nurture, and development, in: *Perception and Cognition at Century's End*, J. Hochberg, ed., Academic Press, San Diego, pp. 333–371.

Spetch, M. L., Cheng, K., and MacDonald, S. E. 1996, Learning the configuration of a landmark array: I. Touch-screen studies with pigeons and humans, *J. Comp. Psychol.* **110**:55–68.

Spetch, M. L., Cheng, K., MacDonald, S. E., Linkenhoker, B. A., Kelly, D. M., and Doerkson, S. 1997, Learning the configuration of a landmark array in pigeons and humans. II. Generality across search tasks, *J. Comp. Psychol.* **111**:14–24.

Sugita, Y. 1999, Grouping of image fragments in primary visual cortex, *Nature* **401**:269–272.

Sutton, J. E., Olthof, A., and Roberts, W. A. 2000, Landmark use by squirrel monkeys: *Saimiri sciures, Anim. Learn. and Behav.* **28**:28–42.

Suzuki, S., Augerinos, G., and Black, A. H. 1980, Stimulus control of spatial behavior on the eight-arm maze rats, *Learn. Motiv.* **11**:1–18.

Thinus-Blanc, C. 1996, *Animal Spatial Cognition: Behavioural and Neural Approaches*, World Scientific Press, Singapore.

Tinklepaugh, O. L. 1928, An experimental study of representative factors in monkeys, *J. Comp. Psychol.* **8**:197–236.

Tommasi, L., Vallortigara, G., and Zanforlin, M. 1997, Young chickens learn to localize the centre of a spatial environment, *J. Comp. Physiol.* **180**:567–572.

Tommasi, L., and Vallortigara, G. 2000, Searching for the centre: Spatial cognition in the domestic chick, *J. Exp. Psychol.: Anim. Behav. Process* **26**:477–486.

Tommasi, L., and Vallortigara, G. 2001, Encoding of geometric and landmark information in the left and right hemispheres of the avian brain, *Behav. Neurosci.* **115**:602–613.

Tommasi, L., Bressan, P., and Vallortigara, G. 1995, Solving occlusion indeterminacy in chromatically homogeneous patterns, *Perception* **24**:391–403.

Towe, A. L. 1954, A study of figural equivalence in the pigeon, *J. Comp. Physiol. Psychol.* **47**:283–287.

Ullman, S. 1979, *The Interpretation of Visual Motion*, MIT Press, Cambridge, MA.

Vallortigara, G. 1992, Affiliation and aggression as related to gender in domestic chicks (*Gallus gallus*), *J. Comp. Psychol.* **106**:53–57.

Vallortigara, G. 2000, Comparative neuropsychology of the dual brain: A stroll through left and right animals' perceptual worlds, *Brain Lang.* **73**:189–219.

Vallortigara, G., and Andrew, R. J. 1991, Lateralization of response by chicks to change in a model partner, *Anim. Behav.* **41**:187–194.

Vallortigara, G., and Regolin, L. 2002, Facing an obstacle: Lateralization of object and spatial cognition, in: *Comparative Vertebrate Lateralization*, R. J. Andrew and L. J. Rogers, eds., Cambridge University Press, Cambridge, pp. 383–444.

Vallortigara, G., and Tommasi, L. 2001, Minimization of modal contours: An instance of an evolutionary internalized geometric regularity? *Brain Behav. Sci.* **24**:706–707.

Vallortigara, G., Rogers, L. J., and Bisazza, A. 1999, Possible evolutionary origins of cognitive brain lateralization, *Brain Res. Rev.* **30**:164–175.

Vallortigara, G., Zanforlin, M., and Pasti, G. 1990, Geometric modules in animal's spatial representation: A test with chicks, *J. Comp. Psychol.* **104**:248–254.

Vallortigara, G., Regolin, L., Rigoni, M., and Zanforlin, M. 1998, Delayed search for a concealed imprinted object in the domestic chick, *Anim. Cogn.* **1**:17–24.

Vallortigara, G., Cozzutti, C., Tommasi, L., and Rogers, L. J. 2001, How birds use their eyes: Opposite left–right specialisation for the lateral and frontal visual hemifield in the domestic chick, *Curr. Biol.* 11:29–33.

Wasserman, E. A., Kirkpatrick-Steger, K., Van Hamme, L. J., and Biedermann, I. 1993, Pigeons are sensitive to the spatial organization of complex visual stimuli, *Psychol. Sci.* **4**:336–341.

Watanabe, S., and Ito, Y. 1991, Discrimination of individuals in pigeons, *Bird Behav.* **9**:20–29.

Wallach, H., and O'Connell, D. N. 1953, The kinetic depth effect, *J. Exp. Psychol.* **45**:205–217.

Wu, H. M., Sackett, G. P., and Gunderson, V. M. 1986, Social stimuli as incentives for delayed response performance by infant pigtailed macaques, *Primates* **27**:229–236.

Zanforlin, M. 1981, Visual perception of complex forms (anomalous surfaces) in chicks, *Ital. J. Psychol.* **8**:1–16.

PART TWO

Social Learning

CHAPTER THREE

Socially Mediated Learning among Monkeys and Apes: Some Comparative Perspectives

Hilary O. Box and Anne E. Russon

INTRODUCTION

As an order, the group primates is markedly diverse for its size of some 300 species. The group includes adult male gorillas that weigh up to 5–6,000 times that of the smallest primates, the dwarf and mouse lemurs of Madagascar. There is also much diversity in locomotion and in diet, encompassing small carnivorous species such as tarsiers, and much larger folivores such as howler monkeys. A critical distinction between these taxa, as within the order as a whole, is that of nocturnal and diurnal living and their respective sensory adaptations and social organization. Such diversity clearly cautions against simplistic generalizations about the order as a whole. We should be aware that taxonomic relatedness—including the fact that we look like many other primate species—may lead to unsubstantiated assumptions; by our

Hilary O. Box • Department of Psychology, University of Reading, Reading, UK. **Anne E. Russon** • Department of Psychology, Glendon College of York University, Toronto Ontario, Canada.

Comparative Vertebrate Cognition, edited by Lesley J. Rogers and Gisela Kaplan. Kluwer Academic/Plenum Publishers, 2004.

relatedness, we may be tempted to consider characteristics and propensities with biological, social, and political implications, but we should be careful about making generalizations on the basis of taxonomic relatedness alone (Rowell, 1999). Rowell has argued, for example, that generalizations about behavior on the basis that species are primates would, in some cases, be less useful than comparisons among groups of animals that are niche equivalents (Rowell, 1999).

Significantly, however, simian primates (monkeys and apes) have often been considered to have a special status among non-human animals by the interests of the people who have studied them. As our nearest living relatives for instance, the primates are of interest to us in a number of ways, as in studies of human origins, language, and culture (e.g., de Waal, 2001; Gibson and Ingold, 1993). They are also of interest in issues of welfare and conservation (e.g., Byrne, 1999a; Cowlishaw and Dunbar, 2000). Hence, a persistent perspective concerning a variety of behavioral studies of monkeys and apes has been generated with reference to human mental abilities and social propensities (Box, 1994; Rowell, 1999). A good example refers to socially mediated learning that has figured prominently in arguments for human superiority in areas such as language, culture, and imitation.

In this chapter, we use socially mediated learning as the context in which to discuss some aspects of comparison between species of monkeys and the great apes, as well as to lesser extents, between monkeys and apes, and other groups of animals, in order to clarify some comparative perspectives. More generally, the domain of socially mediated learning is particularly important for our understanding of ecological, social, and cognitive influences that contribute to meeting ecological challenges in comparative and evolutionary contexts. We begin with some information about socially mediated learning in general.

SOCIALLY MEDIATED LEARNING

Socially mediated learning (hereafter called social learning for brevity) refers to the acquisition of behavior that is influenced by activities of other individuals, either directly or indirectly (Box, 1984; Heyes, 1994). These other individuals are frequently members of the same species but not necessarily so. Social learning contributes to a wide range of behavior as in learning what to eat, where to find food and how to process it, how to interact with members of one's social unit, and how to choose mates, as well as learning to identify and respond appropriately to predators. It may also generate novel behavior that is not part of the typical repertoire of the species.

Hence, the functions of social learning primarily concern the acquisition of skills and information that enable individuals to adjust competently to the demands of their environments; social and physical alike. Effects of social learning may include reducing time, effort, and risk, and increasing rates of behavioral change within a group (Giraldeau, 1997). Social learning influences survival and fitness among individuals, although it is possible, of course, that socially maintained behaviors might not always be advantageous (Laland and Williams, 1998).

Further, the social acquisition of information may be enabled by various mechanisms (e.g., Galef, 1988; Heyes, 1994) as by local enhancement, stimulus enhancement, and social facilitation, during which the attention of individuals is focused toward locations, stimuli, or activities that facilitate behavior. In fact, the catalog of mechanisms to support social learning is much longer (Galef, 1998; Whiten and Ham, 1992) but most attention has focused on local and stimulus enhancement, and response facilitation.

By some contrast, and despite assumptions as in everyday speech, there is a long tradition of treating imitation as a separate mechanism that underpins relatively rare examples of social learning in which motor activities are learned directly by observation (e.g., Galef, 1988; Heyes, 1996; Heyes and Ray, 2000; Janik and Slater, 2000; Thorndike, 1911; Whiten and Ham, 1992).

The term social learning then, is very broad; it encompasses socially creating ecological opportunities in different ways that tap different learning mechanisms. It is part of the evolution of behavioral strategies among a wide diversity of animal groups. Hence, social influences, learning mechanisms, and the bio-behavioral responsiveness that facilitate them, vary with regard to phylogenetic, ecological, sensori-motor, and cognitive adaptations of different taxa. Theoretical discussions include those of Aoki and Feldman (1987), Boyd and Richerson (1988), Laland et al. (1993, 1996), Sibly (1999), Byrne (1999b), Byrne and Russon (1998), and Visalberghi and Fragaszy (2002) that address evolutionary, genetic, cognitive, and environmental contexts that make social learning possible and advantageous.

Although social learning is still an under-subscribed area in behavioral biology, and one that has never been adequately represented, there has been substantive interest in the influence of social learning in different animal groups. Primates are prominent among them, but so, to a lesser extent, are marine mammals (Boran and Heimlich, 1999; Herman, 2002; Janik and Slater, 2000; Rendell and Whitehead, 2001), bats (Wilkinson, 1992; Wilkinson and Boughman, 1999), rodents (Galef, 1996; Galef and Allen, 1995; Galef and Wigmore, 1983; Heyes, 1996; Laland, 1999; Terkel, 1996; Zohar and Terkel, 1992) and various species of birds (Atkins and Zentall, 1996; Galef et al., 1986; Lefebvre, 1986; Moore, 1992, 1996).

Custance et al. (2002a) remark that most "instigated" studies of social learning, including imitation, have focused on mechanisms rather than function (see also Box, 1994; Box and Gibson, 1999). Custance et al. (2002a) reviewed all such studies on non-human primates since 1950 (84 publications, 109 data sets). Most of the data sets, whether from ecologically valid tests or not, revealed positive effects of social input, particularly enumerating benefits in the rate of the acquisition of behavior. However, those that incorporated the controls considered essential to identifying the social mechanisms involved, have been conducted in highly artificial conditions with little consideration for ecological validity, including social as well as physical ecology. This means that we know relatively little about how social learning in its different forms operates in natural conditions (see also Box, 1994).

In fact, although much of the work has been in captivity with controlled experimental techniques, there are also significant experimental studies of social learning in the natural environment, as in bats by Wilkinson (1992), white-throated magpie jays by Langen (1996), and in pigeons by Lefebvre (1986). Moreover, recent experiments in captivity extend our perspectives and understanding of the mechanisms involved in social learning. It is important, for example, to consider methods by which we may extend the longstanding and pervasive methodological technique in which trained individuals act as demonstrators for naive individuals, as in obtaining food. Subsequently, the naive observers are given the opportunity to perform the task alone. In some cases, groups for comparison may be challenged with the same task, but without a successful demonstrator or, in which there is a demonstrator, but a different solution to the task is used. There are clear advantages of using such techniques; the outcomes and social contexts may be relatively unambiguous, but studies of "transmission chains" address additional and more natural perspectives (Galef and Allen, 1995; Laland, 1999; Laland and Williams, 1997). For example, observer rats whose foraging performance is enhanced through social learning from a demonstrator rat, may subsequently act as a demonstrator for the next observer in a line of transmission.

Variations on this experimental theme are combined with explorations of the additive effects of different aspects of social influence, such as being exposed to the breath of another animal after eating, as well as being able to search at a site that has the scent marks of other animals. Hence, multiple transmission learning events are advocated. Such experiments give information on the fixation of patterns of behavior within groups by way of information transmission among individuals. The questions are relevant, for example, to the potential survival of learned traditions in animal populations when the original animals are no longer present, and to the development of innovative behavior.

In another very different example, work by Hudson et al. (1999) shows that in an extreme social system with limited maternal care, and no direct assistance in the transmission to a sudden weaning, pups of the European rabbit (*Oryctolagus cuniculus*) may acquire information, as about the diet of their mothers—in utero, during nursing, and by fecal pellets in the nest. Anatomical, physiological, and behavioral evidence shows that pups perceive and process social information that is critical to their survival even as newborns, as well as later in the context of facilitating their transition to independence. Mothers provide a wide range of olfactory information.

In recent years also, and for various methodological and conceptual reasons, a number of people have been more concerned with the social influences, and the biobehavioral propensities that support learning among species, and individuals within species, rather than upon learning as such (Box, 1994, 1999; King, 1994, 1999; Visalberghi and Fragaszy, 2002). Studies that address social relationships (Boesch, 1993a, 1993b; Huffman, 1984; Huffman and Hirata, 2003; Russon and Galdikas, 1995), developmental parameters (Inoue-Nakamura and Matsuzawa, 1997; Russon, 2003), and individual differences (Box, 1999) are good cases in point. Social learning involves the strategic adjustments of individuals of different species in dynamic social and ecological contexts; it involves the extents to which individuals cope with, maintain, and create social and other environmental opportunities. Critically, animals of significantly different dispositions react with their environments and learn about their environments in different ways that, in turn, have significant consequences for their life strategies (Box, 1999). Hence, this suggests an interactive approach that involves emotion, attention, activity, and mental ability, and emphasizes social learning as processes of communication, and not as the acquisition of "acts" or "packages" (see discussions by Box [1999] and King

[1999]). There is also a longstanding interest in cognitive processes that underpin transmission processes such as imitation and culture (Byrne and Russon, 1998; Galef, 1988, 1992; Russon, 1999b; Whiten, 2000).

Further, stimulating discussions in social learning have involved consideration of the influence of changes that animals make to their environments, as in dams and burrows used across generations. (e.g., see Giraldeau and Caraco, 2000; Jones et al., 1997, for "information ecology" and Laland et al. 2000, for "niche construction"). In the latter context, for instance, responses to problems posed by the environment are not exclusive contributors to natural selection. In modifying their environments by their behavior, animals may contribute to the process of selection. Hence, adaptation is no longer seen as a one-way process, but a two-way process in which populations of animals set, as well as solve problems.

It is also pertinent to the development of research in social learning that there is more interest amongst field biologists to ask questions across species about the characteristics and propensities of learning opportunities in social contexts. Studies about social learning on previously rarely investigated species of mammals, for example, are being brought into the domain, and are at a stimulating stage of opening up questions that relate to the biology of the different species (Box and Gibson, 1999). These include macropod marsupials by Higginbottom and Croft (1999); caribou, musk ox and arctic hare by Klein (1999); naked mole rats by Faulkes (1999); and canids by Nel (1999).

Importantly, however, and with some notable exceptions (e.g., Byrne and Byrne, 1993; Inoue-Nakamura and Matsuzawa, 1997; Russon and Galdikas, 1995; Whitehead, 1986), field studies are rarely conducive to discerning the mechanisms of social learning, but they do provide information on how social learning operates in natural populations in terms of the problems, enabling conditions, social dynamics, and functions involved. Ideally, we should consider studies in the field, and controlled experimental studies as complementary. Terkel (1996) provides an excellent example in which field observations and experimental studies with Black rats (*Rattus rattus*) clearly show that the ability to strip pine cones is acquired by immature rats in close proximity to a female that is experienced in stripping cones—a technique that is necessary for life in its specific habitat.

Although field studies do sometimes allow studies of mechanisms, the point is that they ask different questions that are ultimately critical to our understanding of the functions of social learning. For example, studies of unrelated taxa such as musk oxen, caribou, and artic hare, living in the same Artic habitat can help clarify how species with significantly different characteristics and behavioral propensities deal with similar environmental challenges, including opportunities for social learning (e.g., Klein, 1999). Conversely, studies of species that are closely related, as among felids (e.g., Kitchener, 1999) and canids (e.g., Nel, 1999) can help determine how taxa that have similar anatomical, sensory, motor, and mental capacities adapt to the demands of different environments.

We should examine carefully assumptions about the relevance of social learning in the lifestyles of different groups of animals. For example, carnivores are of interest because in many cases they need to acquire skills in dealing with their food that herbivores do not. In some instances they may incur considerable risks, and predatory skills are only honed after much practice. Adults of some species such as among felids, killer whales, bat-eared foxes, and some dolphins provide practice opportunities for their young. In some of these species, adults may actively demonstrate methods of capturing prey (e.g., Caro, 1994; Guinet and Bouvier, 1995; Nel, 1999). Such examples may inadvertently give the impression that social learning is functionally more important for some carnivores than for herbivores. More information is required on the extent to which social learning may play a role in the development of plant processing techniques in other mammals (Box and Gibson, 1999) as well as in plant selection (Prins, 1996).

Central questions of field studies include: What do animals of different age, sex, and species need to know to adjust competently to their environments? What characteristics and propensities do they have to acquire such skill and information? and What opportunities do they have to learn in socially mediated ways? Hence, field studies may engender questions, and lead to experiments about social dynamics (Cambefort, 1981) and about sex differences in social learning as described by Lee and Moss (1999) for African savannah elephants; by Boesch and Boesch-Achermann (2000) for chimpanzees, and by Russon (2003) for orangutans. Studies of physical, developmental, and social interactions are thereby encouraged.

It is also important that social learning may not always be necessary to acquire complex behavior in animals such as the primates (Milton, 1993), but natural contexts facilitate behavior, and behavior may be honed in natural contexts.

Given that, with notable exceptions (e.g., Byrne, 1999c; Byrne and Byrne, 1993; Whitehead, 1986), studies of social learning in real life frequently present severe methodological constraints, King's (1994, 1999) distinction between social information acquisition and social information donation provides a significant advance in this regard. The former includes any behavior that helps, or potentially helps, immature animals gain information from adults. Social information donation refers to cases in which adults direct an action or behavior at immature animals. The technique may be applied to any age group of animals, in captivity and in the field, and in different functional contexts. The methodological importance of the distinction is that it provides a means to delineate reliably the relative contributions of individuals in which they might extract information from others.

Moreover, with reference to the literature on the acquisition of foraging skills among species of monkeys, the predominant pattern is that of social information acquisition. In this context, it is of interest that there is evidence for both social information acquisition and social information donation among New World marmosets and tamarins (Callitrichidae). In these monkeys, food sharing by both spontaneous offering and in response to begging is a conspicuous part of infant life (Feistner and McGrew, 1989). Rapaport (1999), for instance, has shown that in golden lion tamarins (*Leontopithecus rosalia*), adults transferred foods to immatures that were familiar to the adults but novel to the young, as well as more frequently, food that was unfamiliar to them all, than food that was familiar to both immatures and adults. Rapaport suggests that adults change their behavior in ways that facilitate learning about food by immatures.

Such cases are important because information donation is rare among monkeys and apes (certainly in feeding contexts). Information donation is often discussed within the controversial topic of "teaching" and its human cognitive orientations. Information donation also raises questions about the influence of lifestyle. For example, marmosets and tamarins are highly cooperative in many aspects of their lives and are socially very tolerant (e.g., Caine, 1993). We need to know more about ecological influences upon information donation, of course, but social influences in these animals certainly fit with the facilitation of their transfer of information. Importantly in this context, for example, marmosets and tamarins are cooperative breeders and Snowdon (2001) has recently argued that the unusual social dynamics of cooperative breeding systems may lead to more efficient cognitive skills including social learning, information donation, and imitation than among species with different social systems. This hypothesis opens up new emphases of interest in social and cognitive influences in social learning among primate taxa that are not known to be cognitively astute, and make interesting comparisons with more traditional taxonomic divides, in which mental abilities are assumed to be of paramount importance.

There is also good evidence for social information donation in a diversity of other mammals. Among cetaceans, for example, there are well-known cases for the demonstration and acquisition of predatory techniques (see Boran and Heimlich, 1999, for a general review), as in the capture of seal pups by whales (Hoelzel, 1991), and the unique cooperative interaction between bottlenose dolphins and human fishermen on the coast of Brazil (Pryor et al., 1990). Among canids, Nel (1999) describes how bat-eared fox (*Otocyon megalotis*) males initiate their pups into foraging, and Caro's (1994) classic studies with cheetahs provide examples in which females provide opportunities for their cubs to learn to deal with prey, to refine their innate killing skills, and to take advantage of prey species that are relatively available in particular areas. Once again we can begin to ask questions in different taxa about the interactions of cognitive, social, and ecological influences in social learning, which vary considerably in these taxa.

Overall, information that we now have shows that social learning is ecologically relevant to a wide diversity of animals, and not necessarily more important functionally to groups such as monkeys and apes that have often been assumed to have a "special position," typically in terms of their mental abilities (Box, 1994; Laland et al., 1993). It is important that we no longer make simplistic phylogenetic assumptions about mechanisms and functions of social learning. We should consider the scope of the phenomena; study in what taxa, and under what conditions, social learning occurs, and in relation to life strategies of individuals of different taxa (Box and Gibson, 1999). Impediments to progress in these comparative regards, as among mammals, have been the tendency to overestimate and overemphasize assumed capacities for social learning among monkeys and apes with reference to humans, and the implied tendencies by many behavioral biologists to underestimate the importance of social learning in supporting ecological strategies among a wide diversity of animal groups. Hence, although important conceptual and methodological advances have been made, many comparative questions as in mammalian social learning are unanswered (Box and Gibson, 1999), and the relative "status" of primate taxa in various domains of social learning remain open. Further, questions of such comparative status not only include primate and non-primate taxa, but comparisons of social learning between primate taxa. In fact, although there is little information for prosimian species (such as lemurs, lorises, and bushbabies) and many species of monkeys, that which we do have shows stimulating quantitative differences between monkeys and apes that emphasize behavioral diversity within the order, and raise questions about the relative "special status" of the great apes. For many reasons, comparison between monkeys and great apes is a major domain of comparative research in primate social learning. Great apes, for example, are considered to learn more complex skills than monkeys in some aspects of their lives, and there is clear evidence for an ape/monkey divide in various domains of behavioral biology (e.g., Byrne, 1995).

Hence, the aim of the following sections of this chapter is to discuss current perspectives of social learning among monkeys and great apes. As previously acknowledged, however, there is extensive diversity within these groups. Moreover, we have selected topics of core interest in social learning among monkeys and apes, especially imitation and culture. We begin with studies of imitation.

IMITATION

A persistent and contentious domain of study within social learning has been that of imitation namely, doing an act from seeing it done (Thorndike, 1911), or response learning by observation (Heyes and Ray, 2000). Imitation contributes in various ways to the development of human skills and social interactions and, some claim, is a facility that humans have from birth (Meltzoff, 1996). Byrne (2000) provides an instructive discussion of the advantages and constraints of ways in which different academic disciplines have studied imitation.

Further, imitation is generally thought of as the most demanding category of social learning in cognitive terms in that a translation of observed novel input into a matching output may involve more complex central processing than other mechanisms of social learning (Heyes, 1998; Whiten and Byrne, 1988). Comparative studies of other primates have been of interest. The central question has been: do non-human primate species have such ability?

Imitation in Monkeys

A paper by Watson in 1908 foreshadowed results that have been much emphasized in monkey research ever since. His work stemmed from the apparent contradiction between Thorndike's (1911) experiments, which did not support imitation as playing an important role in primate behavior, and Hobhouse's (1915) experiments, which found that various animal species, but monkeys, in particular, learn by "perception of result" and imitation.

Watson's experiments (some of which emulated those of Hobhouse, 1915) with four captive monkeys of three different species (two rhesus monkeys, one cebus, and one baboon) included problems of the "perception of a simple relation," and others "of the manipulation type." In relational problems, monkeys observed Watson as the demonstrator perform tasks involving pulling in food with a rake, drawing in food with a cloth, obtaining food from the bottom of a bottle by means of a fork, and pushing food out from the middle of a fixed, horizontal glass cylinder by using a light stick. In manipulative problems, the monkeys mainly observed another monkey, as in opening a latch, and a box, by various mechanical means. After much patient study, Watson found that "there was never the slightest evidence of inferential imitation manifested in the actions of any of these animals. There was never imitation either of my movements or the movements of the animal that was successfully manipulating the mechanism" (p. 172).

In the results of the perception of relational tasks there was also no evidence of an appreciation of a relationship between two objects, such as a rake, or a cloth and the food. The glass cylinder, for example, was provisioned with a piece of banana at the center that was slowly pushed out by the demonstrator with a stick. In the tests, the stick was placed nearby. Watson found that even when, in later trials, the stick was actually placed in the cylinder before the tests, the monkeys did not use it appropriately; the stick was removed in favor of using the hands!

Watson also described a series of spontaneous activities in which one animal would begin an activity that was subsequently repeated by the other animal. He noted that his anecdotal material would compare favorably with findings of Romanes (1884/1977) a strong proponent of imitation in monkeys, but that close scrutiny "of such acts, especially during the period of their genesis, does not lead me to think that the higher forms of imitation are present in them" (p. 178). Today, this sort of behavioral copying might be interpreted as a form of response facilitation: the first animal's actions simply prime or trigger similar actions that the observing animal already knows (Byrne and Russon, 1998).

Subsequently, there have been many more varied and rigorous experiments that reach generally similar conclusions. An influential review by Visalberghi and Fragaszy (1990), updated in 2002, emphasizes their well-known conclusion that there is no evidence, either from experimental work, or from observations in the field (Galef, 1990, 1992) that monkeys demonstrate an ability to imitate. Much of their own empirical work has involved New World capuchin monkeys (*Cebus apella*). These monkeys have a number of characteristics and propensities that make them excellent subjects to test in this domain. Whereas capuchins are comparable with other species of monkeys in tasks that assess their attentional, conceptual, and memory abilities (Tomasello and Call, 1997), they stand out as socially very tolerant (all age groups are interested in one another's activities), as highly manipulative and exploratory, and as cognitively more astute than most other monkeys known in this context (Anderson, 1996). They are the most versatile known among monkeys in their use of tools (Anderson, 2000; Visalberghi, 1993), for example, and they spontaneously show new behaviors in their manipulative activities, including their use of tools (Fragaszy and Adams-Curtis, 1991; Visalberghi, 1987).

Visalberghi and Fragaszy find no evidence of imitative abilities in these monkeys. There are many examples, including instances in which monkeys learn to use tools in simple actions to obtain a desirable food item. In her well-known paradigm, Visalberghi (1993) found that monkeys who had been unsuccessful at solving a tool task, pushing food from a transparent tube with a stick, were still unable to solve the task after observing proficient monkeys solving it, even after repeated demonstrations. They did not improve their technique in orienting the tool toward the tube, and they did not selectively focus on any aspects of the task that were relevant to solving it.

Visalberghi and Fragaszy (2002) also reviewed a variety of experimental methods that they, and others, have used to study imitation in capuchins. These include a two-action model, a method that has been used with a wide variety of animals, in which there are two functionally equivalent, but different methods of solving a task (Dawson and Foss, 1965). Different individuals only observe one of these and are tested subsequently on whether their performance matched that of the method demonstrated (see Custance et al., 1999). We may accept this as one of the most powerful methods available for testing imitation (Shettleworth, 1999). Under these test conditions, capuchins did not match such demonstrated novel actions.

In other experiments (Perucchini et al., 1997) pairs of monkeys were given identical sets of objects, and observations made about the extents to which they spontaneously matched each other's activities. They did not watch each other a great deal and did not match each other's activities spontaneously. Again, investigations into whether capuchins would match the actions of a human demonstrator in a program to ultimately teach specific behavior, and sequences of behavior, that would be of help to quadriplegic people (e.g., Hemery et al., 1998; Fragaszy unpublished) concluded, for instance, that although the monkeys predictably contact objects that have been acted upon by the demonstrator and, to lesser extents they would "move an object to achieve (or toward) a demonstrated movement or new position of the object," they did not match the actions that were performed (Herve and Deputte, 1993).

Some few other species have been studied in imitative learning. For example, Custance et al. (2002b) presented puzzle boxes to pigtailed macaques (*Macaca nemestrina*) and 2-year-old children. A human demonstrator showed two alternative methods for manipulating objects that were designed to test whether the monkeys would learn about the movement of the objects (emulation) or about which part of the objects to manipulate (stimulus enhancement). The monkeys watched for a reasonable proportion of the demonstrations, but showed little evidence of social learning. Alternatively, the young children watched much more, and showed clear evidence of stimulus enhancement, emulation, and imitation.

There are occasional positive reports of imitation in species of monkeys. In a two-action experiment with common marmosets (*Callithrix jacchus*), two of the five monkeys showed strong signs of matching the way in which a conspecific model opened a doorway to retrieve food (Bugnyar and Huber, 1997). A microanalysis of the successful monkey's behavior supported the conclusion. In other work by Custance et al. (1999), capuchins and chimpanzees were both shown a human model opening an artificial fruit in one of two alternative ways. The capuchins did not copy some of the features that the chimpanzees had copied, but they did show signs of the overall approach they had observed, and more than is consistent with only stimulus enhancement. Custance et al. (1999) suggest that the performance of these monkeys can be considered as emulation or as a simple form of imitation (see Whiten [2000] for a discussion of the "the gray area" between these concepts).

The conclusion from wide-ranging experiments has been, and still is, in many respects, that there is a lack of convincing evidence for imitation learning per se in monkeys. However, the evidence does now include additional perspectives that deserve experimental attention. The case is certainly not closed. In fact, the lack of robust evidence for monkeys in this area has encouraged different perspectives of study critical to biologically realistic developments in social learning, leading, for instance, to methodological discussions. There are those that question the social relationship of human demonstrator and primate observer. (Note an excellent and prophetic review of social learning in monkeys by Hall as long ago as 1963.) One current common sense view has expressed doubt about the appropriateness of using human demonstrators for nonhuman species, given that humans use their own species as demonstrators but may then compare results between the performance of monkeys and children on the basis of this imbalance in design.

Imitation in Great Apes

Imitation research has spotlighted the great apes, even more so than monkeys, because of their evolutionary position between Old World monkeys and humans. For this reason, they may have unique imitative abilities among nonhuman primates.

Indeed, great apes consistently perform better than monkeys on tests of imitation. On Watson's imitation tasks that monkeys consistently fail (e.g., rake in food, open a latch or box by mechanical means), great apes regularly succeed. Recent studies on great ape imitation have used three types of tasks, Watson's "perceiving simple relations," his "manipulative tasks," and "arbitrary actions." While the basic pattern of great ape success and monkey failure stands, the pattern that is emerging from this research is raising new debates on how to understand their imitative abilities.

Studies on imitating "simple relations" typically use the classic demonstration, tool use, because using a tool requires manipulating the relation between the tool and other objects (Russon, 1999a). From his experiments on imitating raking food with a tool, Tomasello concluded that great apes learn something about general functional relations between tool and food from observing demonstrations (e.g., Call and Tomasello, 1994; Tomasello et al., 1987, 1993). Other studies have shown that great apes can imitate the *particular* relation demonstrated, for example, raking food with a stick tool, as demonstrated, rather than pushing or banging it (Byrne and Russon, 1998; Meinel, 1995; Myowa-Yamakoshi and Matsuzawa, 1999; Russon, 1999a; Russon and Galdikas, 1995). Visalberghi conducted a comparative study on imitating tool use in chimpanzees and capuchins using the tube task, which is a modern version of Watson's glass cylinder task. Chimpanzees learned from observing a proficient demonstrator solve this tool task by focusing selectively on the aspects of the task relevant to solving it, although capuchins did not (Visalberghi, 1993; Visalberghi and Limongelli, 1996).

Great apes also succeed in imitating "manipulative" tasks. Field studies of gorillas and orangutans suggest that imitation plays a role in acquiring manipulative skills, such as how to pick and fold stinging nettles so that they can be eaten safely, or, how to get juice from sugar cane (Bryne, 1999c; Byrne and Byrne, 1993; Russon and Galdikas, 1993). Experiments using the two-action method with an artificial fruit task confirmed this suggestion. To open the artificial fruit, gorilla and chimpanzee observers imitated the manipulations they saw demonstrated (Stoinski et al., 2001; Whiten and Custance, 1996); orangutans were less successful, but their experiment suffered methodological problems (Custance et al., 2001). Other studies with the artificial fruit have shown the same difference in imitation between great apes and monkeys (for marmosets, see Caldwell and Whiten [1999]; for capuchins, see Custance et al. [1999]; and for baboons, see Marshall et al. [1999]).

Chimpanzees can also imitate arbitrary nonfunctional actions, like gestures, as several experiments using the two-action method have shown (Custance et al., 1995; Myowa-Yamakoshi and Matsuzawa, 1999; Tomasello et al., 1993). Similar methods show comparable abilities in orangutans (Miles et al., 1996).

Great apes succeed about equally in imitating manipulative and simple relational tasks, perhaps because the two types of tasks are not effectively different. The manipulative and tool tasks typically used are both functional tasks, both require recreating simple relations between one object and another, and both probably employ the same kind of means-ends understanding (Byrne and Russon, 1998; Russon 1998; Stoinski et al., 2001). This may also explain similarities in great apes' limitations on both types of tasks: typically, they imitated only parts of what they observed. By contrast, great apes may be better imitators of relational tasks than of arbitrary actions (Myowa-Yamakoshi and Matsuzawa, 1999, 2000). Myowa-Yamakoshi and Matsuzawa suggested that functional tasks might be easier to imitate than arbitrary actions because in functional tasks, the outcome of the demonstrated behavior is salient and it constrains, or even defines, the behavioral means. Specialists in cognitive development consider relations between objects to be more difficult to understand than individual actions (e.g., Case, 1985), however. That great apes imitate simple relational tasks better than individual actions may mean that they encode demonstrations at higher, not lower, cognitive levels, congruent with their advanced cognitive abilities.

This mosaic pattern of successes and failures in great ape imitation has drawn research attention to the many sources of information provided by a demonstration and to the question of which sources great apes attend to and which they select to imitate (Call, 1999; Custance et al., 2001; Myowa-Yamakoshi and Matsuzawa, 1999, 2000; Russon and Galdikas, 1995). In behavior that can be hierarchically decomposed, for instance, matching could occur at many levels (Byrne and Russon, 1998; Russon, 1999a).

Studies have tested whether great apes imitate at the level of outcomes, motor actions, action sequences, object–object relations, and behavioral programs (Byrne and Russon, 1998; Call and Tomasello, 1994; Russon, 1999a; Tomasello, 1996; Whiten, 1998; Whiten et al., 1996). Some findings suggest that chimpanzees attend to and imitate outcomes (Call and Tomasello, 1998; Myowa-Yamakoshi and Matsuzawa, 2000; Povinelli, 1991): they imitate the result of a demonstrator's actions (e.g., open a bolt lock) but use their own instead of the demonstrated motor actions (e.g., pull versus twist the bolt). These findings should be interpreted with care. Studies typically demonstrate functional tasks, like tool use, in which outcomes constrain and sometimes even define which actions are needed, and in which simple relations are focal features (e.g., Byrne and Russon, 1998; Russon, 1999a; Smith and Bryson, 1994; Stoinski et al., 2001). To remove a screw from a board, for instance, twisting is among the few actions possible. In such tasks, motor actions and outcomes are not as distinct as implied. Debates aside, this type of research illustrates that imitation is more complicated than most studies on non-human species have recognized, and probably a multilevel ability; that great apes' imitative abilities touch on some but not all of the features demonstrated; and that several biobehavioral factors probably contribute to great apes' imitative pattern.

Great apes' imitative success has mainly been attributed to a "cognitive gap" that separates them from monkeys, generated by some special cognitive capacity like mental representation or rudimentary symbolism (e.g., Byrne, 1995; Russon et al., 1996). Their imitative weaknesses may likewise reflect cognitive limitations (Russon et al., 1998). Great apes appear to have relatively few cognitive building blocks for encoding and re-enacting demonstrations. They may be able to combine only a few elements in one behavioral program and to encode the relations they observe only in undifferentiated form (e.g., join sticks by butting versus screwing them together). For all these reasons, their imitation produces rough, low fidelity copies. Their cognitive structures may resist modification, so they may persist with familiar behaviors rather than try new ones.

The focus on cognition has led to the relative neglect of other biobehavioral factors that govern or influence imitation, such as function and social dynamics. Development is a third factor of fundamental importance, for which, however, little comparable evidence is available on non-primate species. It is important to note that rich social learning conditions may be critical to developing imitative abilities in great apes although probably not, as some have claimed, special "human" enculturation (Parker and McKinney, 1999; Russon, 1999b; Suddendorf and Whiten, 2001).

Relatively little is known about the function of great ape imitation because research has emphasized imitation elicited under controlled experimental conditions over spontaneous imitation in species-normal contexts (Custance et al., 2002a; Miklósi, 1999). Experimentalists insist that imitation cannot be established conclusively from spontaneous behavior in the field, but few experimental studies have considered ecological validity and most employ highly contrived artificial tasks. This is unfortunate because function may be important comparatively, if indeed chimpanzees imitate some tasks better than others (e.g., simple functional relations versus arbitrary actions).

Imitation can serve two distinct functions, communicative and instrumental (Yando et al., 1978). Accordingly, it may assume two different forms: a communicative form, for copying detailed motor acts (action-level imitation, mimicry), and an instrumental form, for copying the organizational structure of behavior and filling in details independently (program-level imitation, method learning) (Byrne, 2002; Byrne and Russon, 1998). The former is better for communication because it ensures clarity of meaning and behavioral meshing between observer and model. The latter is better for instrumental behavior, like tool use, because effective solutions are "probabilistic" and require case-by-case revision at the detail level (Langer, 2000). Instrumental problems are virtually guaranteed to vary in detail from one instance to the next, so the principles demonstrated are what count, not the details.

Imitation can serve both functions in great apes but if they are better at program- than action-level imitation, the instrumental function may be the more biologically important. This suggestion is not new. Great ape imitation has been linked with acquiring complex manipulative and tool-assisted foraging techniques, which are required by a diet that relies seasonally on foods that are difficult to obtain (Byrne, 1995, 1997; Byrne and Russon, 1998; Parker and Gibson, 1979; Russon, 1998). Great apes' reliance on difficult foods is probably linked to their dietary breadth (over 200 species in some populations; Rodman, 2000), which in turn is probably linked to the seasonal food scarcities caused by their fruit preferences and their exceptionally high body mass (Waterman, 1984). These dietary conditions reflect two ecological factors that have been offered to explain the distribution of imitation across taxa: opportunistic-generalist versus conservative-specialist lifestyles (Klopfer, 1961) and foraging pressures that select for complex versus simple motor techniques (Russon et al., 1998). These factors do not occur in the same fashion in monkeys. Some monkeys rely seasonally on difficult foods but none reach the large body mass of the great apes, and monkeys' greater capacity for folivory and their different ranging and grouping patterns may favor other dietary options. Monkeys, then, may face ecological pressures that are less demanding of instrumental imitation.

In this light, action-level and program-level imitation may not simply be prominent nodes along a continuum of increasingly complex levels of imitation, but rather distinct forms, with distinct patterns of occurrence, which evolved in response to very different needs. This would be consistent with suggestions that the two forms of imitation are served by different mechanisms: kinesthetic-visual matching for action-level imitation, hierarchical plan construction for program-level imitation.

Social dynamics can also influence social learning (e.g., Coussi-Korbel and Fragaszy, 1995). Insofar as social dynamics differ between great apes, other primates, and other mammals, their respective imitative abilities may also differ. Primate societies differ from those of other mammals because of their stable, mixed-sex groups (e.g., van Schaik, 1996). Primate groups have a high potential for social complexity because they support long-term and dynamic social relationships within and between sexes. Some of this complexity is seen in primates' sophisticated social cognition (e.g., individual recognition, rank- and relationship-sensitive conflict management, alliances, tactical deception, exchanges of services like grooming and agonistic support). Great apes societies are distinctive within non-human primates for their fission–fusion tendencies and described by some as more flexible and more egalitarian social structures (Fuentes, 2000; van Schaik et al., in press). These social dynamics could differentially favor imitation: they broaden the social pathways along which individuals can share expertise because they enhance social tolerance.

A second facet of social dynamics that may differentially favor imitation in great apes over other non-human primates is the social use of gaze. Imitation relies on observers being able to watch demonstrators in sustained fashion. Prolonged gaze, or staring, typically functions as a form of threat in non-human species (Kaplan and Rogers, 2002; Redican, 1975). In great apes, however, staring and direct eye contact do not always make recipients withdraw or submit; indeed, subordinates often stare at dominants, not the reverse. These forms of gaze may serve other social purposes in great apes like initiating play or copulation, inviting reconciliation, intervening, greeting, and begging or scrounging (Russon, personal observation; Yamagiwa, 1992, in press). Tolerance for staring could facilitate imitation by enabling prolonged social observation. More flexible use of gaze could be a product of cognitive enhancement, insofar as cognitive takeovers of behavior tend to enhance flexibility. Alternatively, it could be a product of more flexible and more egalitarian societies, in which dominance is less rigidly defined and enforced.

Primate Imitation in Broader Perspective

Research on non-primate species has brought to light further biological considerations that are important in the study of imitation. Studies on non-primate species have often assessed vocal imitation (e.g., see Kaplan, this volume), for instance. Although vocal imitation is not considered directly comparable to visuo-motor imitation, such studies nevertheless point to biological factors likely to affect imitative propensities.

Species that have passed experimental tests of visuo-motor imitation include rats (Heyes and Dawson, 1990; Heyes et al., 1992, 1994), budgerigars (Dawson and Foss, 1965; Galef et al., 1986), a Gray parrot (Moore, 1992, 1996), pigeons (Zentall et al., 1996), Japanese quails (Atkins and Zentall, 1996), Carib grackles (Lefebvre et al., 1996, 1997), ravens (Fritz and Kotrschal, 1999), and dolphins (e.g., Marten et al., 1996; Xitco et al., 1998). Based on naturalistic tests, dolphins and Asian elephants may imitate tool-based or manipulative instrumental behaviors such as using leafy branches as fly switches (Hart et al., 2001; Rendell and Whitehead, 2001; see also Nissani, this volume).

In particular, studies of non-primate taxa have brought to light further methodological pitfalls in the comparative study of imitation. Cues from other sensory modalities may confound tests of visuo-motor imitation. For instance, the original claims that rats imitate were withdrawn because odor cues left by demonstrators were major contributors to behavioral matching (Gardner, 1997). For species that can imitate, variation may occur in the quality of imitation. Dolphins and great apes stand out as imitative generalists, capable of imitating a wide range of both motor and vocal behaviors, whereas other species have imitated a narrower range of actions (Herman, 2002; Herman and Pack, 2001; Myoma-Yamakoshi and Matsuzawa, 2000; Rendell and Whitehead, 2001). Dolphins and great apes also tend to imitate in "real time," whereas the parrot imitated only after multiple demonstrations and "incubation" periods several months long (Moore, 1992).

Confounding variables have also come to light through comparing imitation across a broader range of species. Motivational and procedural confounds are common risks because different species respond differentially to food deprivation, task and social stimuli, etc. Variables likely to confound imitation tests include opportunism, tolerance of humans, social preferences, development, attentional or perceptual biases, memory capacity, and degrees of neophobia (Miklósi, 1999; Russon et al., 1998). Again, failure to control for species differences in sociality is likely to have confounded many experimental tests of primate imitation, thereby compromising their validity (Boesch, 1993a, 1993b; Russon, 2003; Russon and Galdikas, 1995). Methods for removing such problems are not yet well established for variables likely to confound imitation tests. In this light, we consider the effects of cognitive, social, and functional variables on imitation.

Many of the non-primate taxa for which imitation has been claimed are reputedly capable of high-level cognition, including ravens, Gray parrots, elephants, and dolphins. Yet, imitation has also been claimed for species not known for sophisticated cognition, such as rats and avian non-mimids. However, much of the evidence for imitation in the latter species is strongly disputed, mainly on methodological grounds, and includes the case in which the claim for imitation has been withdrawn (e.g., Gardner, 1997). A capacity for high-level cognition then may still be a critical factor in potentiating imitation, although the possibility remains that it is not the master key and that species not noted for their cognitive sophistication may yet demonstrate imitative capabilities. What is clear is that imitation almost certainly depends on additional factors, like affect, motivation, social relations, memory, and perception.

The primary function of an imitative capacity for a species is therefore important to consider, as some have (e.g., Janik and Slater, 2000; Moore, 1996; Russon et al., 1998). It could well be communicative, coordinating activities and maintaining social cohesion. The most commonly cited adaptive advantages of imitative behavior in the non-human world are social, not instrumental: enhancing communication and group harmony through standardization. It is harder to envisage how imitation contributes to instrumental skills, which must be fine-tuned to physical versus social conditions and which outdate rapidly in changeable environments (e.g., Galef, 1992).

In support of this view, social communication is an important adaptive dimension in species potentially capable of imitation and social dynamics are important in motivating imitation (Boesch, 1993a, 1993b; Custance and Bard, 1994; Lefebvre et al., 1997; Moore, 1992; Pepperberg, 1988; Russon and Galdikas, 1995; Sayigh et al., 1993; Tyack, 1986). A wide diversity of species, non-human primates included, hone innately guided communicative signals to standard forms and uses (e.g., Seyfarth et al., 1980; Tomasello et al., 1985). A case in point is vocal imitation, which may be used by wild bottlenose dolphins in acquiring signature whistles (Sayigh et al., 1993) and by avian mimids in replicating vocalizations to handle social encounters (Moore, 1992; Pepperberg, 1988). Visuo-motor imitation may be used by some avian species in coping with particular types of social competition (Lefebvre et al., 1996, 1997). Imitative capacities may therefore be associated with a suite of social conditions that occur in a number of families of highly social taxa. Typically, these taxa are also long-lived. Imitative capacities may have evolved primarily in lineages where sociality and social learning, with the attendant value of social communication and of matching others' behavior, were critical precursors.

Some evidence nonetheless suggests that imitation may have an instrumental function in some non-primate species. It helps dolphins, perhaps grackles, and ravens acquire instrumental skills (Fritz and Kotrschal, 1999, 2002; Moore, 1992; Sayigh et al., 1993; Tayler and Saayman, 1973). Byrne and Russon (1998) offered the program-level model of imitation to address the problem of how imitation could contribute to instrumental skills entailing complex manipulative behaviors. Program-level imitation depends on highly sophisticated cognitive processes, so imitation that serves instrumental functions may be restricted to species with exceptionally high intellectual capacities. The picture that emerges is that imitation may have evolved to serve different functions in different taxa.

Both communication and instrumentality probably contribute to imitation in many taxa, given the biological factors that have been linked with imitation: long life-span, high-level cognition, and prolonged parental care (Roper, 1986), relatively egalitarian, fission–fusion societies (van Schaik et al., in press), opportunistic-generalist versus conservative-specialist lifestyle (Klopfer, 1961), scramble versus interference competition, and foraging pressures that select for complex versus simple motor techniques (e.g., embedded foods versus leaves or grains) (Dolman et al., 1996; Lefebvre et al., 1996; Russon et al., 1998). Individual factors probably apply differentially across species. The Gray parrot's capacity for visuo-motor imitation, for instance, may owe to an evolutionary path characterized by increasingly sophisticated mimetic learning (Moore, 1992). Nonvocal mimicry (copying sounds by movement versus vocalizing, e.g., banging objects with the beak to mimic percussive sounds) occurs in two species that are among the best vocal mimics, Gray parrots and starlings, and motor imitation occurs only in the Gray parrot. Several social and ecological factors may contribute to imitation in ravens, including high adaptability, neophobia and neophilia, and a degree of fission–fusion (Fritz and Kotrschal, 1999). Within cetaceans, evidence for imitation includes both communication and instrumental behavior so imitation may owe to both sets of factors in these species.

To turn back to primates, strictly social functions for imitation are unlikely (Russon et al., 1998). Effectively, primate imitation is great ape imitation. As such, it is likely to be a manifestation of higher cognition and probably closely linked with the pressures of exceptionally broad and difficult feeding niches, the extremely large body mass, and the extended life histories that distinguish the great apes from monkeys (e.g., Parker, 1996). Accordingly, although social functions are undoubtedly involved, great ape imitation may function equally, or even primarily, to boost immature great apes' acquisition of the complex foraging techniques they need to achieve independence in this feeding niche.

CULTURE

The study of culture has been considered from many perspectives. Anthropologists, for example, study culture as a human phenomenon, some think uniquely so. Tyler's definition of 1871 proposes that "Culture—is that complex whole which includes knowledge, belief, art, law, morals, custom, and other capabilities and habits acquired by man as a member of society" (p. 1), but this classic definition exists alongside several hundred others. Other definitions specify the transmission processes involved, typically symbolism, teaching, and language. Again, imitation has been emphasized as a cornerstone of culture (e.g., Bonner, 1980; Donald, 1991; Parker and Russon, 1996; Tomasello et al., 1993). Yet others consider that effectively sharing social information is the crux of culture, however it is achieved (e.g., Fritz and Kotrschal, 1999; King, 1994; McGrew, 1998). In these latter terms culture may be described generally in terms of sets of skills and knowledge that are shared by members of a community, and transmitted across generations through nongenetic channels. It is in these terms that many studies of culture among non-human animal species are considered.

For non-human primates, then, culture typically includes the social transfer of expertise, operating across generations and at the group level (Kummer, 1971; McGrew, 1998; Nishida, 1987). Moreover, primate cultures are commonly described in terms of behavioral traditions, broadly construed as shared practices that are relatively long-lasting, shared among group members, and acquired in part via social influences on learning (Fragaszy and Perry, 2003) or, more narrowly, as behaviors that occur habitually, repeatedly by several group members, or customarily by all group members of appropriate age, sex, and status (McGrew, 1998).

In broad biological perspective, culture is construed as one of the two processes (the other being regarded as genetic) that can transmit behavior across generations and so shape evolutionary change (Nishida, 1987). In this sense, cultural differences (traditions) are well-established phenomena in animals that are sustained by various social transmission mechanisms (Whiten and Ham, 1992). Commonly accepted criteria for behavioral traditions include the spread of novel behavior, and observations of differences in the expression of the same behavior among populations within a species.

Behavioral Traditions among Monkeys

One of the most famous examples in the literature on the behavior of monkeys is that of a young female Japanese macaque (Imo) in washing provisioned sweet potatoes on the island of Koshima in Japan (Kawai, 1965), and the subsequent adoption of this, and other behaviors by many other troop members. This was an early reported study of behavioral traditions among non-human primates and has consistently attracted a great deal of discussion.

Studies of the Koshima monkeys began in 1953 and have continued to the present. Hence, it is of significant interest that data are now available for seven generations of monkeys to document the history of the community. Critically, it has been possible to document which animal started the behavior, and how, and when, it was transmitted to other monkeys, as well as how the behavior changed during its transmission from one generation to the next.

There have been criticisms. An influential one was Galef's (1990), that the slow and steady rate of diffusion of the behavior was not in accord with a prediction of accelerated behavior if the mechanism of imitation was involved. However interesting a lack of evidence for imitation is in comparative terms, it is also irrelevant to the generally important point that, whatever the mechanisms involved, the data show evidence for the emergence, modification, and social routes of transmission (for discussion of the Koshima island studies, see de Waal, 2001).

There are other good examples of behavioral traditions among monkeys. For instance, stone handling among Japanese macaques provides a much-discussed case that was first described from its origins by Huffman (see review by Huffman, 1996). Stones are handled by monkeys in a number of ways that are apparently functionless; they are rubbed together; they produce a sound of click clacking, and are subsequently dropped. Stone handling began with a 3-year-old middle-ranking female in 1979, and subsequently spread to all the immatures in one group. Hence, once again, both the origin and the dissemination of the behavior are well documented. Similarly, Tanaka's (1998) detailed studies of removing parasitic louse eggs from the skin of other Japanese monkeys within matrilineal groups showed that when one female used a different technique in this regard, it was subsequently adopted by all her kin.

Matsuzawa (2002) has recently compared cultural phenomena between common chimpanzees and monkeys. He emphasizes three main points, first, that the cultural phenomena of Koshima monkeys are mainly the outcome of human provisioning. This is an interesting point with respect to the discussion by Kummer and Goodall (1985) in which they predict sudden environmental change, such as provisioning, as a major factor that may influence the development of behavioral innovations. Matsuzawa's second point is that the behavioral traditions of the monkeys contain no use of tools. This is in notable contrast with those of behavioral traditions in common chimpanzees, as is discussed in the next section. The third point is that whereas there is an emphasis on observational learning in chimpanzee studies, individual learning among monkeys is more prevalent.

From a different perspective, it is interesting that examples of behavioral traditions are reported for few species of monkeys as yet. There are some other examples, as in hamadryas baboons (Kummer, 1971), but the vast majority of information for monkeys comes from the genus Macaca, and especially the Japanese macaques, that have generated by far the largest catalog of apparent traditions among monkeys. These traditions include not only foraging (washing sweet potatoes, sluicing wheat, washing other foods, eating dead fish that has been washed ashore, capturing small octopus, and preying on frogs and lizards); but hygiene/comfort (bathing in hot springs and removing louse eggs); grooming (using stones in social grooming); and communication (grooming contact calls, courtship gestures and posture.)

Little is known about other species of macaques, and it is difficult to know what to make of the information in comparative terms. It may be relevant to future discussions, however, based upon a larger taxonomic database, that the 20 or so species of macaques live in a wide diversity of habitats across Asia; that there is a diversity of dietary adaptations; there are differences in ranging behavior, habitat use, and foraging behavior, and differences in sociodemographics, sexual, and social behavior (cf. Clarke and Boinski, 1995).

Considering other species for which behavioral traditions might be predicted, the much studied cebus monkeys are potentially interesting. As we noted earlier, they are highly cognitively astute, manipulative, socially tolerant, and generalist-extractive foragers, but evidence for cultural traditions has been very slight. Recent information by Panger and Perry (2002), however, describes observations on white-faced capuchins (*Cebus capucinus*) at three geographically close and ecologically similar study sites in Costa Rica. Of the 61 foods that were compared, 20 were found to be processed differently across the sites, and that most of the differences were either customary or habitual. Social learning processes are also implicated by observations analyzed from two of the sites; individuals who shared more time in proximity with each other, also shared different food processing techniques.

Behavioral Traditions among the Great Apes

By far the largest catalog of great ape traditions is derived from wild common chimpanzees. It spans differential food identification/preferences, many types and styles of tool use, habit nuances or style (e.g., in hunting, choice of prey and how consumed), grooming (hand clasp, social scratch), and vocalizations (pant hoot dialects). Whiten et al. (1999) reported 39 traditions that include tool use, hygiene/comfort, communication, grooming, and other behaviors (e.g., self-tickle).

To give one or two examples, chimpanzees in Gombe National Park famously fish for termites by inserting a twig (selected and perhaps modified) or a blade of grass into termite mounds; the soldier and worker ants that cling to the tool are carefully removed by drawing the tool across the mouth. Mahale chimpanzees more rarely eat termites and, when they do, they use their hands to break up the soil and remove the insects with their fingers. Moreover, they do use stick tools and scale trees to obtain wood boring ants. These ants—together with the same species of termites—are also present in Gombe, but the chimpanzees rarely eat them. In another well-known example, from West Africa, chimpanzees in the Taï Forest use stone tools against anvils to open five species of nutritious nuts but chimpanzees in neighboring communities do not, even though the conditions for stone nut-cracking are equally favorable (Boesch and Boesch-Achermann, 2000).

Much attention has been paid to tool cultures among common chimpanzees, for obvious reasons of interest with regard to ourselves, but other cultural traditions among common chimpanzees have been identified that involve grooming techniques and hunting style. When a kill of a red colobus monkey is made for example, different populations of chimpanzees share food in different ways. At Gombe, sons who kill share with mothers and brothers, but not with rival males; they also share with females with sexual swellings, and with high-ranking females (Stanford, 2001). At Taï there is a reciprocal system in which the male that kills shares meat with other members of the hunt regardless of whether they are relatives or allies (Boesch and Boesch-Achermann, 2000).

For wild orangutans, communicative and tool use traditions have been identified (Peters, 2001; van Schaik and Knott, 2001). The discovery of tool use traditions in the community of Sumatran orangutans at Suaq Balimbing is especially interesting because previously, tool use had very rarely been reported in orangutans in the wild. In this community, several types of tools are customary, including probes and hammers for extracting insects from tree holes, and probes for extracting seeds from Neesia fruit (Fox et al., 1999). Another wild orangutan community is known to eat Neesia, at Gunung Palung in Borneo, but none of its members has ever been seen using tools to obtain it (van Schaik and Knott, 2001). Gunung Palung orangutans have other traditions. They hold leaves against their mouths when they kiss-squeak (a species-typical orangutan vocalization), apparently to amplify the sound. This leafy technique is habitual in Gunung Palung and unknown elsewhere (Peters, 2001). Numerous other potential traditions were identified at a recent orangutan cultures workshop (van Schaik et al., 2003). Communities of forest-living rehabilitant orangutans also share practices, probably because they are unusually sociable and heavily focused on learning, and have offered evidence of shared habitual foraging and hygiene-related skills within cliques of immatures (Russon, 2002).

For wild gorillas, one traditional difference has been proposed between two western lowland populations in food identification or preference (weaver ants versus termites). Wild bonobos have provided virtually no evidence of traditions. Some zoo groups of lowland gorillas and bonobos have behavioral traditions (e.g., using sticks to rake in food, clapping hands to express pleasure), so both species probably share the potential for behavioral traditions despite limited evidence from the wild to date.

Facilitating Influences on Behavioral Traditions among Great Apes

We may now consider facilitating influences on the potential for cultural traditions among simian primates, especially for comparisons between monkeys and great apes.

If we start from the "uniquely human" view of culture, which requires grounding in symbolism, there are comparative points to be made relevant to the great apes. In the majority of cases, great apes use traditional tool skills for straightforward practical purposes like foraging or hygiene. In a few cases in chimpanzees, however, objects are used to create communicative displays that may rely on rudimentary symbolism. An example is using plant materials to communicate a desire to mate. Male chimpanzees in Mahale clip leaves in front of a desirable female, noisily tearing off leaves from a picked stem with fingers and teeth (Nishida, 1987). Males in Gombe do not leaf clip (Goodall, 1986), but they send the same message by rapidly shaking a small branch or bush. Taï chimpanzees also leaf clip but they changed how they use this display over time: leaf clipping was once a part of adult males' tree drumming sequence, then abruptly it appeared during resting or sexual frustration, and finally it disappeared from tree drumming (Boesch, 1996). Two communities, then, used different behaviors to signify the same outcome, and one community altered the function of a single behavior. The implication is that in chimpanzees, the relationship between a behavior and its meaning can be arbitrary and this is the essence of rudimentary symbolism. Agreement has not been reached on whether, technically, traditions like this qualify as symbolic. Such examples nonetheless raise the possibility that great ape traditions have some grounding in symbolism, although symbolism is neither extensive nor powerful in their case. To date, there is no suggestion of anything similar in monkey societies.

Differences between the behavioral traditions in great apes and monkeys are consistent with views that culture and cognition are interdependent in primates (e.g., Donald, 1991). Monkeys rarely if ever generate tool-using traditions but almost all great apes do; monkeys' traditional manipulative skills are less complex than those of great apes; and monkeys' communicative traditions have never shown the rudimentary symbolic features found in great apes. Such differences between primate species in the complexity of their cultures, both the traditions acquired and the means of transmitting them, may be tied to differences in their cognition, especially the social learning abilities that govern cultural transmission (e.g., Parker and Russon, 1996; Whiten, 2000). The stronger imitative abilities seen in great apes than in monkeys are probable contributors.

The functions of behavioural traditions for a species should track the functions of its social learning, notably, that great ape imitation may be geared primarily to acquiring complex instrumental skills. Virtually half (19/39) of the chimpanzee traditions identified by Whiten et al. (1999) are tool skills used in foraging and another 7/39 are tool or manipulative skills used elsewhere (hygiene-comfort). Next most common were communicative traditions (10/39, including 4 leaf clipping variants). The remaining 3/39 items are difficult to classify: grooming hand clasp, self-tickle, aimed throwing. If pant hoot dialects and social scratch are added from McGrew (1998), communication and grooming counts increase. While the paucity of evidence on other great apes precludes similar analysis, this list of traditions suggests that chimpanzee traditions serve transmission of foraging skills first, and then communication and hygiene skills.

Macaque traditions, which by default mainly represent monkeys, span a relatively similar range: foraging skills, other functional skills (hygiene/comfort, grooming, one tool use item), and communication. In terms of frequency, such functional skills outweigh communication as they do in great apes.

Differences suggested are less complex traditions in macaques than great apes (almost no tool skills, simpler foraging techniques) and a smaller range of traditions in macaques than chimpanzees.

Taken at face value, these patterns suggest that traditions in monkeys and great apes are more strongly geared to sustaining instrumental functional skills than communication. Given the patchiness of the evidence and the likelihood that findings are biased toward researchers' interests, however, these patterns may not be representative of the distribution of primate traditions.

Primate traditions depend on social processes beyond the mental mechanisms that enable social learning. Rehabilitant orangutans are astute social learners, for instance, but their socially acquired behaviors are not known to spread community-wide (McGrew, 1998; Russon and Galdikas, 1993). Effectively, the concept of social learning does not capture the role of the community in sustaining traditions. Several social parameters within a community must probably be favorable for cultural transmission to occur, with respect to such factors as relationships between learners and experts, availability of multiple sources of social input, pathways for social learning, how traditional expertise is distributed through the community, indirect social influences, and bases for social tolerance. Exactly what social conditions are needed to enable behavior to spread widely remains unclear, although some candidates have been proposed.

One probable requirement is high social tolerance, which may be necessary for sharing manipulative skills (Coussi-Korbel and Fragaszy, 1995). In primates, social tolerance may affect social transmission through the relationships it facilitates. Relationships influence the pathways along which social learning is channeled because they affect who can be near whom when behavior is performed, and where attention is likely to be directed (see also Box, 1999). Species variation in social tolerance, then, could affect social transmission by altering how, and how rigidly, relationship networks are defined. A second likely requirement is tolerance and relationship patterns that enable strong horizontal and oblique pathways of social influence (van Schaik and Knott, 2001). In primates, early social influence is typically vertical, within mother–offspring dyads. Vertical influence predominates through infancy in simian primates and continues into juvenility in great apes, after which it wanes. Accordingly, it mainly supports basic expertise (given learners' physical and cognitive immaturity) and narrow diffusion. Wide behavior spread may then depend on whether a primate species offers strong horizontal or oblique social tolerance beyond infancy.

Three features of great ape sociality may provide these requirements. (a) Their tendencies to fission–fusion, flexibility, and egalitarianism may enhance horizontal and oblique diffusion by broadening social tolerance. (b) Their life history profiles favor horizontal and oblique diffusion in the period between puberty and adulthood. Dispersal, sexually based gregariousness, and sophisticated physical and cognitive capabilities occur together during this period (Russon, 2003). Dispersal opens opportunities for horizontal and oblique diffusion via sexual liaisons, because immigrants typically gain admission to new groups on the basis of sexual attractiveness and sexual alliances with group members (Feistner and McGrew, 1989; van Schaik et al., in press; Yamagiwa, 1992, in press). That these social changes coincide with advanced cognitive and physical capabilities may not be accidental. Powerful capabilities support rapid sharing of complex expertise, and it is critical to extend adult-level expertise during this period because of the new needs created by changing social roles and ranges (Russon, 2002). (c) Their considerable food sharing (van Schaik et al., in press; Yamagiwa, in press) would facilitate sharing food-related expertise within relationships where food is commonly shared.

Individually, these social features do not necessarily differentiate great apes from other non-human primates. Cebus, as already noted, are socially very tolerant. Atelines (spider monkeys, muriquis) live in fission–fusion societies that closely resemble those of the great apes (van Schaik et al., in press). None of the other simian primates shares the constellation of social parameters found in great apes, however, or this constellation combined with more powerful cognition. Accordingly, great ape sociality may offer different cultural potentials. The processes involved have been articulated in Japanese macaques (Huffman and Hirata, 2003) and chimpanzees (Boesch and Boesch-Achermann, 2000; Inoue-Nakamura and Matsuzawa, 1997), by studies that have tracked how behaviors spread through a group from their inception and how naïve individuals acquire them. The probability is that these factors operate together, interactively, to support behavioral traditions.

If primate processes of behavioral traditions are integrated in this fashion, then it is the particular constellation of features, with all their changing qualities and interactions, that shape behavioral traditions, and processes and not any one feature acting alone. What builds stronger cultural potential in great apes than monkeys is the particular mix of qualities.

As a final perspective in this area, it is important to consider that behavioral traditions, sustained by a variety of social learning mechanisms, should be common in many animals. We require much more information (e.g., Box and Gibson, 1999) although recent work emphasizes that opportunities for social learning are available and may be important in a wide diversity of mammalian species (Box and Gibson, 1999) and are known to exist for many altricial avian species.

Evidence for sharing practices at the community level and across generations is more difficult to obtain, but recent work offers interesting possibilities. Fritz and Kotrschal (1999) have argued that ravens' socio-ecology fosters the formation of behavioral traditions. Ravens are reputedly cognitively sophisticated, ecologically highly adaptable, both neophobic and neophilic, and some age-sex classes form fission–fusion groups. Further, Rendell and Whitehead (2001) have presented evidence for the sharing of practices between mother and offspring and for group-specific behavioral traditions in some cetaceans. These cetaceans have sophisticated social systems, presenting opportunities for individuals to practice behavioral traditions, and versatile and complex cognition (e.g., mental representation, to support broad and flexible repertoires of foraging techniques).

Evidence for behavioral traditions in non-primate taxa is currently suggestive, but it emphasizes that we probably fail to appreciate the potential of many non-primate taxa for behavioral traditions. This means that primate achievements in this regard are not unique: to a considerable degree, they probably owe to biological factors, especially socio-ecological ones, that affect other taxa as well. If there are distinctions to be made, they are probably distinctions of degree or distinctions reflecting the particular way in which the constellation of potentiating factors is mixed.

REFERENCES

Anderson, J. R., 1996, Chimpanzees and capuchin monkeys: Comparative cognition, in: *Reaching into Thought: The Minds of the Great Apes*, A. E. Russon, K. A. Bard, and S. T. Parker, eds., Cambridge University Press, Cambridge, pp. 23–56.

Anderson, J. R., 2000, Tool-use, manipulation and cognition in capuchin monkeys (Cebus), in: *New Perspectives in Primate Evolution and Behaviour*, C. Harcourt, ed., Westbury Publishing, Otley, pp. 91–110.

Aoki, K., and Feldman, M. W., 1987, Towards a theory for the evolution of cultural communication: Coevolution of signal and reception, *Proc. Natl. Acad. Sci. USA*, **84**:7164–7168.

Atkins, C. K., and Zentall, T. R., 1996, Imitative learning in male Japanese quail (*Coturnix japonica*) using the two-action method, *J. Comp. Psy.* **110**(3):316–320.

Boesch, C., and Boesch-Achermann, H., 2000, *The Chimpanzees of the Taï Forest: Behavioural Ecology and Evolution*, Oxford University Press, Oxford.

Boesch, C., 1993a, Towards a new image of culture in wild chimpanzees? *Behav. Brain Sci.* **16**:514–515.

Boesch, C., 1993b, Aspects of transmission of tool-use in wild chimpanzees, in: *Tools, Language, and Cognition in Human Evolution*, K. R. Gibson, and T. Ingold, eds., Cambridge University Press, Cambridge, pp. 171–182.

Boesch, C., 1996, Three approaches to assessing chimpanzee culture, in: *Reaching into Thought: The Minds of the Great Apes*, A. E. Russon, K. A. Bard, and S. T. Parker, eds., Cambridge University Press, Cambridge, pp. 404–429.

Bonner, J. T., 1980, *The Evolution of Culture in Animals*, Princeton University Press, Princeton.

Boran, J. R., and Heimlich, S., 1999, Social learning in cetaceans: Hunting, hearing and hierarchies, in: *Mammalian Social Learning: Comparative and Ecological Perspectives*, O. H. Box and K. R. Gibson, eds., Cambridge University Press, Cambridge, pp. 282–307.

Box, H. O., 1984, *Primate Behaviour and Social Ecology*, Chapman & Hall, London and New York.

Box, H. O., 1994, Comparative perspectives in primate social learning: New lessons for old traditions, in: Current Primatology, J. J. Roeder, B. Thierry, J. R. Anderson, and N. Herrenschmidt, eds., Université Louis Pasteur, Strasbourg, pp. 321–327.

Box, H. O., 1999, Temperament and socially mediated learning among primates, in: *Mammalian Social Learning: Comparative and Ecological Perspectives*, H. R. Box, and K. R. Gibson, eds., Cambridge University Press, Cambridge, pp. 33–56.

Box, H. O., and Gibson, K. R., eds., 1999, *Mammalian Social Learning: Comparative and Ecological Perspectives*, Cambridge University Press, Cambridge.

Boyd, R., and Richerson, P. J., 1988, An evolutionary model of social learning: The effects of spatial and temporal variation, in: *Social Learning: Psychological and Biological Perspectives*, T. R. Zentall and G. B. Galef, eds., Lawrence Erlbaum Associates, Hillsdale, New Jersey, pp. 29–48.

Bugnyar, T., and Huber, L., 1997, Push or pull: An experimental study on imitation in marmosets, *Anim. Behav.* **54**:817–831.

Byrne, R. W., 1995, *The Thinking Ape: Evolutionary Origins of Intelligence*, Oxford University Press, Oxford.

Byrne, R. W., 1997, The Technical Intelligence hypothesis: An alternative evolutionary stimulus to intelligence? in: Machiavellian Intelligence, 2nd ed. R. W. Byrne, and A. Whiten, eds., Cambridge University Press, Cambridge, pp. 289–311.

Byrne, R. W., 1999a, Primate cognition: Evidence for the ethical treatment of primates, in: Attitudes to Animals: Views in Animal Welfare, F. L. Dolins ed., Cambridge University Press, Cambridge, pp. 114–125.

Byrne, R. W., 1999b, Imitation without intentionality: Using string parsing theory to copy the organisation of behaviour, *Anim. Cogn.* **2**:63–72.

Bryne, R. W., 1999c, Cognition in great ape ecology: Skill-learning ability opens up foraging opportunities, in: *Mammalian Social Learning: Comparative and Ecological Perspectives*, H. O. Box and K. R. Gibson, eds., Cambridge University Press, Cambridge, pp. 333–350.

Bryne, R. W., 2000, Changing views on imitation in primates, in: *Primate Encounters*, S. C. Strum and L. M. Fedigan, eds., University of Chicago Press, pp. 296–309.

Byrne, R. W., 2002, The primate origins of human intelligence, in: *The Evolution of Intelligence*, R. J. Sternberg and J. C. Kaufman, eds., Lawrence Erlbaum Associates, Mahwah, pp. 79–95.

Byrne, R. W., and Byrne, J. E. M., 1993, Complex leaf-gathering skills of mountain gorillas (*Gorilla g. berengei*): Variability and standardization, *Am. J. Primatol.* **31**:241–261.

Byrne, R. W., and Russon, A. E., 1998, Learning by imitation: A hierarchical approach, *Brain Behav. Sci.* **21**:667–721.

Caine, N. G., 1993, Flexibility and co-operation as unifying themes in Saguinus social organisation and behaviour: The role of predation pressures, in: *Marmosets and Tamarins: Systematics, Behaviour and Ecology*, A. B. Rylands, ed., Oxford University Press, Oxford, pp. 200–219.

Caldwell, C. A., and Whiten, A., 1999, Observational learning in the marmoset monkey (*Callithrix jacchus*). *Proc. AISB Convention, Edinburgh: Symp Imitation Animals and Artefacts*, pp. 27–31.

Call, J., and Tomasello, M., 1994, The social learning of tool use by orangutans (*Pongo pygmaeus*), *Hum. Evol.* **9**:297–313.

Call, J., and Tomasello, M., 1998, Distinguishing intentional from accidental actions in orangutans (*Pongo pygmaeus*), chimpanzees (*Pan troglodytes*), and human children (*Homo sapiens*), *J. Comp. Psy.* **112**:192–206.

Call, J., 1999, Levels of imitation and cognitive mechanisms in orangutans, in: *The Mentalities of Gorillas and Orangutans*, S. T. Parker, R. W. Mitchell, and H. L. Miles eds., Cambridge University Press, Cambridge, pp. 316–341.

Cambefort, J. P., 1981, The comparative study of culturally transmitted patterns of feeding habits in the chacma baboons, *Papio ursinus* and the vervet monkeys *Cercopithecus aethiops, Folia Primatol.* **36**:243–263.

Caro, T. M., 1994, *Cheetahs of the Serengeti Plains*, University of Chicago Press, Chicago.

Case, R., 1985, *Intellectual Development: Birth to Adulthood*, Academic Press, New York.

Clarke, A. S., and Boinski, S., 1995, Temperament in non-human primates, *Am. J. Primatol.* **37**:103–125.

Coussi-Korbel, S., and Fragaszy, D., 1995, On the relation between social dynamics and social learning, *Anim. Behav.* **50**:1441–1453.

Cowlishaw, G., and Dunbar, R. I. M., 2000, *Primate Conservation Biology*, University of Chicago Press, Chicago.

Custance, D. M., and Bard, K. A., 1994, The comparative and developmental study of self-recognition and imitation: The importance of social factors, in: *Self-awareness in Animals and Humans: Developmental Perspectives*, S. T. Parker, R. W. Mitchell, and M. L. Boccia, eds., Cambridge University Press, Cambridge, pp. 207–226.

Custance, D. M., Whiten, A., and Bard, K. A., 1995, Can young chimpanzees imitate arbitrary actions? Hayes and Hayes (1952) revisited. *Behaviour* **132**:839–858.

Custance, D. M., Whiten, A., and Fredman, T., 1999, Social learning of an artificial fruit task in capuchin monkeys (*Cebus apella*), *J. Comp. Psy.* **113**:1–11.

Custance, D. M., Whiten, A., Sambrook, T., and Galdikas, B. M. F., 2001, Testing for social learning in the "artificial fruit" processing of wildborn orangutans (*Pongo pygmaeus*), Tanjung Puting, Indonesia, *Anim. Cogn.* **4**:305–313.

Custance, D. M., Whiten, A., and Fredman, T., 2002a, Social learning and primate reintroduction. *Int. J. Primatol.* **23**(3):479–499.

Custance, D., Rigamont., Previde, E. P., and Spiezio, C., 2002b, Testing for social learning on four puzzle boxes in pig-tailed macaques (*Macaca nemestrina*) and young human infants, *Abstract of the X1Xth Congress of the International Primatological Society*, Beijing, China, p. 237.

Dawson, B. V., and Foss, B. M., 1965, Observational learning in budgerigars, *Anim. Behav.* **13**:470–474.

De Waal, F., 2001, *The Ape and the Sushi Master*, Basic Books, New York.

Dolman, C., Templeton, J., and Lefebvre, L., 1996, Mode of foraging competition is related to tutor preference in Zenaida aurita, *J. Comp. Psy.* **110**(1):45–54.

Donald, M., 1991, *Origins of the Modern Mind*, Harvard University Press, Cambridge.

Faulkes, C. G., 1999, Social transmission of information in a eusocial rodent, the naked mole-rat (*Heterocephalus glaber*), in: *Mammalian Social Learning: Comparative and Ecological Perspectives*, H. O. Box and K. R. Gibson, eds., Cambridge University Press, Cambridge, pp. 205–219.

Feistner, A. T. C., and McGrew, W. C., 1989, Food sharing in primates: A critical review, in: *Perspectives in Primate Biology*, Vol. 3 P. K. Seth and S. Seth, eds., Today and Tomorrows Printers and Publishers, New Delhi, pp. 21–36.

Fox, E. A., Sitompul, A. F., and van Schaik, C. P., 1999, Intelligent tool use in wild Sumatran orangutans, in: *The Mentalities of Gorillas and Orangutans: Comparative Perspectives*, S. T. Parker, R. W. Mitchell, and H. L. Miles, eds., Cambridge University Press, Cambridge, pp. 99–116.

Fragaszy, D. M., and Adams-Curtis, L. E., 1991, Environmental challenges in groups of capuchins, in: *Primate Responses to Environmental Change*, H. O. Box, ed., Chapman & Hall, London, pp. 239–264.

Fragaszy, D. M., and Perry, S., eds., 2003, *Towards a Biology of Traditions: Models and Evidence*, Cambridge University Press, Cambridge.

Fritz, J., and Kotrschal, K., 1999, Social learning in common ravens, *Corvus corvax*, *Anim. Behav.* **57**:785–793.

Fritz, J., and Kotrschal, K., 2002, On avian imitation: Cognitive and ethological perspectives, in: *Imitation in Animals and Artifacts*, K. Dautenhahn and C. L. Nehaniv, eds., MIT Press, Cambridge MA and London, pp. 133–155.

Fuentes, A., 2000, Hylobatid communities: Changing views of pair bonding and social organization in hominoids, *Yrbk Phys. Anth.* **43**(Suppl. 31 to *Am. J Phys. Anthropol.*): 33–60.

Galef, B. G., Jr., 1988, Imitation in animals: History, definition, and interpretation of data from the psychological laboratory, in: *Social Learning: Psychological and Biological Perspectives*, T. R. Zentall and B. G. Galef, Jr., eds., Lawrence Erlbaum Associates, Hillsdale, pp. 3–28.

Galef, B. G., Jr., 1990, Tradition in animals: Field observations and laboratory analyses, in: *Interpretations and Explanations in the Study of Behaviour: Comparative Perspectives*, M. Bekoff and D. Jamieson, eds., Westview Press, Boulder, CO., pp. 74–95.

Galef, B. G., Jr., 1992, The question of animal culture, *Hum. Nat.* **3**:157–178.

Galef, B. G., Jr., 1996, Social enhancement of food preferences in Norway rats: A brief review, in: *Social Learning in Animals: The Roots of Culture*, C. M. Heyes and B. G. Galef, Jr., eds., Academic Press, New York, pp. 49–64.

Galef, B. G., Jr., 1998, Recent progress in studies of imitation and social learning in animals, in: *Advances in Psychological Science*, Vol. 2, *Biological and Cognitive Aspects.* M. Sabourin, F. Craik, and M. Robert, eds., Psychology Press, pp. 275–299.

Galef, B. G., Jr., and Allen, C., 1995, A model system for studying animal traditions, *Anim. Behav.* **50**:705–717.

Galef, B. G., Jr., Manzig, L. A., and Field, R. M., 1986, Imitation learning in budgerigars: Dawson and Foss (1965) revisited, *Behav. Proc.* **13**(1–2):191–202.

Galef, B. G., Jr., and Wigmore, S. W., 1983, Transfer of information concerning distant foods: A laboratory investigation of the "information-centre" hypothesis, *Anim. Behav.* **31**:748–758.

Gardner, M. R., 1997, *Imitation: The Methodological Adequacy of Directional Control Tests*, Unpublished doctoral thesis, University of London, London.

Gibson, K. R., and Ingold, T., eds., 1993, *Tools, Language and Cognition in Human Evolution*, Cambridge University Press, Cambridge.

Giraldeau, L. A., 1997, The ecology of information use, in: *Behavioural Ecology: An Evolutionary Approach*, Fourth edition, N. Davies and J. Krebs, eds., Blackwell Science, Oxford, pp. 42–68.

Giraldeau, L. A., and Caraco, T., 2000, *Social Foraging Theory*, Princeton University Press, Princeton.

Goodall, J., 1986, *The Chimpanzees of Gombe: Patterns of Behavior*, The Belknap Press, Cambridge.

Guinet, C., and Bouvier, J., 1995, Development of intentional stranding hunting techniques in killer whales (*Orcinus orca*) calves at Crozet Archipelago, *Can. J. Zool.* **73**:27–33.

Hall, K. R. L., 1963, Observational learning in monkeys and apes, *Brit. J. Psy.* **54**:201–226.

Hart, B. J., Hart, L. A., McCoy, M., and Sarath, C. R., 2001, Cognitive behaviour in Asian elephants: Use and modification of branches for fly switching, *Anim. Behav.* **62**:839–847.

Hemery, C., Fragaszy, D. M., and Deputte, B. L., 1998, Human-socialised capuchins match objects but not actions, *Paper Presented at the XV11th Congress of the International Primatological Society*, Antananarivo, Madagascar.

Herman, L. M., and Pack, A. A., 2001, Laboratory evidence for cultural transmission mechanisms, *Behav. Brain Sci.* **24**(2):335–336.

Herman, L. M., 2002, Vocal, social and self-imitation by bottlenosed dolphins, in: *Imitation in Animals and Artifacts*, K. Dautenhahn and C. Nehaniv eds., MIT Press, Cambridge, pp. 63–108.

Herve, N., and Deputte, B. L., 1993, Social influence in manipulations of a capuchin monkey raised in a human environment: A preliminary case study, *Primates* **34**(2):227–232.

Heyes, C. M., 1994, Social learning in animals: Categories and mechanisms, *Bio. Rev.* **69**:207–231.

Heyes, C. M., 1996, Genuine imitation? in: *Social Learning in Animals*, C. M. Heyes and B. G. Galef, Jr., eds., Academic Press, New York, pp. 371–389.

Heyes, C. M., 1998, Theory of mind in non-human primates, *Behav. Brain Sci.* **21**(1):101–148.

Heyes, C. M., and Dawson, G. R., 1990, A demonstration of observational learning in rats using a bidirectional control, *Quart. J. Exp. Psy.* **42B**:59–71.

Heyes, C. M., Dawson, G. R., and Nokes, T., 1992, Imitation in rats: Initial responding and transfer evidence, *Quart. J. Exp. Psy.* **45B**:229–240.

Heyes, C. M., Jaldow, E., Nokes, T., and Dawson, G. R., 1994, Imitation in rats (*Rattus norvegicus*): The role of demonstrator action, *Behav. Proc.* **32**:173–182.

Heyes, C. M., and Ray, E., 2000, What is the significance of imitation in animals? *Adv. Study Behav.* **29**:215–245.

Higginbottom, K., and Croft, D. B., 1999, Social learning in marsupials, in: *Mammalian Social Learning: Comparative and Ecological Perspectives*, H. O. Box and K. R. Gibson, eds., Cambridge University Press, Cambridge, pp. 80–101.

Hobhouse, L. T., 1915, *Mind in Evolution*. Macmillan, London.

Hoelzel, A. R., 1991, Killer whale predation on marine mammals at Punta Norte, Argentina: Food sharing, provisioning and foraging strategy, *Behav. Ecol. Sociol* **129**:1–8.

Hudson, R., Schaal, B., and Bilko, A., 1999, Transmission of olfactory information from mother to young in the European rabbit, in: *Mammalian Social Learning: Comparative and Ecological Perspectives*, H. O. Box and K. R. Gibson, eds., Cambridge University Press, Cambridge, pp. 141–157.

Huffman, M. A., 1984, Stone-play of Macaca fuscata in Arashiyama B. troop: Transmission of a non-adaptive behavior, *J. Hum. Evol.* **13**:725–735.

Huffman, M. A., 1996, Acquisition of innovative cultural behaviors in non-human primates: A case study of stone handling, a socially transmitted behavior in Japanese macaques, in: *Social Learning in Animals: The Roots of Culture*, C. M. Heyes and B. G. Galef, Jr., eds., Academic Press, London, pp. 267–289.

Huffman, M. A., and Hirata, S., 2003, Biological and ecological foundations of primate behavioral traditions, in: *Towards a Biology of Traditions: Models and Evidence*, D. M. Fragaszy and S. Perry, eds., Cambridge University Press, Cambridge, pp. 267–296.

Inoue-Nakamura, N., and Matsuzawa, T., 1997, Development of stone tool use by wild chimpanzees (*Pan troglodytes*), *J. Comp. Psy.* **111**:159–173.

Janik, V., and Slater, P. J. B., 2000, The different roles of social learning in vocal communication, *Anim. Behav.* **60**:1–11.

Jones, C., Lawton, J., and Shachak, M., 1997, Positive and negative effects of organisms as ecosystems engineers, *Ecology* **78**:1946–1957.

Kaplan, G., and Rogers, L. J., 2002, Patterns of gazing in orangutans (*Pongo pygmaeus*), *Int. J. Primatol.* **23**(3):501–526.

Kawai, M., 1965, Newly acquired pre-cultural behavior of a natural troop of Japanese monkeys on Koshima Island, *Primates* **6**:1–30.

King B. J., 1994, *The Information Continuum: Evolution of Social Information Transfer in Monkeys, Apes, and Hominids*, School of American Research Press, Santa Fe.

King, B. J., 1999, New directions in the study of primate learning, in: *Mammalian Social Learning: Comparative and Ecological Perspectives*, H. O. Box and K. R. Gibson, eds., Cambridge University Press, Cambridge, pp. 17–32.

Kitchener, A. C., 1999, Watch with mother a review of social learning in the Felidae, in: *Mammalian Social Learning: Comparative and Ecological Perspectives*, H. O. Box and K. R. Gibson, eds., Cambridge University Press, Cambridge, pp. 236–258.

Klein, D. R., 1999, Comparative social learning among artic herbivores: The caribou, muskox and artic hare, in: *Mammalian Social Learning: Comparative and Ecological*

Perspectives, H. O. Box and K. R. Gibson, eds., Cambridge University Press, Cambridge, pp. 126–140.

Klopfer, P. H., 1961, Observational learning in birds: The establishment of behavioural modes, *Behaviour* **17**:71–80.

Kummer, H., 1971, *Primate Societies: Group Techniques of Ecological Adaptation*, Aldine, Chicago.

Kummer, H., and Goodall, J., 1985, Conditions of innovative behaviour in primates, *Phil. Trans. R. Soc. Lond. B* **308**:203–214.

Laland, K. N., 1999. Exploring the dynamics of social transmission with rats, in: *Mammalian Social Learning: Comparative and Ecological Perspectives*, H. O. Box and K. R. Gibson, eds., Cambridge University Press, Cambridge, pp. 174–187.

Laland, K., Odling-Smee, J., and Feldman, M., 2000, Niche construction, biological evolution, and cultural change, *Behav. Brain Sci.* **23**(1):131–146.

Laland, K. N., Richerson, P. J., and Boyd, R., 1993, Animal social learning: Toward a new theoretical approach, *Pers. Ethol.* **10**:249–277.

Laland, K. N., Richerson, R. J., and Boyd, R., 1996, Developing a theory of animal social learning, in: *Social Learning in Animals: The Roots of Culture*, C. M. Heyes and B. G. Galef, Jr., eds., Academic Press, New York, pp. 129–154.

Laland, K. N., and Williams, K., 1997, Shoaling generates social learning of foraging information in guppies, *Anim. Behav.* **53**:1161–1169.

Laland, K. N., and Williams, K., 1998, Social transmission of maladaptive information in the guppy, *Behav. Ecol.* **9**:493–499.

Langen, T. A., 1996, Social learning of a novel foraging skill by White-throated Magpie-jays (*Calocitta formosa*): A field experiment, *Ethology* **102**:157–166.

Langer, J., 2000, The descent of cognitive development, (with peer commentaries), *Dev. Sci.* **3**(4):361–388.

Lee, P. C., and Moss, C. J., 1999, The social context for learning and behavioural development among African elephants, in: *Mammalian Social Learning: Comparative and Ecological Perspectives*, H. O. Box and K. R. Gibson, eds., Cambridge University Press, Cambridge, pp. 102–125.

Lefebvre, L., 1986, Cultural diffusion of a novel food-finding behaviour in urban pigeons: An experimental field test, *Ethology* **71**:292–304.

Lefebvre, L., Koelle, M., Brown, K., and Templeton, J., 1997, Carib grackles imitate conspecifics and Zenaida dove tutors, *Behaviour* **134**(13–14):1003–1017.

Lefebvre, L., Palameta, B., and Hatch, K. K., 1996, Is group-living associated with social learning? A comparative test of a gregarious and a territorial Columbid, *Behaviour* **133**(3–4):241–261.

Marshall, S. R. J., Whiten, A., and Caldwell, C. (1999). Observational learning in baboons, *Presented at the Association for the Study of Animal Behaviour Conference on Evolution of Mind*, London.

Marten, K., Shariff, K., Parakos, S., and White, J., 1996, Ring bubbles of dolphins, *Sci. Am.* **275**(2):83–87.

Matsuzawa, T., 2002, Monkey culture and chimpanzee culture, *Abstract of the X1Xth Congress of the International Primatological Society*, Beijing, China.

McGrew, W. C., 1998, Culture in nonhuman primates? *Ann. Rev. Anth.* **27**:301–328.

Meinel, M., 1995, *Eliciting True Imitation of Object Use in Captive Orangutans*, Unpublished BA Thesis, Glendon College, York University, Toronto, Canada.

Meltzoff, A. N., 1996, The human infant as imitative generalist: A 20-year progress report on infant imitation with implications for comparative psychology, in: *Social Learning in Animals: The Roots of Culture*, C. M. Heyes and B. G. Galef, Jr., eds., Academic Press, London, pp. 347–370.

Miklósi, A., 1999, The ethological analysis of imitation, *Biol. Rev.* **74**:347–374.

Miles, H. L., Mitchell, R. W., and Harper, S., 1996, Imitation, pretense and self-awareness in a signing orangutan, in: *Reaching into Thought: The Minds of the Great Apes*, A. E. Russon, K. A. Bard, and S. T. Parker, eds., Cambridge University Press, Cambridge, pp. 278–299.

Milton, K., 1993, Diet and social organisation of a free-ranging spider monkey population: The development of species-typical behaviour in the absence of adults, in: *Juvenile Primates: Life History, Development and Behaviour*, M. E. Pereira and L. A. Fairbanks, eds., Oxford University Press, Oxford, pp. 173–181.

Moore, B., 1996, The evolution of imitative learning, in: *Social Learning in Animals: The Roots of Culture*, C. M. Heyes and B. G. Galef, Jr., eds., Academic Press, London, pp. 245–265.

Moore, B. R., 1992, Avian movement imitation and a new form of mimicry: Tracing the evolution of a complex form of learning. *Behaviour* **122**:231–263.

Myowa-Yamakoshi, M., and Matsuzawa, T., 1999, Factors influencing imitation of manipulatory actions in chimpanzees (*Pan troglodytes*), *J. Comp. Psy.* **113**: 128–136.

Myowa-Yamakoshi, M., and Matsuzawa, T., 2000, Imitation of intentional manipulatory actions in chimpanzees (*Pan troglodytes*), *J. Comp. Psy.* **114**:381–391.

Nel, J., 1999, Social learning in canids: An ecological perspective, in: *Mammalian Social Learning: Comparative and Ecological Perspectives*, H. O. Box and K. R. Gibson, eds., Cambridge University Press, Cambridge, pp. 250–278.

Nishida, T., 1987, Local traditions and cultural transmission, in: *Primate Societies*, S. Smuts, D. L. Cheney, R. M. Seyfarth, R. W. Wrangham, and T. T. Struhsaker, eds., University of Chicago Press, Chicago, pp. 467–474.

Panger, M., and Perry, S., 2002, The role of social influences on cross-site differences in food processing techniques used by free-ranging white-faced capuchins (*Cebus capucinus*) in Costa Rica, *Abstract of the X1Xth Congress of the International Primatological Society*, Beijing, China.

Parker, S. T., 1996, Apprenticeship in tool-mediated extractive foraging: Imitation, teaching and self-awareness in great apes, in: *Reaching into Thought: The Minds of the Great Apes*, A. E. Russon, K. A. Bard, and S. T. Parker, eds., Cambridge University Press, Cambridge, pp. 348–370.

Parker, S. T., and Gibson, K. R., 1979, A developmental model for the evolution of language and intelligence in early hominids, *Behav. Brain Sci.* **2**:367–408.

Parker, S. T., and McKinney, M. L., 1999, *Origins of Intelligence: The Evolution of Cognitive Development in Monkeys, Apes, and Humans*, The Johns Hopkins University Press, Baltimore.

Parker, S. T., and Russon, A. E., 1996, On the wild side of culture and cognition in the great apes, in: *Reaching into Thought: The Minds of the Great Apes*, A. E. Russon, K. A. Bard, and S. T. Parker, eds., Cambridge University Press, Cambridge, pp. 430–450.

Pepperberg, I. M., 1988, The importance of social interaction and observation in the acquisition of communicative competence: Possible parallels between avian and human learning, in: *Social Learning Psychological and Biological Perspectives*, T. R. Zentall and B. G. Galef, Jr., eds., Lawrence Erlbaum Associates, Hillsdale, pp. 279–300.

Perucchini, P., Bellagamba, F., Visalberghi, E., and Camaioni, L., 1997, Influenza del comportamento di un conspecifico sulla manipolazione nel cebo, *Eta Evolutiva* **58**:50–58.

Peters, H., 2001, Tool use to modify vocalisations by wild orang-utans, *Folia Primatol.* **72**(4):242–244.

Povinelli, D. J., 1991, *Social Intelligence in Monkeys and Apes*, Unpublished doctoral dissertation, Yale University, New Haven.

Prins, H. H. T., 1996, *Ecology and Behaviour of the African Buffalo: Social Inequality and Decision Making*, Chapman & Hall, London.

Pryor, K., Lindberg, J., Lindberg, S., and Milano, R., 1990, A dolphin-human fishing co-operative in Brazil, *Mar. Mammal. Sci.* **6**:77–82.

Rapaport, L. G., 1999, Provisioning of young Golden lion tamarins (Callitrichidae, Leontopithecus rosalia): A test of the information hypothesis, *Ecology* **105**(7): 619–636.

Redican, W. K., 1975, Facial expressions in nonhuman primates, in: *Primate Behavior*, L. A. Rosenblum, ed., Academic Press, New York, pp. 103–194.

Rendell, L., and Whitehead, H., 2001, Culture in whales and dolphins, *Behav. Brain Sci.* **24**:309–382.

Rodman, P. S., 2000, Great ape models for the evolution of human diet, www. cast. uark.edu/local/icaes/conferences/wburg/posters/psrodman/GAMHD.htm, Dec.

Romanes, G. J., 1884/1977, *Mental Evolution in Animals*, AMS Press, New York.

Roper, T. J., 1986, Cultural evolution of feeding behaviour in mammals, *Sci. Prog.* **70**:571–583.

Rowell, T. E., 1999, The myth of peculiar primates, in: *Mammalian Social Learning: Comparative and Ecological Perspectives*, H. O. Box and K. R. Gibson, eds., Cambridge University Press, Cambridge, pp. 6–16.

Russon, A. E., 1998, The nature and evolution of orangutan intelligence, *Primates* **39**:485–503.

Russon, A. E., 1999a, Orangutans' imitation of tool use: A cognitive interpretation, in: *Mentalities of Gorillas and Orangutans*, S. T. Parker, R. W. Mitchell, and H. L. Miles, eds., Cambridge University Press, Cambridge, pp. 119–145.

Russon, A. E., 1999b, Naturalistic approaches to orangutan intelligence and the question of enculturation, *Int. J. Comp. Psy.* **12**(4):181–202.

Russon, A. E., 2002, Comparative developmental perspectives on culture: The great apes, in: *Between Biology and Culture: Perspectives on Ontogenetic Development*, H. Keller, Y. H. Poortinga, and A. Schoelmerich, eds., Cambridge University Press, Cambridge, pp. 30–56.

Russon, A. E., 2003, Developmental perspectives on great ape traditions, in: *Towards a Biology of Traditions: Models and Evidence*, D. Fragaszy and S. Perry, eds., Cambridge University Press, Cambridge, pp. 329–364.

Russon, A. E., Bard, K. A., and Parker, S. T., eds., 1996, *Reaching into Thought: The Minds of the Great Apes*, Cambridge University Press, Cambridge.

Russon, A. E., and Galdikas, B. M. F., 1993, Imitation in free-ranging rehabilitant orangutans, *J. Comp. Psy.* **107**(2):147–161.

Russon, A. E., and Galdikas, B. M. F., 1995, Constraints on great ape imitation: Model and action selectivity in rehabilitant orangutans (*Pongo pygmaeus*), *J. Comp. Psy.* **109**(1):5–17.

Russon, A. E., Mitchell, R. W., Lefebvre, L., and Abravanel, E., 1998, The comparative evolution of imitation, in: *Piaget, Evolution, and Development*, J. Langer and M. Killen, eds., Lawrence Erlbaum Associates, Mahwah, pp. 103–143.

Sayigh, L. S., Tyack, P. L., and Wells, R. S., 1993, Signature whistle development in bottlenose dolphins is affected by early experience, *23rd International Ethological Conference*, Torremolinos, Sept. 1–9.

Seyfarth, R. M., Cheney, D. L., and Marler, P., 1980, Monkey responses to three different alarm calls: Evidence for predator classification and semantic communication, *Science* **210**:801–803.

Shettleworth, S. J., 1999, *Cognition, Evolution, and Behavior*, Oxford University Press, Oxford.

Sibly, R. M., 1999, Evolutionary biology of skill and information transfer, in: *Mammalian Social Learning: Comparative and Ecological Perspectives*, H. O. Box and K. R. Gibson, eds., Cambridge University Press, Cambridge, pp. 57–71.

Smith, I. M., and Bryson, S. E., 1994, Imitation and action in autism: A critical review, *Psy. Bull.* **116**:259–273.

Snowdon, C. T., 2001, Social processes in communication and cognition in callitrichid monkeys: A review, *Anim. Cogn.* **4**(3–4):247–257.

Stanford, C., 2001, *Significant Others*, Basic Books, New York.

Stoinski, T. S., Wrate, J. L., Ure, N., and Whiten, A., 2001, Imitative learning by captive western lowland gorillas (*Gorilla gorilla gorilla*) in a simulated food processing task, *J. Comp. Psy.* **115**(3):272–281.

Suddendorf, T., and Whiten, A., 2001, Mental evolution and development: Evidence for secondary representation in children, great apes, and other animals, *Psy. Bull.* **127**(5):629–650.

Tanaka, I., 1998, Social diffusion of modified louse egg-handling techniques during grooming in free-ranging Japanese macaques, *Am. J. Phys. Anthropol.* **98**:197–201.

Tayler, C. K., and Saayman, G. S., 1973, Imitative behaviour by Indian Ocean bottlenose dolphins (*Tursiops aduncus*) in captivity, *Behaviour* **44**:286–298.

Terkel, J., 1996, Cultural transmission of feeding behaviour in the black rat (*Rattus rattus*), in: *Social Learning in Animals*, C. M. Heyes and B. G. Galef, Jr., eds., Academic Press, London, pp. 17–47.

Thorndike, E. L., 1911, *Animal Intelligence*, Macmillan, New York.

Tomasello, M., 1996, Do apes ape? in: *Social Learning in Animals: The Roots of Culture*, C. M. Heyes and B. G. Galef, Jr., eds., Academic Press, London, pp. 319–349.

Tomasello, M., and Call, J., 1997, *Primate Cognition*, Oxford University Press, New York.

Tomasello, M., Davis-Dasilva, M., Camak, L., and Bard, K., 1987, Observational learning of tool use by young chimpanzees, *Hum. Evol.* **2**:175–185.

Tomasello, M., George, B., Kruger, A., Farrar, M., and Evans, A., 1985, The development of gestural communication in young chimpanzees, *J. Hum. Evol.* **14**:175–186.

Tomasello, M., Savage-Rumbaugh, E. S., and Kruger, A. C., 1993, Imitative learning of actions on objects by children, chimpanzees, and enculturated chimpanzees, *Child Dev.* **64**:1688–1705.

Tyack, P., 1986, Population biology, social behavior and communication in whales and dolphins, *Trends Ecol. Evol.* **1**:175–188.

Tyler, E. B., 1871, *Primitive Culture*, Murray, London.

Van Schaik, C. P., 1996, Social evolution in primates: The role of ecological factors and male behaviour, *Proc. Brit. Acad.* **88**:9–31.

Van Schaik, C. P., Ancrenaz, M., Beaver, G., Galdikas, G. M. F., Knott, C., Priatna, D. et al., 2003, Orangutan cultures and the evolution of material culture, *Science*, **299**:102–105.

Van Schaik, C. P., and Knott, C., 2001, Geographic variation in tool use in Neesia fruits in orangutans, *Am. J. Phys. Anthropol.* **114**:331–342.

Van Schaik, C. P., Preuschoft, S., and Watts, D. P., in press, Great ape social systems, in: *The Evolution of Great Ape Intelligence*, A. E., Russon, and D. R. Begun, eds., Cambridge University Press, Cambridge.

Visalberghi, E., 1987, Acquisition of nut-cracking behaviour by 2 capuchin monkeys (*Cebus apella*). *Folia Primatol.* **49**:168–181.

Visalberghi, E., 1993, Tool use in a South American monkey species, An overview of characteristics and limits of tool use in Cebus apella, in: *The Use of Tools by Humans and Nonhuman Primates*, A. Berthelet and J. Chavaillon, eds., Oxford University Press, Oxford, pp. 118–131.

Visalbergi, E., and Fragaszy, D. M., 1990, Do monkeys ape? in: *"Language" and Intelligence in Monkeys and Apes*, S. T., Parker and K. R. Gibson, eds., Cambridge University Press, Cambridge, pp. 247–273.

Visalberghi, E., and Fragaszy, D. M., 2002, Do monkeys ape? Ten years after, in: *Imitation in Animals and Artifacts*, K., Dautenhahn, and C. Nehaniv, eds., MIT Press, Cambridge. pp. 471–499.

Visalberghi, E., and Limongelli, L., 1996, Acting and understanding: Tool use revisited through the minds of capuchin monkeys, in: *Reaching into Thought: The Minds of the Great Apes*, A. E., Russon, K. A., Bard, and S. T. Parker, eds., Cambridge University Press, Cambridge, pp. 57–79.

Waterman, P. G., 1984, Food acquisition and processing as a function of plant chemistry, in: *Food Acquisition and Processing in Primates*, D. J. Chivers, B. A., Wood, and A. Bilsborough, eds., Plenum, New York, pp. 177–211.

Watson, J. B., 1908, Imitation in monkeys, *Psych. Bull.* **V**(6):159–179.

Whitehead, J. M., 1986, Development of feeding selectivity in mantled howling monkeys (*Alouatta palliata*), in: *Primate Ontogeny, Cognition and Social Behaviour*, J. Else and P. C. Lee, eds., Cambridge University Press, Cambridge, pp. 105–117.

Whiten, A., 1998, Imitation of the sequential structure of actions by chimpanzees (*Pan troglodytes*), *J. Comp. Psy.* **112**:270–281.

Whiten, A., 2000, Primate culture and social learning, *Cogn. Sci.* **24**(3):477–508.

Whiten, A., and Byrne, R. W., 1988, Tactical deception in primates. *Behav. and Brain Sciences,* **11**:233–733.

Whiten, A., and Custance, D. M., 1996, Studies of imitation in chimpanzees and children. in: *Social Learning in Animals, The Roots of Culture*, C. M., Heyes and B. G. Galef, Jr., eds., Academic Press, London, pp. 291–318.

Whiten, A., Custance, D. M., Gomez, J. C., Teixidor, P., and Bard, K. A., 1996, Imitative learning of artificial fruit processing in children (*Homo sapiens*) and chimpanzees (*Pan troglodytes*), *J. Comp. Psy.* **110**:3–14.

Whiten, A., Goodall, J., McGrew, W. C., Nishida, T., Reynolds, V., Sugiyama, Y., et al., 1999, Culture in chimpanzees, *Nature* **399**:682–685.

Whiten, A., and Ham, R., 1992, On the nature and evolution of imitation in the animal kingdom: Reappraisal of a century of research, in: *Advances in the Study of Behavior*, P. J. B., Slater, J. S., Rosenblatt, C., Beer, and C., Milinski, eds., Vol. 21. Academic Press, New York, pp. 238–283.

Wilkinson, G. S., 1992, Information transfer at evening bat colonies, *Anim. Behav.* **44**:501–518.

Wilkinson. G. S., and Boughman, J. W., 1999, Social influences on foraging in bats, in: *Mammalian Social Learning: Comparative and Ecological Perspectives*, H. O., Box, and K. R. Gibson, eds., Cambridge University Press, Cambridge, pp. 188–204.

Xitco, M. J., Jr., Harley, H. E., and Brill, R., 1998, Action level imitation by bottlenose dolphins, *Napoli Social Learning Conference*, Naples, Italy.

Yamagiwa, J., 1992, Functional analysis of social staring behavior in an all-male group of mountain gorillas, *Primates* **33**:523–544.

Yamagiwa, J. (in press). Diet and foraging of the great apes: Ecological constraints on their social organizations and implications for their divergence, in: *The Evolutionary Origins of Great Ape Intelligence*, A. E. Russon and D. R. Begun, eds., Cambridge University Press, Cambridge.

Yando, R., Seitz, V., and Zigler, E., 1978, *Imitation: A Developmental Perspective*, Lawrence Erlbaum Associates, Hillsdale.

Zentall, T. R., Sutton, J. E., and Sherburne, L. M., 1996, True imitative learning in pigeons, *Psychol. Sci.* 7(6):343–346.

Zohar, O. and Terkel, J., 1992, Acquisition of pine cone stripping behaviour in black rats (*Rattus rattus*). *Int. J. Comp. Psychol.* **5**:1–6.

CHAPTER FOUR

Social Learning, Innovation, and Intelligence in Fish

Yfke van Bergen, Kevin N. Laland, and William Hoppitt

INTRODUCTION

Social learning refers to learning that results from exposure to the behavior of other individuals. It enables animals to acquire locally adaptive information about beneficial or dangerous situations from more experienced conspecifics and heterospecifics, thus avoiding some of the risks associated with trial and error learning (Galef, 1995; Giraldeau and Caraco, 2000; Laland et al., 1993, 1996). Many animals are capable of social learning, and empirical and theoretical findings suggest that in most cases it does not require advanced cognitive abilities (Galef, 1988; Heyes, 1994; Whiten and Ham, 1992). In fact, most cases of social learning in animals appear to be mediated by simple processes such as local or stimulus enhancement, in which the behavior of another animal draws an individual's attention to a location or stimulus, about which the observer subsequently learns something.

Yfke van Bergen, Kevin N. Laland, and William Hoppitt • Sub-Department of Animal Behaviour, University of Cambridge, Madingley, UK.

Comparative Vertebrate Cognition, edited by Lesley J. Rogers and Gisela Kaplan. Kluwer Academic/Plenum Publishers, 2004.

As researchers who regularly track the literature on social learning in fish, birds, and primates, in our judgment there is a greater tendency among animal behaviorists to provide higher order cognitive explanations for behavior in primates than other taxa. Social learning in primates is often considered to be mediated by more "complex" processes such as observational learning, emulation, and imitation, rather than the simpler processes of local and stimulus enhancement. Primates are more closely related to humans than are other orders, are generally long-lived, have a long period of immaturity with much parental investment, possess large brains and live in complex social groups. In short, they possess characteristics that many researchers suppose make it likely that social learning is more important in primates than in other orders, and more likely that social learning in primates is mediated by processes reflecting complex cognitive abilities. Humans have a tendency, therefore, to attribute a special character to social learning in primates (Fragaszy and Visalberghi, 1996). In this chapter, we question whether this assumption of more extensive and sophisticated social learning in primates is justified. We argue that, as far as possible, identical explanations should be utilized when the same or similar behavior is exhibited by different species, irrespective of their degree of relatedness to humans, or assumed cognitive capabilities. Our argument applies in comparisons of social learning in fish and primates. The fact that primates may be capable of sophisticated forms of cognition, such as imitation, emulation, or the attribution of mental states to others, itself a contentious and unresolved issue (Tomasello and Call, 1997; Visalberghi, 1992; Whiten, 2000), while fish are not—although this is also not established—is to our minds irrelevant. The behaviors of fish, birds, and primates are to be interpreted according to the same criteria, and on a level playing field. Simple explanations and hypotheses that have been devised to account for the behavior of other taxa, including fish, may be regarded as null hypotheses when the same or similar behavior is observed in primates, with cognitively demanding explanations invoked only when there is evidence to reject these.

If we are to understand the ecological and evolutionary significance of social learning, we need to consider a framework of social learning that is task- and species-independent (Coussi-Korbel and Fragaszy, 1995). In the studies outlined in this chapter, we focus on functional perspectives on social learning and innovation, leaving aside for the moment psychological processes that facilitate learning.

Fish have comparatively small brains and are not renowned for their intelligence or learning abilities. Yet, empirical evidence suggests that they can be surprisingly good at learning from conspecifics. This chapter will outline a series of laboratory experiments on innovation and social learning in the guppy (*Poecilia reticulata*), a small tropical fish endemic to South America. Together, these studies indicate that the relatively small brains of these fish do not pose an impediment to learning, and that many phenomena of interest to primatologists can be demonstrated in fish. We argue that there is comparatively little that is special or sophisticated about social learning processes in non-human primates, and that perhaps the differences between non-human primates and other animals are more quantitative than qualitative (Bshary et al., 2002; Rowell, 1999).

To the extent that primatologists study their animals because they are interested in the processes that underlie complex behavior, we encourage them to consider how simpler study species may be just as productive. The studies outlined in this chapter illustrate the practical benefits of using fish as a study species, avoiding many of the problems primatologists face, such as small sample sizes or demographic constraints. Fish are generally easy and inexpensive to keep in the laboratory, as they usually require little space and have simple feeding requirements. Consequently, it is much easier to obtain large numbers of experimental populations, enabling population-level analyses. Moreover, working with fish allows use of experimental designs that are simply not feasible or economically viable with primates: it is much easier to set up replicate populations or transmission chain studies, for instance. In addition, it is relatively easy to vary social and ecological conditions between populations to model environmental variation in a controlled manner. Most fish also have faster rates of reproduction than primates, and so the fitness consequences of social learning over generations can be investigated over relatively short time spans, whereas a similar study in primates would take decades.

In addition to these practical benefits, guppies are a good model system for social learning studies for a number of reasons. Guppies live in social groups, prefer to shoal with familiar individuals, and are capable of individual recognition (Griffiths and Magurran, 1997a, 1997b, 1998; Magurran et al., 1994). Consequently, the basis for more complex social behavior is certainly present in this species. Furthermore, there is now strong empirical evidence that guppies are able to learn socially from conspecifics (Lachlan et al., 1998; Laland and Williams, 1997, 1998). Below, we first outline the findings from a series of studies of social learning and innovation in the guppy conducted in our laboratory, and then discuss how these can help us to interpret similar phenomena in primates.

TRADITIONS AND SOCIAL LEARNING IN GUPPIES

For most animals, locating and processing food are among the greatest challenges faced on a daily basis. It is thus generally accepted that many important cognitive skills in animals evolved in a foraging context (Tomasello and Call, 1997). Social learning is likely to play an important role in foraging for many species, and social influences on foraging behavior have been studied extensively in a broad range of species (Galef and Giraldeau, 2001). Animals can learn when, where, what, and how to forage from more experienced conspecifics in social groups. There is evidence from several fish species including goldfish (*Carassius* spp.), guppies (*Poecilia reticulata*), and minnows (*Phoxinus phoxinus*) that social grouping increases foraging efficiency in fish (Day et al., 2001; Morgan, 1988; Morgan and Colgan, 1987; Pitcher et al., 1982; Pitcher and House, 1987; Ryer and Olla, 1991, 1992). The studies on social learning and innovation in this chapter are mostly set in an ecologically valid foraging context because we are interested in social learning as a biologically significant phenomenon.

In two experimental studies, Laland and Williams (1997) demonstrated that guppies can learn the route to a hidden food source from knowledgeable conspecifics. In the first experiment, naïve observer fish were allowed to swim with demonstrator fish trained to use one of two equivalent routes to a feeder. A shaping procedure was used to train the demonstrators to swim through one of two differently colored holes in a partition to gain access to a floating feeder on the other side of the tank. During training sessions, one of the holes was closed off and the floating feeder placed close to the target hole. As the demonstrators became more familiar with the task, the feeder was progressively moved away from the starting position, toward a position equidistant to each of the holes. Subsequently these trained fish acted as route "demonstrators" to the experimental subjects, or "observers," with both routes now open. After 5 days of swimming with trained demonstrators, the observer fish were tested singly, and preferentially used the route their demonstrators had taken. This experiment provides clear evidence for the social learning of foraging information in fish. Simply by shoaling with knowledgeable conspecifics, guppies can learn the route to a food source. The experiment also indicated that frequency-dependent learning may be important in guppies. Different numbers of demonstrators were used in the first experiment, and performance of the observers increased significantly with the number of demonstrators. It would seem that single demonstrators may be less reliable than multiple demonstrators, or larger numbers of demonstrators may provide more visible or salient demonstrations for naïve individuals.

Guppies do not respond well to being tested in isolation. A second experiment therefore extended these findings using a more suitable design for social species: a transmission chain. Small founder populations of four individuals were trained to take one of the two routes, after which founder members were gradually replaced by naïve individuals. Three days after all the original founder members had been removed, the group still showed a strong preference for the route that their founders had been trained to use, demonstrating that a tradition for an arbitrary route to a feeder can be maintained in guppy populations. We suggest that equivalent processes may underlie the traditional use of mating sites, schooling sites, and migration routes characteristic of many fish (Helfman and Schultz, 1984; Warner, 1988).

Using essentially the same experimental design, Laland and Williams took the study one step further, providing the first empirical evidence that maladaptive information can be socially transmitted through animal populations. Founder groups of fish were trained to take either an energetically costly circuitous route or a less costly short route to a feeder. A pilot study had established that there was a significant energetic cost associated with the longer route compared with the short route. A transmission chain was set up, replacing founders with naïve individuals. Three days after all the founders had been removed, the groups of untrained fish still took the circuitous founder routes. Moreover, fish foraging alone learned to take the short route much faster than naïve fish that had shoaled with demonstrators trained to take the long route, indicating that socially learned information can inhibit learning of optimal behavior patterns. Thus, guppies can learn routes to food from conspecifics and maintain behavioral traditions, even if the information that is being transmitted is maladaptive.

The next step in this series of studies was to investigate the social and ecological factors that influence reliance on social learning in this species. Given that shoaling tendencies are likely to facilitate social learning, Lachlan et al. (1998) varied characteristics of shoals to determine which factors influence shoaling preferences and the transmission of information. Concerning shoaling preferences, they found that guppies (a) prefer to join a shoal rather than a single fish, (b) assort by size (small fish preferring to shoal with similar sized individuals), (c) prefer to shoal with fish that have had repeated successful foraging experiences over those that have not had these experiences, and (d) prefer to shoal with familiar individuals. Guppies were found to be more likely to adopt the behavior of a shoal of demonstrators than alternative behavior shown by one demonstrator. Moreover, following a shoal to a food site frequently resulted in the naïve individual learning a route to food after just a single trial. Taken together, these results imply that learned behavior may diffuse through a fish population in a nonrandom manner. The findings also have interesting implications for understanding of the social transmission of adaptive and maladaptive information. If guppies can use behavioral cues to assess whether other individuals have had recent foraging success, and follow them preferentially, maladaptive information would not be transmitted for long in the wild. This ability to pick out successful demonstrators may be one factor that makes social learning an adaptive strategy for animals.

Swaney et al. (2001) investigated whether familiarity and proficiency of demonstrators affect the performance of observers. Demonstrators were trained to swim to a food source, after which shoals of demonstrators were introduced to shoals of naïve observers, and the time taken for each fish in the group to reach the food source was recorded over 15 trials. The demonstrators were either familiar or unfamiliar to the observers and either well trained or poorly trained. The results are shown in Figure 1.

Observers performed significantly better when the demonstrators were familiar. Over all trials, the proficiency of the demonstrators did not affect the performance of the observers. However, initially (e.g., on the first trial) the observers exposed to well-trained demonstrators performed significantly better than those exposed to poorly trained demonstrators, while observers with poorly trained demonstrators learned the route to the feeder faster. This latter finding is perhaps counterintuitive. Analysis of the data suggests that well-trained demonstrators move too quickly for the observers to follow, and only provide observers with a local-enhancement-type tip-off about the route. Poorly trained demonstrators move more slowly, giving the observers more time to shoal with and follow them. This suggests that the optimal conditions for social learning may be situations where demonstrators are only slightly superior to observers in performance. Furthermore, two alternative forms of

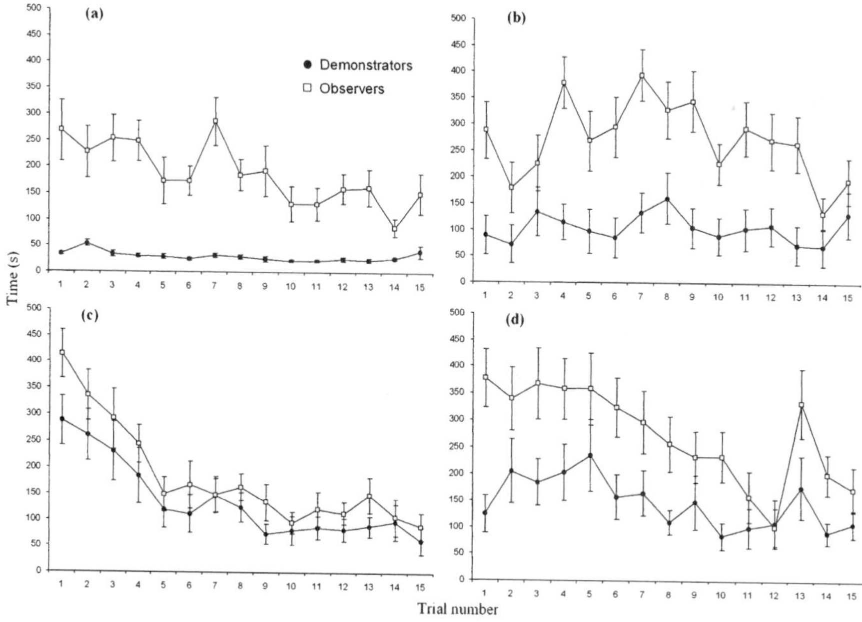

Figure 1. Latency to complete the foraging task by both observers and demonstrators over trials, split into the four groups. Observers had (a) well-trained, familiar demonstrators, (b) well-trained, unfamiliar demonstrators, (c) poorly trained, familiar demonstrators, and (d) poorly trained, unfamiliar demonstrators. (Reprinted from Swaney, W., Kendal, J. R., Capon, H., Brown, C., and Laland, K. N., 2001, Familiarity facilitates social learning of foraging behaviour in the guppy, *Anim. Behav.* **62**:591–598, by permission of Academic Press, an imprint of Elsevier Science.)

social learning would seem to be operating in this simple system: individuals are learning by following demonstrators through the maze, or by having their attention drawn to the entrance.

These studies have demonstrated that fish are able to learn socially, and that traditions can be maintained in fish populations, even if maladaptive information is being transmitted. There is some indication that the processes involved in social learning in fish are simple local and stimulus enhancement mechanisms. These experiments have also given us some insight into which features of demonstrators have important influences on social learning. Learning appears to be frequency-dependent or conformist, as fish learn better with more demonstrators. Fish prefer to shoal with and learn better from familiar individuals and may even be able to select successful demonstrators.

Demonstrator competence is also important—under some circumstances, poor demonstrators may be better to learn from, perhaps because learning from the mistakes of others is an important factor in social learning, as has been demonstrated in birds (Mason, 1988; Templeton, 1998) and human children (Want and Harris, 2001).

INNOVATION IN GUPPIES—IS NECESSITY THE MOTHER OF INVENTION?

Animals often respond to novel social and ecological conditions with new behavioral patterns, or "innovations" (Kummer and Goodall, 1985; Lee, 1991). Innovation is an important component of behavioral plasticity, for instance, it is probably vital in enabling individuals of generalist or opportunistic species to survive in changing environments (Lefebvre et al., 1997). Responses to novelty are of interest to behavioral ecologists working in the fields of foraging behavior, optimality, decision-making and behavioral plasticity, and are particularly important to the study of social learning and the transmission of information, since virtually all transmission of novel behavioral patterns depends on innovation on the part of the initiator of the novel behavior.

Most of what is known about animal innovation stems from observations of natural populations. These observations suggest that particular classes of individuals may be more predisposed to innovate than others, although many of these observations have an anecdotal quality. For instance, low-ranking females are often considered to be the innovators in primate populations (Reader and Laland, 2001), perhaps due to a number of high profile case studies in the literature, such as Imo, the famous potato washing macaque (Kawai, 1965), and because researchers have expressed expectations as to which classes of individual should innovate. However, these are all assumptions and they are not supported by empirical evidence (Reader and Laland, 2001). We need controlled empirical studies on animal innovation. It is not clear whether innovators are primarily particularly creative, intelligent, or nonconformist individuals, individuals exposed to particular environmental factors or individuals in certain motivational states. Below, we outline several laboratory studies on innovation in guppies that investigate the characteristics of innovators and the spread of innovations through guppy populations.

Laland and Reader (1999a) exposed small populations of guppies, composed of individuals varying in sex, hunger level, and body size, to novel foraging tasks, in the form of simple maze partitions through which the fish had to swim to gain access to a food source, and recorded the category of the first fish to reach the novel food source. A subsequent study has found a strong correlation in guppies between time taken to complete a maze for the first time and number of trials to learn the maze according to a trials-to-criterion measure (Reader and Laland, 2000), legitimizing the use of the first individual in a population to swim the maze as a reliable measure of innovation. Laland and Reader found that females were more likely to innovate than males, food-deprived fish were more likely to innovate than nonfood-deprived fish, and smaller fish were more likely to innovate than larger fish. It appears that differences in innovatory tendencies in guppies are best accounted for by differences in motivational state. Innovators were neither the most active fish nor those with the fastest swimming speed. Moreover, the observed patterns disappear when the experiments are repeated with no food in the mazes. Here, the most parsimonious explanation for the observed individual differences in problem solving is that innovators do not need to be particularly intelligent or creative, but are driven to find novel solutions to foraging problems by hunger, or by the metabolic costs of growth or pregnancy.

These findings did not rule out the possibility that over and above these state-dependent factors are "personality" differences that affect an individual's propensity to innovate. To investigate this, Laland and Reader exposed populations of fish to three novel foraging tasks, recording whether fish that had completed the first two tasks fastest performed faster in the third task than fish that had not innovated in the first two tasks. Past innovators were found to be more likely to innovate than past non-innovators. As the design took steps to rule out a number of potential confounding variables, this constituted evidence that there are genuine personality differences in innovative tendency in guppies. It is interesting that there is evidence for innovative individuals in a species not particularly renowned for its intelligence or problem-solving capabilities.

To investigate further how motivational state affects innovation, Laland and Reader (1999b) explored the relationship between past foraging success and foraging innovation. Weight changes of individuals in two mixed-sex populations of guppies were monitored over a 2-week period, and the competitive foraging ability of each individual was measured by recording the number of food items eaten. These populations were then exposed to three novel maze tasks, and the time for each fish to complete the task was recorded. The prediction was that poor competitors—fish that had gained the least weight and obtained the least food items during scramble competition—would be more likely to innovate when presented with the novel foraging tasks. In male guppies, the latency to complete the foraging tasks was indeed found to correlate with both weight gain and the number of food items consumed during scramble competition. However, there was no such correlation in female guppies (see Figure 2). Females appeared more motivated to solve the foraging tasks than males, regardless of how they had fared during the scramble competition.

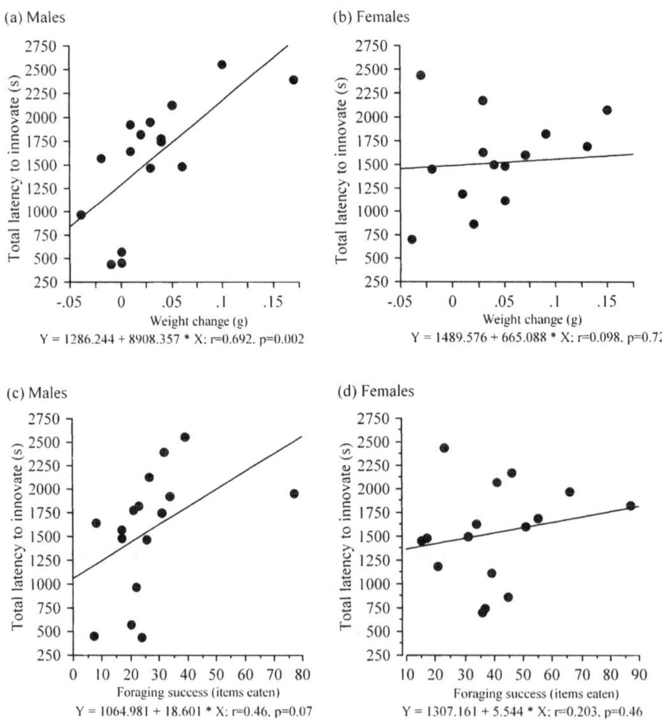

Figure 2. A strong correlation is found between weight change and total time to innovate in (a) males, but no relationship is found in (b) females. Results are similar for the relationship between past foraging success and total time to innovate in (c) males and (d) females. (Reprinted from Laland, K. N., and Reader, S. M., 1999, Foraging innovation is inversely related to competitive ability in male but not in female guppies, *Behav. Ecol.*, **10**:270–274, by permission of Oxford University Press.)

What could explain the finding that female guppies are more likely to show foraging innovations than male guppies, irrespective of past foraging success? Laland and Reader (1999b) reasoned that these sex differences could be accounted for by considering parental investment patterns. In many vertebrate species in which female parental investment exceeds that of males, male reproductive success is most effectively maximized by prioritizing mating and is limited by access to receptive females (Trivers, 1972). In contrast, female reproductive success is limited by access to food resources, particularly in guppies, a species in which the females can store sperm. Guppies are viviparous, thus female parental investment is much greater than that of males. Females also have indeterminate growth, and there is a direct correlation between energy intake and fecundity in females, whereas males stop growing at sexual maturity (Reznick and Yang, 1993; Sargent and Gross, 1993). Consequently, finding high-quality food has greater marginal fitness value for females than for males, and this may explain why females should be more investigative than males and are constantly searching for new food sources.

Having established that individual guppies vary in their propensity to innovate, Reader and Laland (2000) turned to the ways in which innovations are socially transmitted through animal populations. Mixed-sex populations of guppies were presented with three novel foraging tasks, and time to complete the task was recorded for each individual over 15 trials (see Figure 3).

In experiment 1, the populations were made up of equal numbers of food-deprived and nonfood-deprived individuals, while experiment 2 compared small, young fish with large, older fish. Food-deprived fish were faster than nonfood-deprived fish at completing the tasks. Although there was no overall effect of size, there was a significant interaction between sex and size. Adult females completed the tasks much faster than adult males, but no sex difference was found in juveniles. In both experiments, there was a significant sex difference, with novel foraging information spreading faster through female subgroups than through male subgroups. Females were also found to learn at a faster rate than males. These findings are most likely due to motivational differences between the sexes, corroborating the earlier findings of Laland and Reader (1999a, 1999b). The absence of a sex difference in younger fish is also consistent with the parental investment explanation given above, since young fish are not expected to show investment asymmetries.

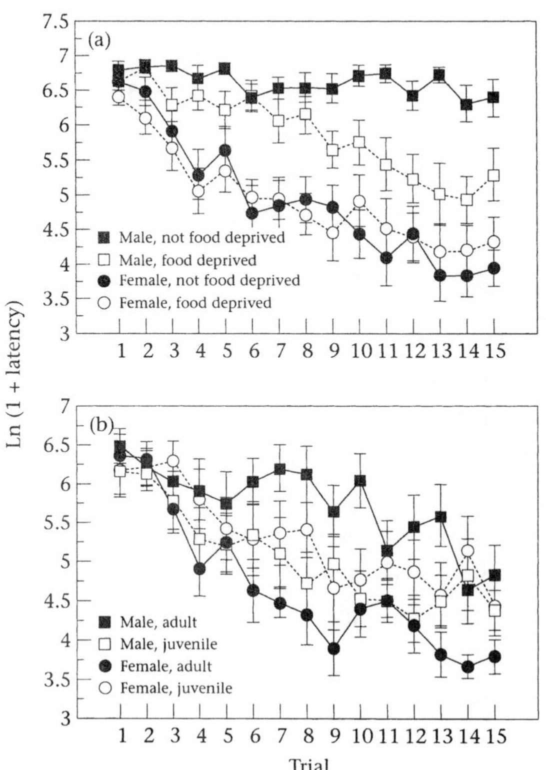

Figure 3. Log-transformed mean latency(ies) to enter goal zone against trial for (a) food-deprived and nonfood-deprived fish and (b) adult and juvenile fish. In (a) $N = 60$, made up of three groups of 20 fish each; in (b) $N = 54$, made up of three groups of 20, 18, and 16 fish. Vertical bars indicate SE. (Reprinted from Reader, S. M., and Laland, K. N., 2000, Diffusion of foraging innovations in the guppy, *Anim. Behav.* **60**:175–180, by permission of Academic Press, an imprint of Elsevier Science.)

CONFORMITY AND SOCIAL RELEASE

Conformity is a common feature of social species, meaning that the behavior of the majority causes other individuals in a social group to adopt the same behavior. It is therefore likely to affect both social learning processes and the likelihood of innovation. Lachlan et al. (1998) proposed that the tendency exhibited by guppies to shoal with the largest number of fish might generate positive frequency-dependent social learning, from here on referred to as conformity. The most common behavior in a shoal would be adopted more rapidly if individuals adopt the behavior of the majority, which would act against the transmission of novel alternative behavior patterns exhibited by only a few individuals (Lachlan et al., 1998). Consequently, conformity may inhibit the spread of innovations through social groups.

Day et al. (2001) found that conformity may both increase and reduce opportunities for social learning, depending on environmental conditions. In the first experiment of an investigation of how shoal size affects foraging efficiency in guppies, they presented a food source to shoals of guppies in open water, and large shoals were found to locate food faster than small shoals. The finding that individuals in large shoals typically find food faster than individuals in small shoals has been reported in many species of fish (Morgan, 1988; Morgan and Colgan, 1987; Pitcher et al., 1982; Pitcher and House, 1987; Ryer and Olla 1991, 1992). However, in experiment 2, the fish had to swim through an opaque maze partition to get to a food source, and in this situation the exact opposite result was found, namely that smaller shoals located food faster than larger shoals. The third experiment was designed to clarify the contradictory results of the first two experiments. Experiment 2 was repeated, using a transparent rather than an opaque maze partition. Fish in larger shoals were once again found to locate the food source faster in these conditions. Figure 4 shows these results.

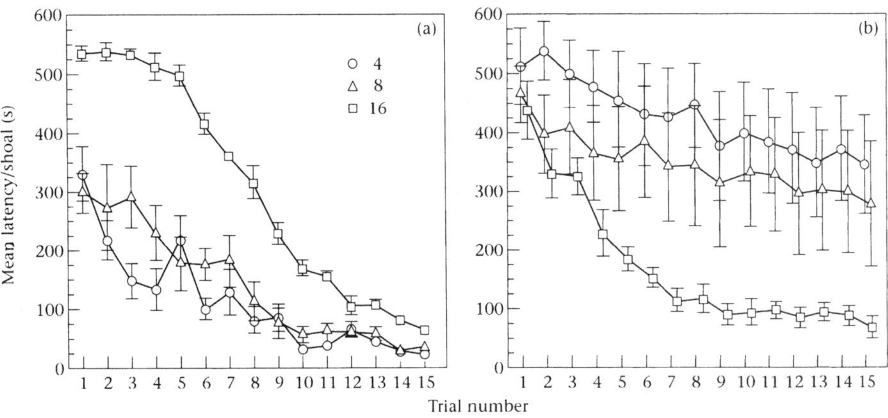

Figure 4. Mean shoal latencies for fish in each shoal size to (a) pass through the opaque partition (experiment 2) and (b) pass through the transparent partition (experiment 3) to enter the goal zone. (Reprinted from Day, R. L., Macdonald, T., Brown, C., Laland, K. N., and Reader, S. M., 2001, Interactions between shoal size and conformity in guppy social foraging, *Anim. Behav.* **62**:917–925, by permission of Academic Press, an imprint of Elsevier Science.)

How does conformity help to explain these findings? Day et al. reasoned that fish in large shoals have more shoal mates from whom to acquire information, and therefore have greater opportunity to use social cues from successful conspecifics to locate a novel food source. The process of local enhancement will also be more important in large shoals, since large numbers at a food site will attract conspecifics more rapidly than small aggregations. As individuals adopt the behavior demonstrated by the majority of the shoal, conformity explains the findings of experiment 1, which in this case facilitates social learning. As mentioned earlier, guppies have a preference to join large over small shoals, and by reverse logic, one might expect that individuals would be more willing to leave smaller than larger shoals. In the second experiment, fish had to swim through an opaque partition to locate a novel food source. Since this involved breaking visual contact with the rest of the shoal, this could be interpreted as leaving the shoal. Thus, those foraging innovations requiring individual fish to move away from the shoal and discover a novel food source may be less likely in large than small shoals because fish could be expected to be more reluctant to break visual contact with the rest of the shoal. Here, social learning opportunities are hindered by conformity. In experiment 3, visual contact between fish was maintained because the partition was transparent, so fish passing through the maze partition were not leaving the shoal. Under these conditions, social transmission of foraging information is not impeded by conformity.

Conformity may be a common feature of the social learning of many social species and may frequently inhibit the generation and spread of novel behavioral variants, yet there is evidence from both field and laboratory studies that innovations regularly spread through animal populations. How can these apparently conflicting findings be reconciled? Brown and Laland (2002) propose a “social release” hypothesis, suggesting that certain circumstances will release animals from conformity. Changes in the social or ecological environment (such as dispersal to unfamiliar environments or food shortage following a sudden climate change) may result in the loss of a clear pattern of majority behavior. In these situations, individuals may be released from the tendency to conform, aspects of traditional behavior may break down, and novel behavioral variants may spread. Brown and Laland investigated the transmission of information about an escape route from an artificial predator from trained demonstrator fish to naïve observers. The artificial predator was a trawl apparatus with two differently colored holes, through which the fish had to swim to avoid being trapped in a small gap between the trawl and the end of the tank. The fish appear to be highly motivated to escape from the trawl net, but are not harmed in any way by the procedure, making this avoidance task a useful simulation of a situation in which individuals can learn an adaptive escape response from more knowledgeable conspecifics. Five shoals of fish incorporating untrained (sham) demonstrators were assigned as controls. Of the six experimental populations, three shoals had demonstrators trained to take a red outlined hole in the trawl and three had demonstrators trained to take a blue outlined hole in the trawl. Over 15 trials exposing each shoal to the trawl apparatus, the escape latency and color of escape hole taken was recorded for each fish in each of these shoals. The demonstrators were then removed and the remaining observers were again exposed to the trawl for 15 trials, recording escape latency and color of escape hole taken. In the presence of demonstrators, naïve observer fish exhibited a strong preference for the escape route that their demonstrators had been trained to take. They also escaped more quickly than fish in control populations. Once the demonstrators had been removed, however, the observer fish no longer exhibited a preference for the demonstrator route, being equally likely to escape via either escape route.

One interpretation of these results is that the observer fish had not learned to escape when demonstrators were present, but simply shoaled with their escaping demonstrators. However, two findings refute this. First, after their demonstrators had been removed, the observer fish in the experimental shoals still escaped more quickly and frequently than control fish. Second, there was no deterioration in performance in either the latency or frequency of escape among observers in experimental shoals during the trials when demonstrators were absent. There is clear evidence, then, that observers did improve their escape responses by interacting with demonstrators during the trials when demonstrators were present. The interesting conclusion of this study is that while arbitrary features of the escape response, such as the color of the escape route chosen, were rapidly lost in the absence of demonstrators, functional aspects of the response, such as the enhanced latency to escape, were conserved.

In summary, experimental studies have established that under restricted circumstances, suboptimal information can be transmitted among shoals of fish, and that this can result in arbitrary and even maladaptive behavioral traditions (Laland and Williams, 1997, 1998). Conformity can help to ensure that animals in a social group exhibit similar behavior with its antipredator benefits, and can facilitate social learning so that animals acquire locally adaptive behavior. However, in some conditions, conformity may in fact hinder social learning by blocking the diffusion of innovations. The studies outlined in this section also indicate that under certain circumstances animals may be released from conformist social learning, which is how innovations may spread through animal populations.

PRIMATE SUPREMACY RECONSIDERED

Can the findings of the studies on guppies reviewed in this chapter inform research on social learning and innovation in primates? To answer this, we must first consider why primates have been assumed to be unusually sophisticated when it comes to social learning. Rowell (1999) has described in no uncertain terms why primates are considered special. First, researchers have been forced to study primates in a different way, due to features of primate life history, such as long life spans and slow rates of development, and the fact that they are usually only available in small, not readily accessible groups in the wild. In spite of its intuitive appeal, we know of no evidence for the hypothesis that these life history characteristics co-vary with reliance on social learning. The second factor is that different expectations of primates have led to different questions being asked of them. Many researchers expected to find enhanced cognitive ability and social sophistication in primates due to their evolutionary relatedness with humans, and designed and conducted research with this in mind (Rowell, 1999). Until recently, non-primate researchers had simply not asked the same questions as primatologists. However, now that comparative data are becoming increasingly available, Rowell notes that there is no empirical evidence to suggest an obvious distinction in the quality of social behavior between primates and other mammals. As we stated in the introduction, differences between primates and other animals may be more quantitative than qualitative. If this is the case, we need criteria for comparing social systems that are not constrained by taxonomy (Coussi-Korbel and Fragaszy, 1995; Rowell, 1999). A functional perspective on social learning may be of value.

We agree with Coussi-Korbel and Fragaszy (1995) that the overwhelming focus on psychological mechanisms in the field of social learning has somewhat impeded the study of social learning as a biologically significant phenomenon. If one takes a functional approach, then taxonomy and questions about cognitive complexity are no longer an issue (Caro and Hauser, 1992). We are not yet in a position to determine whether the mechanisms underlying social learning processes in primates differ fundamentally from those of other orders. As far as we are aware, the only empirical evidence of a qualitative difference in the psychological mechanisms underlying the social learning of primates and other orders is imitation, which many researchers believe to have been demonstrated in great apes but not in other taxa (Whiten and Custance, 1996). However, there is little consensus as to which examples of ape social learning constitute reliable evidence for imitation, and there is some evidence for imitation in cetaceans and birds (Dorrance and Zentall, 2001; Rendell and Whitehead 2001a, 2001b; Zentall et al., 1996). In the face of this uncertainty, it may pay to focus on two points that have been well established by research into social learning: (a) simple processes such as local and stimulus enhancement can lead to social learning, and (b) cases of apparent imitation can frequently be explained by simple mechanisms such as local and stimulus enhancement. Given these considerations, studies of species that are traditionally seen as cognitively less sophisticated than primates could be profitable. By using simple model systems, as well as taking a functional approach, characteristics of social learning processes may emerge which are universal, or at least widespread. We believe that on a functional level, the features of guppy social learning and innovation reviewed in this chapter can all be related to similar phenomena in primates, and useful comparisons made across diverse taxa.

Coussi-Korbel and Fragaszy (1995) postulated that information may diffuse in a directed manner through many populations of social animals because social species associate in a nonrandom manner, restricting the routes by which information is likely to flow. Social learning is therefore dependent on the identity of the demonstrator and observer: animals are likely to attend to, and learn preferentially from, certain individuals, resulting in what Coussi-Korbel and Fragaszy termed "directed social learning." For example, young animals may be more likely to learn from older conspecifics, and innovations may be more likely to spread to individuals with the same social rank as the innovator. Since guppies exhibit shoaling preferences, and motivational differences influence innovatory tendencies of individuals, there is reason to believe that learned information may flow preferentially through certain subsections of guppy populations, reminiscent of the directed social learning hypothesis.

Proximity in social groups is usually necessary for individuals to acquire knowledge about the behavior of conspecifics and for the transmission of information. However, primates often disperse in foraging situations, and aggression toward subordinates when they come too close to dominants is common. Therefore, social tolerance in a potentially competitive foraging context is likely to facilitate social learning (Coussi-Korbel and Fragaszy, 1995). Van Schaik et al. (1999) proposed the social tolerance hypothesis, which states that social learning may occur more readily in groups where individuals are more tolerant of close proximity of conspecifics (i.e., more readily in egalitarian than despotic societies), and explained how this could facilitate social learning. Social tolerance enhances behavioral coordination, bringing animals in close proximity during foraging, making close range observations possible, and reduces anxiety levels, making it easier for animals to attend to the foraging tasks (Van Schaik et al., 1999). There is evidence for social tolerance and directed social learning in guppies. Social learning in guppies is more frequently directed toward groups of demonstrators than to a single demonstrator (Laland and Williams 1997; Lachlan et al., 1998). More attention is given to familiar conspecifics than to unfamiliar conspecifics, and information is transmitted faster through subgroups of female guppies than subgroups of males (Lachlan et al., 1998; Reader and Laland, 2000, Swaney et al., 2001). A comparative study on social learning processes in egalitarian and despotic societies in primates is needed to establish whether social tolerance and directed social learning play an important part in primate social learning.

Let us reconsider one of the most well-known cases of social learning, that of sweet-potato washing in Japanese macaques (Kawai 1965). Imo, the famous potato- and wheat-washing macaque, has become celebrated as a great innovator. For instance, E.O. Wilson (1975) in *Sociobiology* and John Bonner (1980) in *The Evolution of Culture in Animals* both refer to Imo as "a genius," while Jane Goodall (Kummer and Goodall, 1985) describes her as "gifted." In the 1960s, food washing seemed an unlikely natural behavior for monkeys, and hence these behavior patterns were regarded as novel and intelligent. When the food-washing behavior spread, it was described as "preculture" or "protoculture," the clear implication being that it could be regarded as an analog, perhaps even a homolog, of human culture.

With the benefit of hindsight, and in the light of accumulated scientific knowledge, Imo's achievements and those of her troop members appear more modest. It would seem that food washing is a stable feature of macaque behavior, and Imo's innovation involved the application of an established behavior pattern to novel foods (Visalberghi and Fragaszy, 1990). Experimental studies have revealed that food-washing behavior is learned relatively easily by monkeys and can become common in a troop through processes other than the imitation of a rare "creative genius" (Visalberghi and Fragaszy, 1990). There is no evidence that food washing spread through Imo's troop by imitation, teaching, or indeed any unusually sophisticated form of social learning, and there are grounds for concern that they may even have been an artifact of human provisioning (Galef, 1992; Green, 1975). The lauding of Imo as a genius and the use of the terms protoculture or preculture to describe this case of social transmission was widely accepted only because the behavior was reported for a primate. Would the scientific community have accepted the same terminology to describe social learning in guppies?

Now let us consider how our studies of social learning and innovation in fish compare with the "exemplar" case of potato washing in Japanese macaques. We have carried out a detailed analysis of the factors that underpin foraging innovation in guppies, and found that state-dependent factors account for most of the variation. These and a host of other findings (reviewed in Laland and Reader, 1999a, 1999b; Reader and Laland, 2001) lead us to the view that the adage "necessity is the mother of invention" probably applies to most animal innovation. To our knowledge, no consideration has been given to the possibility that Imo was the first individual to apply food washing to sweet potatoes merely because she was among the hungriest of individuals in the troop, yet this would seem a real possibility. However, our experiments have, in addition, found evidence for personality differences in the problem-solving ability of guppies. Laland and Reader (1999a) found that some fish consistently solved foraging problems, not because they were hungrier, or faster swimmers, or more active, or healthier, but because of some stable personality characteristic. To our knowledge there is no equivalent experimental evidence for personality differences in problem-solving ability in any species of primate. Many primate studies report differences in performance between individuals, but none have ruled out alternative explanations for individual differences (e.g., motivational factors) and hence the variance cannot reliably be interpreted as reflecting differences in personality. The one study (of which we are aware) that sought such evidence, Fragaszy and Visalberghi's (1990) investigation of innovation in capuchin monkeys, failed to find it.

Let us give the single fish that consistently solved our maze problems faster than others the name of Bertha. We do not yet know whether it is appropriate to describe the personality trait that Bertha exhibits as "intelligence" or "creativity," as opposed to, say, "boldness," "perseverance," or "nonconformity," but at least we know that there is some feature of Bertha's make-up that makes her an effective problem solver. Imo has been gifted with highbrow intellectual qualities as opposed to the less salubrious personality traits in an essentially arbitrary manner, indeed in the absence of any evidence that it was her personality to which her success can be attributed. Under such circumstances, Bertha seems to have greater claims on the "genius" plaudit than Imo.

The transmission chain design has proved invaluable in social learning research, especially when testing social species that do not respond well to isolation testing. Using this design, aspects of behavioral traditions maintained by social processes have been investigated in guppy populations (Laland and Williams, 1997, 1998), as well as in rats, birds, and humans (Curio, 1988; Laland et al., 1993; Galef and Allen, 1995). These traditions are a laboratory analog of tool-using and food-processing traditions in primates (e.g., Whiten et al., 1999), but have the advantage that they are easier to study. Studies of guppies have found that arbitrary or maladaptive information can be socially transmitted, which is, after all, a feature of human cultural traditions. The same methods using rats have isolated a number of factors that influence the stability of traditions, and have generated a host of other insights, such as that innovation can emerge through the accumulated efforts of many individuals (Galef and Allen, 1995; Laland, 1999). We do not know of any studies of primate traditions that establish the same findings, yet transmission chain studies carried out on other taxa are virtually ignored in discussions of primate traditions.

Demonstrator competence has been found to be an important factor in social learning in fish, but in a counterintuitive way. Poor demonstrators may be more effective than competent demonstrators, probably because they enable others to learn from their mistakes and because they move more slowly, resulting in more time for local and stimulus enhancement mechanisms. Guppies that were poorly trained to swim through a maze partition, for example, spent more time swimming near the entrance to the maze partition than well-trained demonstrators, which swam through the partition very quickly. This hesitation provided more opportunity for conspecifics to see where the poorly trained demonstrators were and discover the entrance for themselves (Swaney et al., 2001). Similar phenomena are reported in birds and humans (Mason, 1988; Templeton, 1998; Want and Harris, 2001). Perhaps one reason why processes such as imitation have not been demonstrated conclusively in many primates, particularly monkeys, is that proficient demonstrators move too quickly for conspecifics to follow their actions (Young, personal communication).

Studies of behavioral innovation in guppies raise interesting issues when compared with findings of equivalent studies in primates, and similar underlying processes may be involved in both cases. Motivational factors such as hunger appear to be important factors in guppy innovation, and there is reason to believe that the same may be true for innovation in primates. Innovators may be individuals that are poor competitors, driven to search for alternative behavior patterns when traditional strategies prove unproductive. Studies in primates indicate that innovators are frequently low in social rank, or poor competitors on the outskirts of social groups (Kummer and Goodall, 1985; Sigg, 1980; Visalberghi and Fragaszy, 1990). Reader conducted the first extensive investigation into behavioral innovation in primates by an exhaustive literature survey (Reader, 1999; Reader and Laland, 2001). From an analysis of more than one thousand papers from four primate journals, sex, age, and social rank differences in innovation in primates were determined. The incidence of innovation across a wide range of contexts was found to be three times higher in low-ranking than it was in high-ranking individuals.

Reader and Laland (2001) also found that innovation was disproportionately frequent in male compared with female non-human primates. At first sight these findings appear to conflict with the patterns of innovation observed in guppies, in which females were the more innovatory sex. These findings differ from the guppy studies because they cover a range of contexts, and a range of primate species occupying varied niches, not just foraging innovations. Nonetheless, we believe the same basic explanation may apply in both cases.

Laland and Reader's (1999a) account of differences in guppy foraging innovation placed emphasis on the observations that females are the larger sex and have unconstrained growth, whereas males stop growing at sexual maturity. Females, therefore, have greater metabolic costs of growth and pregnancy. However, as female fecundity is directly related to body size, there are accelerating returns from increasing foraging success. Because of size-related differences in fecundity, there can be considerable variability in female reproductive success. Consequently, there may be greater fitness benefits associated with female than male foraging innovation. There is reason to believe that this explanation may equally apply to foraging innovations in other species in which maternal investment is greater than paternal investment, that is, to most mammals, including primates. Here, however, the sex roles are reversed. In most primates it is the males that are the larger sex, with, in many cases, male growth continuing until death. Males exhibit greater variability in mating success than females. In such circumstances it may be beneficial for males, particularly those of low status, to take risks and try out novel solutions that either allow access to the foods they require to fuel the growth they need to be an effective competitor, or constitute alternative strategies to reproductive success. Support for this line of reasoning comes from the observation that the sex difference in primate innovation is significant only in dimorphic species of primates (where sexual selection is deemed to be more active).

Our experiments have provided unequivocal evidence that fish are capable of social learning of foraging information (Laland and Williams, 1997); have demonstrated repeatedly that maze-solving behavior can diffuse through a population (Brown and Laland, 2002; Reader and Laland, 2000), in the process revealing strong evidence for directed social learning, and have shown that alternative routes through a maze can be maintained as arbitrary and even maladaptive traditions (Laland and Williams, 1997, 1998). The same behavior has been found in field experiments with natural populations of guppies in Trinidad (Reader, 1999; Reader et al., in press). In contrast, while we believe it highly likely that sweet-potato washing has spread amongst Japanese macaques through social learning, there is in fact no evidence to this effect. In theory, the macaques could have adopted potato washing independently, and the differences between food-washing and nonfood-washing troops could reflect provisioning, or the differential availability of resources in their environments. Once again, we could argue that if hard evidence is what counts, the maze-solving traditions exhibited by our fish have greater claims on the terms "preculture" or "protoculture" than potato washing in macaques.

Indeed, we could go further and note that, although chimpanzees, with their diverse tool-using traditions (Whiten et al., 1999), are widely regarded as the animal species most deserving of the sacred mantle of "culture," whether the between population differences can truly be attributed to culture remains controversial (Tomasello, 1994). The controversy remains because it is difficult to rule out alternative explanations, such as that the behavioral variation merely reflects differences in the natural ecology of the populations' environments. A strong test of the hypothesis that the behavioral differences between chimpanzee populations are cultural would be to transplant populations. If Gombe-born chimps were found to termite fish and not nut hammer when in Tai, and Tai-born chimps were found to nut hammer and not termite fish when in Gombe, it would be a most cantankerous skeptic who disputed their claim to culture. Of course, such transportation experiments would be unethical (if not impossible) in chimpanzees. Yet, exact equivalent experiments have been carried out with fish populations. Warner (1988) showed that transplanted populations of Bluehead wrasse (*Thalassoma bifasciatum*), a Caribbean coral reef fish, did not adopt the same use of the environment as the populations they replaced, but rather maintained distinct cultural traditions. Helfman and Schultz (1984) transplanted French grunts (*Haemulon flavolineatum*) between populations and found that while those fish placed into established populations adopted the same traditional behavior as the residents, control fish introduced into regions from where the residents had been removed did not adopt the behavior of former residents. How many primatologists will accept that there is currently stronger empirical evidence for culture in fish than in primates?

CONCLUSIONS

We have aimed to show that phenomena thought to be special in primates may also be found in other taxa, even in species not renowned for their intelligence, such as fish. One of the problems one encounters when attempting to compare primate behavior to that of other species is that researchers have simply not asked the same questions of non-primate species. Many researchers still appear to be guided by assumptions rather than evidence. Only some researchers have asked relevant questions of non-mammals (with the positive exception of birds—see Clayton and Emery, this volume). However, perhaps a trend is starting—certainly more comparative data on mammals have become available in recent years. Fortunately, our laboratory is not the only one to see the value of simple model systems for informing issues about primate sociality. Bshary and his colleagues conducted various elegant studies on cleaner wrasse, demonstrating that these fish employ interspecific social strategies at their cleaning stations that are dependent on the identity of the client. This had previously been reported only for mammalian, mainly primate, species (Bshary and Würth, 2001). Moreover, a current review of fish cognition (Bshary et al., 2002) indicates that studies of social intelligence (social strategies, social learning and traditions, and cooperative hunting) and environmental intelligence (special foraging skills, tool use, cognitive maps, anti-predator behavior, and environmental manipulation) in fish can help us understand these same phenomena in primates.

To the extent that they are interested in processes, we ask other behavioral scientists to challenge the assumptions that have so often been made about primates, and consider how simpler study species may be just as productive when considering features of sociality such as social learning and innovation. To the extent that they are interested in primate behavior, we ask them to consider whether the explanatory processes that work for other taxa—yes, even fish—might also work for primates. Indeed, such explanations would serve as valuable null hypotheses, only to be rejected in favor of more cognitively demanding explanations in the face of concrete evidence.

ACKNOWLEDGMENTS

Research supported in part by BBSRC studentships to YVB and WH, and a Royal Society University Research Fellowship to KNL.

REFERENCES

Bonner, J. T., 1980, *The Evolution of Culture in Animals*, Princeton University Press, Princeton, NJ.

Brown, C., and Laland, K. N., 2002, Social learning of a novel avoidance task in the guppy: Conformity and social release, *Anim. Behav.* **64**:41–47.

Bshary, R., and Würth, M., 2001, Cleaner fish *Labroides dimidiatus* manipulate client reef fish by providing tactile stimulation, *Proc. R. Soc. Lond. B.* **268**:1495–1501.

Bshary, R., Wickler, W., and Fricke, H., 2002, Fish cognition: A primate's eye view, *Anim. Cogn.* **5**:1–13.

Caro, T. M., and Hauser, M. D., 1992, Is there teaching in nonhuman animals? *Q. Rev. Biol.* **67**:151–174.

Coussi-Korbel, S., and Fragaszy, D. M., 1995, On the relation between social dynamics and social learning, *Anim. Behav.* **50**:1441–1453.

Curio, E., 1988, Cultural transmission of enemy recognition by birds, in: T. R. Zentall and B. G. Galef, Jr., eds., *Social Learning: Psychological and Biological Perspectives*, Erlbaum, Hillsdale, NJ, pp. 75–97.

Day, R. L., Macdonald, T., Brown, C., Laland, K. N., and Reader, S. M., 2001, Interactions between shoal size and conformity in guppy social foraging, *Anim. Behav.* **62**:917–925.

Dorrance, B. R., and Zentall, T. R., 2001, Imitative learning in Japanese quail (*Coturnix japonica*) depends on the motivational state of the observer quail at the time of observation, *J. Comp. Psychol.* **115**(1):62–67.

Fragaszy, D. M., and Visalberghi, E., 1990, Social processes affecting the appearance of innovative behaviors in capuchin monkeys, *Folia Primatol.* **54**:155–165.

Fragaszy, D. M., and Visalberghi, E., 1996, Social learning in monkeys: Primate "primacy" reconsidered, in: *Social Learning in Animals: The Roots of Culture*, B. G. Galef, Jr., and C. M, Heyes, eds., Academic Press, New York, pp. 65–84.

Galef, B. G., Jr., 1988, Imitation in animals: History, definitions and interpretation of the data from the psychological laboratory, in: *Social Learning: Psychological and Biological Perspectives*, T. Zentall and B. G. Galef, Jr., eds., Erlbaum, Hillsdale, NJ, pp. 3–28.

Galef, B. G., Jr., 1992, The question of animal culture, *Hum. Nat.* **3**:157–178.

Galef, B. G., Jr., 1995, Why behaviour patterns that animals learn socially are locally adaptive, *Anim. Behav.* **49**:1325–1334.

Galef, B. G., Jr., and Allen, C., 1995, A new model system for studying behavioural traditions in animals, *Anim. Behav.* **50**:705–717.

Galef, B. G., Jr., and Giraldeau, L.-A., 2001, Social influences on foraging in vertebrates: Causal mechanisms and adaptive functions, *Anim. Behav.* **61**:3–15.

Giraldeau, L.-A., and Caraco, T., 2000, *Social Foraging Theory*, Princeton University Press, Princeton, NJ.

Green, S., 1975, Dialects in Japanese monkeys: Vocal learning and cultural transmission of locale-specific vocal behavior? *Z. Tierpsychol.* **38**:304–314.

Griffiths, S. W., and Magurran, A. E., 1997a, Familiarity in schooling fish: How long does it take to acquire? *Anim. Behav.* **53**:945–949.

Griffiths, S. W., and Magurran, A. E., 1997b, Schooling preferences for familiar fish vary with group size in a wild guppy population, *Proc. R. Soc. Lond. B.* **264**:547–551.

Griffiths, S. W., and Magurran, A. E., 1998, Sex and schooling behaviour in the Trinidadian guppy, *Anim. Behav.* **56**:689–693.

Helfman, G. S., and Schultz, E. T., 1984, Social tradition of behavioural traditions in a coral reef fish, *Anim. Behav.* **32**:379–384.

Heyes, C. M., 1994, Social learning in animals: Categories and mechanisms. *Biol. Rev.* **69**:207–231.

Kawai, M., 1965, Newly-acquired pre-cultural behaviour of the natural troop of Japanese monkeys on Koshima islet, *Primates,* **6**:1–30.

Kummer, H., and Goodall, J., 1985, Conditions of innovative behaviour in primates, *Phil. Trans. R. Soc. Lond. B.* **308**:203–214.

Lachlan, R. F., Crooks, L., and Laland, K. N., 1998, Who follows whom? Shoaling preferences and social learning of foraging information in guppies, *Anim. Behav.* **56**:181–190.

Laland, K. N., 1999, Exploring the dynamics of social learning with rats, in: *Mammalian Social Learning: Comparative and Ecological Perspectives*, H. O. Box and K. Gibson, eds., Cambridge University Press, Cambridge, pp. 174–187.

Laland, K. N., and Reader, S. M., 1999a, Foraging innovation in the guppy, *Anim. Behav.* **57**:331–340.

Laland, K. N., and Reader, S. M., 1999b, Foraging innovation is inversely related to competitive ability in male but not in female guppies, *Behav. Ecol.* **10**:270–274.

Laland, K. N., and Williams, K., 1997, Shoaling generates social learning of foraging information in guppies, *Anim. Behav.* **53**:1161–1169.

Laland, K. N., and Williams, K., 1998, Social transmission of maladaptive information in the guppy, *Behav. Ecol.* **9**:493–499.

Laland, K. N., Richerson, P. J., and Boyd, R., 1993, Animal social learning: Towards a new theoretical approach, in: *Perspectives in Ethology 10, Behavior and Evolution*, P. P. G. Bateson, P. H. Klopfer, and N. S. Thompson, eds., Plenum Press, New York, pp. 249–277.

Laland, K. N., Richerson, P. J., and Boyd, R., 1996, Developing a theory of animal social learning, in: *Social Learning in Animals: The Roots of Culture*, B. G. Galef, Jr., and C. M. Heyes, eds., Academic Press, New York, pp. 129–154.

Lee, P. C., 1991, Adaptations to environmental change: An evolutionary perspective, in: *Primate Responses to Environmental Change*, H. O. Box, ed., Chapman & Hall, London, pp. 39–56.

Lefebvre, L., Whittle, P., Lascaris, E., and Finkelstein, A., 1997, Feeding innovations and forebrain size in birds, *Anim. Behav.* **53**:549–560.

Magurran, A. E., Seghers, B. H., Shaw, P. W., and Carvalho, G. R., 1994, Schooling preferences for familiar fish in the guppy, *Poecilia reticulata, J. Fish Biol.* **45**:401–406.

Mason, J. R., 1988, Direct and observational learning by red-winged blackbirds (*Agelaius phoneniceus*): The importance of complex stimuli, in: *Social Learning: Psychological and Biological Perspectives*, T. R. Zentall and B. G. Galef, Jr., eds., Lawrence Erlbaum Associates, pp. 99–117.

Morgan, M. J., 1988, The influence of hunger, shoal size and predator presence on foraging in bluntnose minnows, *Anim. Behav.* **36**:1317–1322.

Morgan, M. J., and Colgan, P. W., 1987, The effects of predator presence and shoal size on foraging in bluntnose minnows, *Environ. Biol. Fish.* **20**:105–111.

Pitcher, T. J., and House, A. C., 1987, Foraging rules for group feeders: Area copying depends upon food density in shoaling goldfish, *Ethology*, **76**:161–167.

Pitcher, T. J., Magurran, A. E., and Winfield, I. J., 1982, Fish in larger shoals find food faster, *Behav. Ecol. Sociobiol.* **10**:149–151.

Reader, S. M., 1999, *Social Learning and Innovation: Individual Differences, Diffusion Dynamics and Evolutionary Issues*, Unpublished PhD Thesis, University of Cambridge.

Reader, S. M., and Laland, K. N., 2000, Diffusion of foraging innovations in the guppy, *Anim. Behav.* **60**:175–180.

Reader, S. M., and Laland, K. N., 2001, Primate innovation: Sex, age and social rank differences, *Int. J. Primatol.* **22**(5):787–805.

Reader, S. M., Kendal, J. R., and Laland, K. N., in press, Social learning of foraging sites and escape routes in wild Trinidadian guppies. *Anim. Behav.*

Rendell, L., and Whitehead, H., 2001a, Culture in whales and dolphins, *Behav. Brain Sci.* **24**(2):309–382.

Rendell, L., and Whitehead, H., 2001b. Cetacean culture: Still afloat after the first naval engagement of the culture wars, *Behav. Brain. Sci.* **24**(2):360–382.

Reznick, D., and Yang, A. P., 1993, The influence of fluctuating resources on life-history patterns of allocation and plasticity in female guppies, *Ecology,* **74**:2011–2019.

Rowell, T., 1999, The myth of peculiar primates, in: *Mammalian Social Learning: Comparative and Ecological Perspectives,* in: H. O. Box and K. R. Gibson, eds., Cambridge University Press, Cambridge, pp. 6–16.

Ryer, C. H., and Olla, B. L., 1991, Information transfer and the facilitation and inhibition of feeding in a shoaling fish, *Environ. Biol. Fish.* **30**:317–323.

Ryer, C. H., and Olla, B. L., 1992, Social mechanisms facilitating exploitation of spatially variable ephemeral food patches in a pelagic marine fish, *Anim. Behav.* **44**:69–74.

Sargent, R. C., and Gross, M. R., 1993, Williams' principle: An explanation of parental care in teleost fishes, in: *Behaviour of Teleost Fishes,* Second edition, T. J. Pitcher, ed., Chapman & Hall, London, pp. 333–361.

Sigg, H., 1980, Differentiation of female positions in hamadryas one-male units, *Z. Tierpsychol.* **53**:265–302.

Swaney, W., Kendal, J. R., Capon, H., Brown, C., and Laland, K. N., 2001, Familiarity facilitates social learning of foraging behaviour in the guppy, *Anim. Behav.* **62**:591–598.

Templeton, J. J., 1998, Learning from others' mistakes: A paradox revisited, *Anim. Behav.* **55**:79–85.

Tomasello, M., 1994, The question of chimpanzee culture, in: *Chimpanzee Cultures,* R. Wrangham, W. McGrew, F. de Waal, and P. Heltne, eds., Harvard University Press, Cambridge, MA, pp. 301–317.

Tomasello, M., and Call, J., 1997, *Primate Cognition,* Oxford University Press, New York.

Trivers, R. L., 1972, Parental investment and sexual selection, in: *Sexual Selection and the Descent of Man, 1871–1971,* B. Campbell, ed., Aldine, Chicago, pp. 136–179.

Van Schaik, C. P., Deaner, R. O., and Merrill, M. Y., 1999, The conditions for tool use in primates: Implications for the evolution of material culture, *J. Hum. Evol.* **36**:719–741.

Visalberghi, E., 1992, Tool-use imitation in nonhuman primates and young children, *Int. J. Psychol.* **27**(3–4):427.

Visalberghi, E., and Fragaszy, D., 1990, Do monkeys ape? in: *Language and Intelligence in Monkeys and Apes*, S. T. Parker and K. Gibson, eds., Cambridge University Press, Cambridge, pp. 247–273.

Want, S. C., and Harris, P. L., 2001, Learning from other people's mistakes: Causal understanding in learning to use a tool, *Child Dev.* **72**(2):431–443.

Warner, R. R., 1988, Traditionality of mating-site preferences in a coral reef fish, *Nature,* **335**:719–721.

Whiten, A., 2000, Primate culture and social learning, *Cogn. Sci.* **24**(3):477–508.

Whiten, A., and Custance, D. M., 1996, Studies of imitation in chimpanzees and children, in: *Social Learning in Animals: The Roots of Culture*, B. G. Galef, Jr., and C. M. Heyes, eds., Academic Press, New York, pp. 291–318.

Whiten, A., and Ham, R., 1992, On the nature and evolution of imitation in the animal kingdom: Reappraisal of a century of research, *Adv. Study Behav.* **21**:239–283.

Whiten, A., Goodall, J., McGrew, W. C., Nishida, T., Reynolds, V., Sugiyama, Y. et al., 1999, Cultures in chimpanzees, *Nature,* **399**:682–685.

Wilson, E. O., 1975, *Sociobiology*, Harvard University Press, Cambridge, MA.

Zentall, T. R., Sutton, J. E., and Sherburne, L. M., 1996, True imitative learning in pigeons, *Psychol. Sci.* 7(6):343–346.

PART THREE

Communication

CHAPTER FIVE

The Primate Isolation Call: A Comparison with Precocial Birds and Non-primate Mammals

John D. Newman

INTRODUCTION

Evolution took a major turn when mammals first walked the Earth. Reptilian infants typically develop within the egg to be self-sufficient at hatching. Most mammalian infants, however, require their mother for warmth and sustenance during their earliest developmental period, and will die if they fail to receive these life-sustaining gifts. Infant birds, like reptiles, hatch from eggs, but for most species, they are fed and cared for by one or both parents until they can fend for themselves. As with mammals, infant birds depend on a caregiver to survive during their earliest developmental period. It should not be surprising, then, that infant birds and mammals have evolved vocal signals to alert their caregiver should the infant become separated from its nest or caregiver. With birds, the best developed of these so-called "separation calls" or "isolation calls" appear to

---

John D. Newman • Laboratory of Comparative Ethology, NICHD, National Institutes of Health, Department of Health and Human Services, Poolesville, Maryland, USA.

Comparative Vertebrate Cognition, edited by Lesley J. Rogers and Gisela Kaplan. Kluwer Academic/Plenum Publishers, 2004.

be in precocial species, where the chicks can locomote soon after birth and get their food by foraging with the mother. This sometimes results in occasions when the chicks become separated from the mother, hence the adaptive value for auditory signals to efficiently bring about reunion. In some species, these separation calls continue into adulthood. For example, in the Bobwhite Quail (*Colinus virginianus*), adults use the separation call in the same way as chicks do, namely when a covey becomes separated (Baker and Bailey, 1987a). As is the case for a number of mammalian species (discussed below), Bobwhite Quail can recognize their covey mates based on the auditory information in their separation calls (Baker and Bailey, 1987b). In other species (e.g., the Japanese quail, Coturnix), the neonatal isolation call of chicks matures into "crowing," a call associated with reproduction (Takeuchi et al., 1996).

The present chapter will focus on whether the isolation call of primates has special attributes or, alternatively, can be regarded as representative of isolation calls from mammals in general. I will argue that the evidence supports the latter case, and, additionally, will provide evidence suggesting the existence of shared attributes between the isolation call of primates (and other mammals) and those of precocial birds. The convergent evolution of avian and mammalian isolation calls (and related neural mechanisms) is an interesting subject, but beyond the scope of this chapter. It is true, nevertheless, that some of the same neural structures mediating isolation call production in mammals also have the same function in birds (cf. Panksepp et al., 1988; Takeuchi et al., 1996). More will be said about mechanisms underlying production and perception of the isolation call below.

THE MAMMALIAN ISOLATION CALL

In mammals (including some marsupials: Aitkin et al., 1996), there is widespread use of a vocalization during infancy that enables an infant to maintain, or regain, contact with its mother. Individual variability in vocal acoustic details is often adaptive, particularly in highly social species where there may be many infants crowded together in a limited space and the mothers return from foraging to nurse their infants. Well-documented examples are to be found in several species of bats (e.g., Gould, 1971) and in pinnepeds (e.g., Trillmich, 1981). However, even in less social species such as murid rodents, infants produce a vocalization that is effective in alerting the mother to their location, and stimulates maternal retrieval of the infant (e.g., Koch and Ehret, 1989).

The acoustic details of the vocalizations vary less than might be expected, given the ecological and social differences between species, with most of the differences attributable to the size of the vocalizer. Hence, small infants, such as mouse pups, make high-frequency calls of short duration, whereas larger infants (such as seal pups), make lower frequency calls of longer duration. Despite these differences, there seems to be a basic "mammalian plan" consisting of a tonal vocalization without significant wide-band energy ("noise") or numerous frequency shifts. In addition, there are species differences in many mammals, despite the fact that these calls function mainly during infancy, hence would not be useful as reproductive isolation behaviors. In some cases, at least,

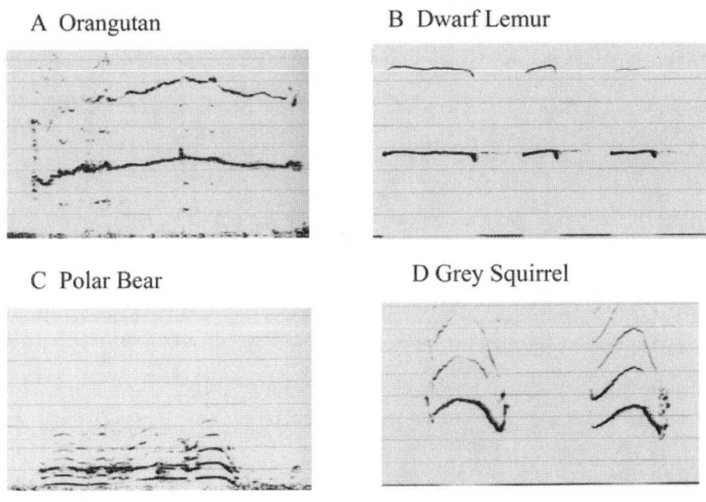

Figure 1. Sound spectrograms of the infant cries (isolation calls) of four mammals. Exact age unknown in each case, but infants were of an age when they would still be nursed by the mother. (A) Orangutan (Primates), (B) Dwarf Lemur (Primates), (C) Polar Bear (Carnivora), and (D) Grey Squirrel (Rodentia). Time scale (horizontal line beneath sound spectograms): A, 800 msec; B, 200 msec, C and D, 400 msec. Frequency scale (horizontal lines running through sound spectrograms), (A) 500 Hz, (B) 2000 Hz, (C) and (D) 1000 Hz. Note that both the dwarf lemur and the grey squirrel are hole-nesting species. The frequency range and duration of the isolation calls of each species approximately reflect the size of the infant, with smaller infants making shorter and higher frequency calls. Note that the smaller infants shown also make multiple calls, whereas the larger infants make single calls. It is unknown at the present time whether this represents a consistent pattern, or merely reflects individual differences within a given species.

these differences appear to reflect other measures of genetic diversity within a group of related species.

Studies of non-primate mammals have documented the existence of infant isolation calls (iics), most of which have the basic tonal structure mentioned above. These include murine rodents (e.g., Colvin, 1973; Hofer, 1996; Hofer et al., 1994; Sales and Smith, 1978; Smith and Sales, 1980), bats of a number of species (e.g., Esser and Schmidt, 1989; Gelfand and McCracken, 1986; Gould, 1971, 1974, 1975, 1979; Gould et al., 1973; Habersetzer and Marimuthu, 1986; Matsumura, 1981; Scherrer and Wilkinson, 1993), guinea pig (e.g., Berryman, 1976), kittens (e.g., Brown et al., 1977; Haskins, 1977, 1979) reindeer (Espmark, 1971), several species of pinnepeds (e.g., Insley, 1992; Trillmich, 1981), and bottlenose dolphins (e.g., Sayigh et al., 1990). Figure 1 shows examples of the isolation calls of four mammalian species, including two primates.

THE PRIMATE ISOLATION CALL

Infant isolation calls are produced by all primate infants, wherever their occurrence has been sought. Acoustically, iics fall into two categories, clicks and tones. Clicks are of short duration and have a broad frequency range. These attributes give clicks good localization properties (cf. Brown et al., 1979), and are found exclusively in certain nocturnal prosimian primates. Presumably, darkness removes visual cues that might assist a mother in locating a lost or separated infant, hence good acoustic localization cues would be an asset to survival. On the other hand, during daylight hours, good acoustic localization cues might also aid a potential predator in finding a vocalizing infant. Perhaps for this reason, infants of diurnal species of primates universally produce "tones" (i.e., tonal calls of limited frequency range and without clicks or other acoustic transients) (cf. Newman, 1985a, 1992). In addition to this basic structural attribute, the isolation calls of some species have been examined in sufficient detail to support several conclusions. First, isolation calls can differ in their acoustic details ("acoustic fine structure") between individuals and between closely related species. Individuality has its obvious survival value, particularly in highly social species, where there may be several infants at the same time. Why there should be species differences in what is primarily an infant vocalization is less clear. However, in one particularly well-studied case, distinctive differences in the acoustic morphology of the isolation calls of two species-groups of squirrel monkeys have been demonstrated (Boinski and Newman, 1988; Newman, 1985b; Newman and Symmes, 1982; Symmes et al., 1979). These differences are present at birth (Lieblich et al., 1980; Newman and Symmes, 1982) and appear to be inherited (Newman, 1985c; Newman and Symmes, 1982). In those species (such as marmosets and tamarins of the primate family Callitrichidae) where iics also come to assume the role of "loud calls" in adults (serving as territorial markers), the differences between species may function as reproductive isolating behaviors, and may also be a behavioral measure of the degree of genetic differentiation during speciation.

ISOLATION CALL DEVELOPMENT

It seems likely that the iic was used only by infants early in mammalian evolution, as is the case in many species. However, some living primates (and perhaps other mammals) continue to use isolation calls into adulthood, where the isolation calls remain relatively unchanged, retaining their basic infantile structure and function. In other living primates (and other mammals), the iic is no longer given once an individual passes through infancy, but the call itself is gradually transformed acoustically and functionally into another vocalization with a completely different function. Examples of the former situation can be found in squirrel monkeys (e.g., Newman and Symmes, 1982) and common marmosets (e.g., Epple, 1968; Norcross and Newman, 1993). Examples of the latter can be found in echolocating bats, where the iic gradually becomes the call used in echolocation (e.g., Gould, 1974). Click-like iics have been less well-studied from this vantage point, but in some prosimian primates, the iic gradually transforms into the adult loud call, which is predominantly tonal (Zimmermann, 1990). These examples suggest that, at some point in mammalian evolution, the brain substrates that mediate isolation call production began to undergo a change, either maintaining their anatomical and functional integrity into adulthood, or were taken over entirely by new neural circuitry.

Primate isolation calls have been used to test the role of individual experience in the acquisition of adult vocal skills. The species used in two of these studies was the rhesus macaque. One study kept individually housed macaque infants in the same room as mother-reared infants and their mothers for the first year of life. The experimental animals developed abnormal isolation calls, indicating that auditory experience is not sufficient for normal vocal development in monkeys (Newman and Symmes, 1974). A more recent study examined the calls of rhesus infants reared in a nursery containing other infants but no adults. In this study, infants were given social experience with age-matched conspecifics, but did not develop the abnormal vocalizations reported by Newman and Symmes in 1974. Their isolation coos were compared with the coos of age-matched infants raised in social groups with their mothers and other adults. Differences were found, but were attributed solely to the larger body size of nursery-reared infants for a given age, which was a result of the enriched formula they were fed (Hammerschmidt et al., 2000). The role of individual experience in isolation call maturation has also been studied in other mammals, such as bats (e.g., Esser and Schmidt, 1989), whereas bats have also been studied from the perspective of heritable isolation call attributes (Scherrer and Wilkinson, 1993).

A WELL-STUDIED PRIMATE

Common marmosets have been particularly well studied because of some interesting attributes of their isolation calls. Infant common marmosets make their species-specific isolation calls ("isolation phee") from birth (cf. Epple, 1968). In addition, they also produce a variety of other vocalizations when separated, but these other calls gradually drop out during development (Newman, 1995). A similar phenomenon has been described for rhesus macaques (Newman, 1995) and prosimian primates (Zimmermann, 1995). During puberty, marmosets start to make phees in the home cage after reaching reproductive maturity and being paired with a partner (Norcross and Newman, 1997; Norcross et al., 1999). Paired adult marmosets make phees in their home cage regularly, as well as phees when separated from their partner. These phee calls are acoustically distinct (as well as differing between males and females) from those made by the same animals when separated (Norcross and Newman, 1993). These acoustic differences are recognized and result in differential responding by conspecific listeners (Norcross et al., 1994). As has been found in squirrel monkeys and rhesus macaques, marmosets show elevated cortisol levels when separated from their home cage and familiar conspecifics (Norcross and Newman, 1999), but the peak elevation is delayed relative to peak production of separation phee calls. Marmosets have also provided evidence that separation phees change their structure over short time spans, in a way that mirrors presumed changes in underlying arousal state over these same time periods (Newman and Goedeking, 1992). Acoustic parameters such as call duration and peak frequency exhibit these changes, while other parameters are more stable, allowing individual differences in phee call structure to continue to communicate the identity of the vocalizer. The fact that phee calls can be shown to consist of sets of parameters that do not all follow the same patterns of variation suggests that some vocalizations may consist of sets of features that are under different physiological control, communicate different information, and may be differentially selected for in terms of their evolutionary history ("attribute hypothesis," Newman and Goedeking, 1992).

NEUROCHEMICAL (PHARMACOLOGICAL) CONTROL OF ISOLATION CALL PRODUCTION

Studies using behavioral pharmacology methods suggest that similar neurochemical substrates mediate isolation call production across a number of primate and non-primate mammals. There is also evidence that opiatergic mechanisms are shared by some birds, as well. A review of the pharmacological control of chick "distress vocalizations" and the role of opiates can be found in Panksepp et al. (1978).

With respect to opiate influence on mammalian isolation calls, there have been studies in guinea pigs (Herman and Panksepp, 1978, 1981), rats (Goodwin and Barr, 1997), dogs (Panksepp et al., 1978), and primates (Harris and Newman, 1988; Kalin et al., 1988; Kalin and Shelton, 1989; Newman, 1988a; Newman et al., 1982; Weiner et al., 1988). All of these studies implicated endogenous opiates in the regulation of isolation call production and other manifestations of separation distress. Opiates have not been examined for their role in regulating production of isolation calls in marmosets, but opiate injections do influence the amount of play behavior initiated by juvenile marmosets (Guard et al., 2002). Other neurochemical systems also appear to play a role in mediating isolation call production, but with greater variability between species (cf. review by Newman, 1991).

NEURAL MECHANISMS OF ISOLATION CALL PRODUCTION

A few studies have investigated the central pathways for isolation call production in non-primate mammals. Herman (Herman and Panksepp, 1981; Panksepp et al., 1988) elicited "distress vocalizations" in the guinea pig similar to separation-induced vocalizations from sites within the medial septum, hypothalamus, midbrain periaqueductal gray (PAG), and adjacent tegmentum. Kyuhou and Gemba (1998) found that electrical stimulation of the PAG in guinea pigs produced two vocal subtypes depending on the part being stimulated; one subtype was the isolation call. Stimulating the brain of cats that were awake results in isolation calls being produced from sites in the hypothalamus (Altafullah et al., 1988; Buchwald et al., 1988).

More extensive studies of the brain substrates mediating isolation call production have been done in non-human primates. Earlier studies using brain stimulation techniques in which calls resembling isolation calls in macaques were produced have been reviewed by MacLean (1990) and MacLean and Newman (1988). The PAG region of the midbrain is known to have widespread importance in vocal expression in amniotes (reptiles, birds, and mammals). Its role in isolation call production has been examined in adult squirrel monkeys (Newman and MacLean, 1982), where bilateral ablations in the PAG and adjacent tegmentum render an animal mute, at least in so far as isolation calls are concerned. More rostral lesions, in the thalamic tegmentum, do not render a monkey mute, but lead to abnormal isolation calls being produced, instead (Newman and MacLean, 1982). Several studies using experimental lesioning methods, in both macaques and squirrel monkeys, have identified the anterior cingulate gyrus and adjacent supplementary motor cortex (Kirzinger and Jürgens, 1982) as structures important for the expression of the isolation calls of adults (MacLean and Newman, 1988; Sutton et al., 1974, 1981, 1985; Trachy et al., 1981). Brain stimulation studies also implicated this region in production of calls with the same basic structure as the isolation call in adult rhesus macaques (Robinson, 1967) and adult squirrel monkeys (Jürgens and Ploog, 1970). Much less is known about the neural substrates underlying expression of the isolation call in infant primates (or infant mammals in general), an obvious shortcoming but not surprising due to the fact that infants generally do not tolerate brain surgery or even surgical levels of anesthesia very well. However, rhesus macaque infants are sufficiently robust to serve as experimental surgical subjects, and some information is available regarding forebrain regulation of their isolation calls. Newman and Bachevalier (1997) found that bilateral lesions of the amygdala in infants led to production of species-typical isolation calls ("coos") that lacked the typical inflective quality (an indication of lower affect in these animals). Newman and Bachevalier (1997; see also Newman et al., 1998) also found in other subjects that bilateral removal of a large portion of the inferior temporal cortex ("area TE") resulted in animals making more calls when separated, compared to age-matched controls, and, in males only, a significant increase in noisy calls during brief separations.

NEURAL MECHANISMS OF ISOLATION CALL PERCEPTION

Numerous studies cited in this chapter have shown that mothers can recognize and respond to the isolation calls of their infants. This clearly suggests that there are mechanisms within the brain that are specialized for detecting and discriminating isolation calls. Unfortunately, identifying these mechanisms and understanding how they function is still at an early stage of knowledge. These issues deserve more detailed examination than is possible in this chapter. However, the reader may find the papers in Newman (1988b) useful as an introduction.

While there have been many studies devoted to central auditory physiology in both birds and mammals, relatively few have dealt with the auditory processing of species-specific vocalizations, and fewer still with the perception of isolation calls or infant cries. One study in birds, using guinea fowl, examined the responses of a major avian auditory center, Field L, to this species' isolation call, the "iambus call" (Scheich et al., 1979). With respect to mammals, few studies outside of non-human primates have examined the neurophysiology of isolation call processing, but Buchwald et al. (1988) examined neural responses in the cat medial geniculate body (MGB, the thalamic auditory nucleus) to artificial sounds as well as to the species-specific isolation call (a single kitten mewing sound). They found that in one defined region of the MGB, nearly 30 percent of the neural responses were only to the isolation call. Several neurophysiology studies have been done using the echolocating bat, *Pteronotus parnellii*, on the processing of vocal sequences from a variety of communication sounds (e.g., Esser et al., 1997). While not studying isolation calls specifically, some of the acoustic profiles used clearly resemble typical isolation calls (cf. Kanwal et al., 1994), so their findings no doubt are relevant to the main theme of this section.

Using very different methods, Ehret (1987) has found evidence for lateralized processing of infant ultrasounds in mice. Mother mice with litters were tested with tone bursts of the same frequency parameter as natural isolation calls (50 kHz) presented through one of two loud speakers, with tone bursts lower than natural isolation calls (20 kHz) being presented through another loud speaker at the opposite end of a runway. Mother mice preferentially chose to approach a speaker presenting the higher frequency tone, suggesting that they possessed a predisposition to approach sounds resembling natural iics. Then, one ear was plugged and the test repeated. Mothers with the right ear plugged failed to show a preference for the 50 kHz tones, whereas mothers with the left ear plugged continued to show this stimulus preference. This suggested that the right hemisphere was more important than the left in performing this choice behavior. The findings were interpreting as indicating the presence of an innate releasing mechanism activated in a communicative context in the left hemisphere of the brain. A study in primates aimed to determine the unit of perception of the isolation call. This study (Ghazanfar et al., 2001), done in captive cotton-top tamarins (*Saguinus oedipus*), used playback methods and the antiphonal calling behavior of listeners to test whether calling was more robust to the individual components of the "combination long call" (comprised of a series of chirps and whistles) or to their combination. The study found that whole calls elicited antiphonal calling significantly more often than either single whistles or single chirps. The authors proposed that this response was mediated by "combination-sensitive" neurons in the auditory system. Such neurons have been found in other mammals, but have not yet been searched for in this tamarin.

No doubt motivated by the relationship to human auditory cortical mechanisms, a significant number of studies have been done on the anatomy and neurophysiology of central auditory processing in non-human primates (cf. Ghazanfar [2002] for a recent overview). Unfortunately, outside of studies on the squirrel monkey (and humans), no studies have focused on the isolation call. Obviously, the human studies have not involved in-depth neuroanatomy or neurophysiology. However, with the availability of "non-invasive" imaging technology, the prospects for studying the processing of our species-specific isolation call (i.e., the infant cry, Newman, 1985a) are being realized. Recent studies using this technique on women with childbearing experience have begun to map the human brain for areas activated by infant cries (Lorberbaum et al., 1999, 2002).

Squirrel monkeys have been used extensively to study auditory communication, including the neurophysiology underlying the perception of isolation calls ("Isolation peep", IP). Many of these studies have been reviewed by Newman (2002). Evidence that the IP of different individuals is unique to that individual ("vocal signature," Lieblich et al., 1980; Symmes et al., 1979), and that mothers could recognize the IPs of their own infants (Symmes and Biben, 1985) was preceded by neurophysiological evidence of this attribute (Newman and Wollberg, 1973), along with an attempt to determine the acoustic features within the IP that foster individual recognition (Symmes and Newman, 1974).

CONCLUSIONS

The isolation call of primates follows the "mammalian plan" seen across a phylogenetically diverse range of mammalian species, hence there is little room to argue that primates are "special" in this respect. The evidence that the iic of some primate infants develops into a functionally different vocalization during adulthood also is a characteristic of the iics of some non-primate mammals, so here too, primates are not special. Thus, the overall conclusion that one can make is that the primate iic, including those of humans, is a highly conserved mammalian trait. It seems plausible that at least some of the underlying neural mechanisms that mediate the production and perception of this functional and structural class of vocalization have also been conserved, and are widely shared. In any case, this is an important area of future research that richly deserves exploring.

REFERENCES

Aitkin, L., Nelson, J., and Shepherd, R., 1996, Development of hearing and vocalization in a marsupial, the Northern Quall, Dasyurus hallucatus, *J. Exp. Zool.* **276**:394–402.

Altafullah, I., Shipley, C., and Buchwald, J. S., 1988, Voiced calls evoked by hypothalamic stimulation in the cat, *Exp. Brain Res.* **71**:21–32.

Baker, J. A., and Bailey, E. D., 1987a, Ontogeny of the separation call in Northern Bobwhite (*Colinus virginianus*), *Can. J. Zool.* **65**:1016–1020.

Baker, J. A., and Bailey, E. D., 1987b, Auditory recognition of covey mates from separation calls in Northern Bobwhite (*Colinus virginianus*), *Can. J. Zool.* **65**:1724–1728.

Berryman, J. C., 1976, Guinea-pig vocalizations: Their structure, causation and function, *Z. Tierpsychol.* **41**:80–106.

Boinski, S., and Newman, J. D., 1988, Preliminary observations on squirrel monkey (*Saimiri oerstedii*) vocalizations in Costa Rica, *Amer. J. Primatol.* **14**:329–343.

Brown, C. H., Beecher, M. D., Moody, D. B., and Stebbins, W. C., 1979, Locatability of vocal signals in Old World monkeys: Design features for the communication of position, *J. Comp. Physiol. Psychol.* **93**:806–819.

Brown, K. A., Buchwald, J. S., Johnson, J. R., and Mikolich, D. J., 1977, Vocalization in the cat and kitten, *Dev. Psychobiol.* **11**:559–570.

Buchwald, J. S., Shipley, C., Altafullah, I., Hinman, C., Harrison, J., and Dickerson, L., 1988, The feline isolation call, in: *The Physiological Control of Mammalian Vocalization*, J. D. Newman, ed., Plenum Press, New York, pp. 119–135.

Colvin, M. A., 1973, Analysis of acoustic structure and function in ultrasounds of neonatal Microtus, *Behaviour* **44**:234–263.

Ehret, G., 1987, Left hemisphere advantage in the mouse brain for recognizing ultrasonic communication calls, *Nature* **325**:249–251.

Epple, G., 1968, Comparative studies on vocalization in marmoset monkeys (Hapalidae), *Folia Primatologia* **8**:1–40.

Espmark, Y., 1971, Individual recognition by voice in reindeer mother–young relationship. Field experiments and playback experiments, *Behaviour* **40**:295–301.

Esser, K. -H., and Schmidt, U., 1989, Mother–infant communication in the lesser spear-nosed bat Phyllostomus discolor (Chiroptera, Phyllostomidae)—evidence for acoustic learning, *Ethology* **82**:156–168.

Esser, K. -H., Condon, C. J., Suga, N., and Kanwal, J. S., 1997, Syntax processing by auditory cortical neurons in the FM-FM area of the mustached bat *Pteronotus parnellii, Proc. Natl. Acad. Sci. USA* **94**:14019–14024.

Gelfand, D. L., and McCracken, G. F., 1986, Individual variation in the isolation calls of Mexican free-tailed bat pups (*Tadarida brasiliensis mexicana*), *Anim. Behav.* **34**:1078–1086.

Ghazanfar, A. A., Flombaum, J. I., Miller, C. T., and Hauser, M. D., 2001, The units of perception in the antiphonal calling behavior of cotton-top tamarins (*Saguinus oedipus*): Playback experiments with long calls, *J. Comp. Physiol. A* **187**:27–35.

Ghazanfar, A. A., ed., 2002, *Primate Audition: Ethology and Neurobiology*, CRC Press, Boca Raton.

Goodwin, G. A., and Barr, G. A., 1997, Evidence for opioid and nonopioid processes mediating adaptive responses of infant rats that are repeatedly isolated, *Dev. Psychobiol.* **31**:217–227.

Gould, E., 1971, Studies of maternal–infant communication and development of vocalizations in the bats Myotis and Eptesicus, *Comm. Behav. Biol.* **5**:263–313.

Gould, E., 1974, Experimental studies of the ontogeny of ultrasonic vocalizations in bats, *Dev. Psychobiol.* **8**(4):333–346.

Gould, E., 1975, Neonatal vocalizations in bats of eight genera, *J. Mammal.* **56**(1):15–29.

Gould, E., 1979, Neonatal vocalizations of ten species of Malaysian bats (Megachiroptera and Microchiroptera), *Am. Zool.* **19**:481–491.

Gould, E., Woolf, N. K., and Turner, D. C., 1973, Double-note communication calls in bats: Occurrence in three families, *J. Mammal.* **54**:998–1001.

Guard, H. J., Newman, J. D., and Roberts, R. L., 2002, Morphine administration selectively facilitates social play in common marmosets, *Dev. Psychobiol.* **41**:37–49.

Habersetzer, J., and Marimuthu, G., 1986, Ontogeny of sounds in the echolocating bat *Hipposideros speoris*, *J. Comp. Physiol. A* **158**:247–257.

Hammerschmidt, K., Newman, J. D., Champoux, M., and Suomi, S. J., 2000, Changes in rhesus macaque coo vocalizations during early development, *Ethology* **106**:873–886.

Harris, J. C., and Newman, J. D., 1988, Primate models for the management of separation anxiety, in: *The Physiological Control of Mammalian Vocalizations*, J. D. Newman, ed., Plenum, New York, pp. 321–330.

Haskins, R., 1977, Effect of kitten vocalizations on maternal behavior, *J. Comp. Physiol. Psychol.* **91**:830–838.

Haskins, R., 1979, A causal analysis of kitten vocalization: An observational and experimental study, *Anim. Behav.* **27**:726–736.

Herman, B. H., and Panksepp, J., 1978, Effects of morphine and naloxone on separation distress and approach attachment: Evidence for opiate mediation of social affect, *Pharmacol. Biochem. Behav.* **9**:213–220.

Herman, B. H., and Panksepp, J., 1981, Ascending endorphin inhibition of distress vocalization, *Science* **211**:1060–1062.

Hofer, M. A., 1996, Multiple regulators of ultrasonic vocalization in the infant rat, *Psychoneuroendocrinology* **21**(2):203–217.

Hofer, M. A., Brunelli S. A., and Shair, H. N., 1994, Potentiation of isolation-induced vocalization by brief exposure of rat pups to maternal cues, *Dev. Psychobiol.* **27**(8):503–517.

Insley, S. J., 1992, Mother–offspring separation and acoustic stereotypy: A comparison of call morphology in two species of pinnepeds, *Behaviour* **120**:103–122.

Jürgens, U., and Ploog, D., 1970, Cerebral representation of vocalization in the squirrel monkey, *Exp. Brain Res.* **10**:532–554.

Kalin, N. H., and Shelton, S. E., 1989, Defensive behaviors in infant rhesus monkeys: Environmental cues and neurochemical regulation, *Science* **243**:1718–1721.

Kalin, N. H., Shelton, S. E., and Barksdale, C. M., 1988, Opiate modulation of separation-induced distress in non-human primates, *Brain Res.* **440**:285–292.

Kanwal, J. S., Matsumura, S., Ohlemiller, K., and Suga, N., 1994, Analysis of acoustic elements and syntax in communication sounds emitted by mustached bats, *J. Acoust. Soc. Am.* **96**:1229–1254.

Kirzinger, A., and Jürgens, U., 1982, Cortical lesion effects and vocalization in the squirrel monkey, *Brain Res.* **233**:299–315.

Koch, M., and Ehret, G., 1989, Estradiol and parental experience, but not prolactin are necessary for ultrasound recognition and pup-retrieval in the mouse, *Physiol. Behav.* **45**:771–776.

Kyuhou, S. -I., and Gemba, H., 1998, Two vocalization-related subregions in the midbrain periaqueductal gray of the guinea pig, *Neuroreport* **9**:1607–1610.

Lieblich, A. K., Symmes, D., Newman, J. D., and Shapiro, M., 1980, Development of the isolation peep in laboratory-bred squirrel monkeys, *Anim. Behav.* **28**:1–9.

Lorberbaum, J. P., Newman, J. D., Dubno, J. R., Horwitz, A. R., Nahas, Z., Teneback, C. C. et al., 1999, The feasibility of using fMRI to study mothers responding to infant cries. *Depression Anxiety* **10**:99–104.

Lorberbaum, J. P., Newman, J. D., Horwitz, A. R., Dubno, J. R., Lydiard, R. B., Hamner, M. B. et al., 2002, A potential role for thalamocingulate circuitry in human maternal behavior, *Biol. Psychiatry* **51**:431–445.

MacLean, P. D., 1990, *The Triune Brain in Evolution*, Plenum Press, New York.

MacLean, P. D., and Newman, J. D., 1988, Role of midline frontolimbic cortex in production of the isolation call of squirrel monkeys, *Brain Res.* **450**:111–123.

Matsumura, S., 1981, Mother–infant communication in a horseshoe bat (*Rhinolophus ferrumequinum nippon*): Vocal communication in three-week-old infants, *J. Mammal.* **62**:20–28.

Newman, J. D., 1985a, The infant cry of primates: An evolutionary perspective, in: *Infant Crying: Theoretical and Research Perspectives*, B. M. Lester and C. F. Z. Boukydis, eds., Plenum Press, New York, pp. 307–323.

Newman, J. D., 1985b, Squirrel monkey communication, in: *Handbook of Squirrel Monkey Research*, L. A. Rosenblum and C. L. Coe, eds., Plenum Press, New York, pp. 99–126.

Newman, J. D., 1985c, Inherited isolation call characteristics in two species of squirrel monkey (Saimiri) and their hybrids, *Int. J. Primatol.* **5**(4):366.

Newman, J. D., 1988a, Ethopharmacology of vocal behavior in primates, in: *Primate Vocal Communication*, D. Todt, P. Goedeking, and D. Symmes, eds., Springer, Berlin, pp. 145–153.

Newman, J. D., ed., 1988b, *The Physiological Control of Mammalian Vocalization*, Plenum Press, New York.

Newman, J. D., 1991, Vocal manifestations of anxiety and their pharmacological control, in: *Psychopharmacology of Anxiolytics and Antidepressants, The International Encyclopedia of Pharmacology and Therapeutics, Section 136*, S. E. File, ed., Pergamon, New York, pp. 251–260.

Newman, J. D., 1992, The primate isolation call and the evolution and physiological control of human speech, in: *Language Origin: A Multidisciplinary Approach*, NATO ASI Series D, Vol. 61. J. Wind, B. Chiarelli, B. Bichakjian, and A. Nocentini, eds., Kluwer, Dordrecht, pp. 301–321.

Newman, J. D., 1995, Vocal ontogeny in macaques and marmosets: Convergent and divergent lines of development, in: *Current Topics in Primate Vocal Communication*, E. Zimmermann, J. Newman, and U. Jürgens, eds., Plenum, New York, pp. 73–97.

Newman, J. D., 2002, Auditory communication and central auditory mechanisms in the squirrel monkey: Past and present, in: *Primate Audition: Ethology and Neurobiology*, A. A. Ghazanfar, ed., CRC Press, Boca Raton, pp. 227–246.

Newman, J. D., and Bachevalier, J., 1997, Neonatal ablations of the amygdala and inferior temporal cortex in rhesus macaques alter vocal response to social separation during the first year of life, *Brain Res.* **758**:180–186.

Newman, J. D., Bachevalier, J., and Mishkin, M., 1998, Infero-temporal cortical ablations affect vocal expression in infant rhesus macaques, *Proc 5th Internet World Congress Biomed Sci '98*, http://www.mcmaster.ca/inabis98/symposia/brudzynski/newman.

Newman, J. D., and Goedeking, P., 1992, Noncategorical vocal communication in subhuman primates: The example of common marmoset phee calls, in: *Nonverbal Vocal Communication: Comparative and Developmental Aspects*, H. Papousek, U. Jürgens, and M. Papousek, eds., Cambridge University Press, Cambridge, pp. 87–101.

Newman, J. D., and MacLean, P. D., 1982, Effects of tegmental lesions on the isolation call of squirrel monkeys, *Brain Res.* **232**:317–329.

Newman, J. D., and Symmes, D., 1974, Vocal pathology in socially deprived monkeys, *Dev. Psychobiol.* 7:351–358.

Newman, J. D., and Symmes, D., 1982, Inheritance and experience in the acquisition of primate acoustic behavior, in: *Primate Communication*, C. T. Snowdon, C. H. Brown, and M. R. Petersen, eds., Cambridge University Press, Cambridge, pp. 259–278.

Newman, J. D., and Wollberg, Z., 1973, Responses of single neurons in the auditory cortex of squirrel monkeys to variants of a single call type, *Exp. Neurol.* **40**:821–824.

Newman, J. D., Murphy, M. R., and Harbaugh, C. R., 1982, Naloxone-reversible suppression of isolation call production after morphine injections in squirrel monkeys, *Soc. Neurosci. Abstr.* **8**:940.

Norcross, J. L., and Newman, J. D., 1993, Context and gender-specific differences in the acoustic structure of common marmoset (*Callithrix jacchus*) phee calls, *Am. J. Primatol.* **301**:37–54.

Norcross, J. L., and Newman, J. D., 1997, Social context affects phee call production by nonreproductive common marmosets (*Callithrix jacchus*), *Am. J. Primatol.* **43**:135–146.

Norcross, J. L., and Newman, J. D., 1999, Effects of separation and novelty on distress vocalizations and cortisol in the common marmoset (*Callithrix jacchus*), *Am. J. Primatol.* **47**:209–222.

Norcross, J. L., Newman, J. D., and CoFrancesco, L. M., 1999, Context and sex differences exist in the acoustic structure of phee calls by newly-paired common marmosets (*Callithrix jacchus*), *Am. J. Primatol.* **49**:165–181.

Norcross, J. L., Newman, J. D., and Fitch, W., 1994, Responses to natural and synthetic phee calls by common marmosets (*Callithrix jacchus*), *Am. J. Primatol.* **33**:15–29.

Panksepp, J., Meeker, R., and Bean, N. J., 1980, The neurochemical control of crying, *Pharmacol. Biochem. Behav.* **12**:437–443.

Panksepp, J., Herman, B., Conner, R., Bishop, P., and Scott, J. P., 1978, The biology of social attachments: Opiates alleviate separation distress, *Biol. Psychiatry* **13**:607–618.

Panksepp, J., Normansell, L., Herman, B., Bishop, P., and Crepeau, L., 1988, Neural and neurochemical control of the separation distress call, in: *The Physiological Control of Mammalian Vocalization*, J. D. Newman, ed., Plenum Press, New York, pp. 263–299.

Robinson, B. W., 1967, Vocalization evoked from forebrain in *Macaca mulatta*, *Physiol. Behav.* **2**:345–354.

Sales, G. D., and Smith, J. C., 1978, Comparative studies of the ultrasonic calls of infant murid rodents, *Dev. Psychobiol.* **11**(6):595–619.

Sayigh, L. S., Tyack, P. L., Wells, R. S., and Scott, M. D., 1990, Signature whistles of free-ranging bottlenose dolphins *Tursiops truncatus*: Stability and mother–offspring comparisons, *Behav. Ecol. Sociobiol.* **26**:247–260.

Scheich, H., Langner, G., and Bonke, D., 1979, Responsiveness of units in the auditory neostriatum of the guinea fowl (*Numida meleagris*) to species-specific calls and synthetic stimuli. II. Discrimination of Iambus-like calls, *J. Comp. Physiol.* **132**:257–276.

Scherrer, J. A., and Wilkinson, G. S., 1993, Evening bat isolation calls provide evidence for heritable signatures, *Anim. Behav.* **46**:847–860.

Smith, J. C., and Sales, G. D., 1980, Ultrasonic behavior and mother–infant interactions in rodents, in: *Maternal Influences and Early Behavior*, R. W. Bell and W. P. Smotherman, eds., Spectrum, New York, pp. 105–133.

Sutton, D., Trachy, R. E., and Lindeman, R. C., 1981, Primate phonation: Unilateral and bilateral cingulate lesion effects, *Behav. Brain Res.* **3**:99–114.

Sutton, D., Larson, C., and Lindeman, R. C., 1974, Neocortical and limbic lesion effects on primate phonation, *Brain Res.* **71**:61–75.

Sutton, D., Trachy, R. E., and Lindeman, R. C., 1985, Discriminative phonation in macaques: Effects of anterior mesial cortex damage, *Exp. Brain Res.* **59**:410–413.

Symmes, D., and Biben, M., 1985, Maternal recognition of individual infant squirrel monkeys from isolation call playback, *Am. J. Primatol.* **9**:39–46.

Symmes, D., and Newman, J. D., 1974, Discrimination of isolation peep variants by squirrel monkeys, *Exp. Brain Res.* **19**:365–376

Symmes, D., Newman, J. D., Riggs, G., and Lieblich, A. K., 1979, Individuality and stability of isolation peeps in squirrel monkeys, *Anim. Behav.* **27**:1142–1152.

Takeuchi, H. -A., Yazaki, Y., Matsushima, T., and Aoki, K., 1996, Expression of Fos-like immunocytochemistry in the brain of quail chick emitting the isolation-induced distress calls, *Neurosci. Lett.* **220**:191–194.

Trachy, R. E., Sutton, D., and Lindeman, R. C., 1981, Primate phonation: Anterior cingulate lesion effects on response rate and acoustical structure, *Am. J. Primatol.* **1**:43–55.

Trillmich, F., 1981, Mutual mother–pup recognition in Galapagos fur seals and sea lions: Cues used and functional significance, *Behaviour* **78**:21–42.

Weiner, S., Coe, C., and Levine, S., 1988, Endocrine and neurochemical sequelae of primate vocalizations, in: *The Physiological Control of Mammalian Vocalization*, J. D. Newman, ed., Plenum Press, New York, pp. 367–394.

Zimmermann, E., 1990, Differentiation of vocalizations in bushbabies Galaginae, Prosimiae, Primates and the significance for assessing phylogenetic relationships, *Z. zoo. Syst. Evolut.-forsch.* **28**:217–239.

Zimmermann, E., 1995, Loud calls in nocturnal prosimians: Structure, evolution and ontogeny, in: *Current Topics in Primate Vocal Communication*, E. Zimmermann, J. D. Newman, and U. Jürgens, eds., Plenum Press, New York, pp. 47–72.

CHAPTER SIX

Meaningful Communication in Primates, Birds, and Other Animals

Gisela Kaplan

INTRODUCTION

Communication plays a role, and sometimes a major role, in almost all aspects of behavior that have to do with higher cognition in animals. Research in this area has usually concentrated on several distinct topics, these being memory, problem solving, learning, using tools, concept formation and categorization, referential signaling, inferring the state of another, and even engaging in hunting. Debates on the cognitive abilities of apes have dominated the last decade and have been the impetus for a substantial volume of research and innovative writing (e.g., Byrne, 1995; Byrne and Whiten, 1988; Carruthers and Smith, 1996; Harcourt, 2000; Heyes, 1998; Heyes and Huber, 2000; Parker and Gibson, 1990; Gibson and Ingold, 1993; Russon et al., 1996; Sternberg and Kaufman, 2002). Many experiments conducted with chimpanzees, gorillas, and orangutans have since underscored the belief that great apes have higher

Gisela Kaplan • Centre for Neuroscience and Animal Behaviour, School of Biological, Biomedical and Molecular Sciences, University of New England, Armidale, NSW, Australia.

Comparative Vertebrate Cognition, edited by Lesley J. Rogers and Gisela Kaplan. Kluwer Academic/Plenum Publishers, 2004.

cognitive abilities, but the view that these might be unique to great apes amongst animals has become increasingly doubtful.

This chapter is focused on two topics: referential signals and communication associated specifically with hunting. The two topics for discussion are chosen advisedly. Finding evidence of referential signaling in any species is deemed to be an important step in demonstrating intentional communication that has to be more than, or different from, expressing affect. Since referential signaling was first discovered in primates, much has been written about the topic. Referential signals, such as alarm calls, are relatively simple vocal forms of communication but they have gained significance in so far as they have been regarded, by some researchers, as rudimentary forms of language (concept formation). Referential signaling can occur between two individuals, whereas the examples of hunting that will be discussed here are group activities of more than two individuals. Although animals can hunt alone or in pairs, the evidence of higher cognition is related particularly to group activities and, of course, hunting requires some form of coordination. By comparison with the recent research interest in referential signals, investigations of the link between hunting and higher cognition has been relatively underrepresented. Nevertheless, there is a continued and abiding interest in hunting and meat eating by chimpanzees and what these might entail in terms of cognition and communication. Also, especially in the literature on whales and dolphins, there is mounting research evidence for behavioral complexity in hunting, involving communication and higher cognition (Boran and Heimlich, 1999). The kind of coordination required in hunting may demand the use of complex communication, although, as we shall see, this is not without controversy.

Great apes, as we have long suspected but now know, have complex communicative abilities. These, together with a range of other abilities, are believed to show higher cognitive functions, some arguably unique. Rather than examine this proposal in the light of neuroscience and the neocortex, a route that Chapters 1 and 9 have taken, I will examine the context of referential signaling and the coordination of group hunting from the social context. The theoretical difficulties that any consideration of communication presents are formidable because, no matter what the cognitive value of a behavior may be, there needs to be a sender and a receiver of a message and there may be different cognitive demands on both of these participants. The question is whether the signal is sent intentionally or merely the consequence of affect; that is, whether it is planned or automatic and innate.

COMMUNICATION FROM THE POINT OF VIEW OF THE RECEIVER

In general, it may be claimed that almost all signals are meaningful at the functional level because a second organism has to respond to the presence of the other. At a very basic level, all living organisms possess some ability to communicate. Even plants are said to communicate. Not all communication requires complex cognitive abilities but may rely on simple signals etc. resulting in either generalized or specific responses. In other words, many types of communication have little, if anything, to do with higher cognition (Zentall, 2000). The complex areas of communication, on the other hand, may well have to do with higher cognition (or intelligence), even if this is at times methodologically difficult to prove. The question is whether primates have a particular place in these discussions by furnishing evidence of unusual abilities which may have lead to characteristics found in human primates or whether we should look instead at factors involving the ecology and social organization of a species in order to understand why higher cognitive abilities are present in some and not other species.

REFERENTIAL SIGNALING

Vocal Signaling in General

We distinguish broadly between the syntax and the semantics of a message. Syntax refers to the structure of a call. Semantics refers to the content or meaning of the message. We use the term "vocabulary" to refer to the semantics, that is, the actual meaning of calls. The two can be intertwined. It is questionable whether vocalizations are ever entirely without meaning. In their simplest aspect, they are expressions of affect (involuntary exclamations in situations of extreme and/or sudden fear, distress, pain). Many vocalizations communicate more than these states, however. Repertoire size or mere length of utterances, by itself, is not an indication of the amount of meaning conveyed. A dog may bark for half an hour or longer and still say only "I am distressed, I have been left alone" or, in the context of wild dogs "stay away, this is my territory." A bird may sing a song at dawn and dusk for hours and equally only indicate "this is my territory." Meaning is far more difficult to assess than simply determining repertoire size. Nevertheless, even determining repertoire size has grown enormously complex over the last few decades as shown in species as different as sea mammals, rodents, and other small prey mammals, birds, and monkeys (see Rogers and Kaplan, 2000).

Moreover, communication is a two-way process and the cognitive demands on the sender may be different from those of the receiver. The latter may decipher a good deal more information than the sender was aware of sending. In questions about higher cognitive function, however, the emphasis is focused not only on the deciphering process by the receiver but also on the possible motivations or intentions of the sender.

Motivational versus Referential Signals Ethology has tended to describe animal behavior in terms of four main motivational systems: aggression, fear, feeding, and sex. These categories are related to physiological processes underlying the behavior. Tinbergen (1953, 1963) argued that behavior was due to relatively invariant and immediate responses to internal and external stimuli, although he believed in development and did recognize that animals learn. Many studies since have shown that vocal behavior in birds, primates, and cetaceans depends on learning which plays an important part in the development of communication. There is a period of plasticity, that is, a period during development when the individual is able to expand its "vocabulary," be this in vocal or gestural range. The period for learning may vary considerably between species. In some species, learning is restricted to the first days or months of life, while in others it may take much longer. In human primates, plasticity is lifelong, and this appears to be the case in bonobos (Savage-Rumbaugh and Lewin, 1994) and in some parrot species (Pepperberg, 1999). This learning is not just a matter of adding to repertoire size. Vocalizations can also become considered responses instead of being simple invariant responses to a set of stimuli.

The earlier distinction that researchers made between "motivational" signals and "referential" signals as two very different paradigms of signal functioning was ultimately not sustainable (Marler et al., 1992). Signals merely reflecting states of affect might exist at one end of the continuum, and signals that are purely "referential," conveying information not derived from the caller's own emotional state, at the other end. Yet many, if not most, signals are somewhere in between these two poles.

In fact, a signal that may be purely "motivational" may be interpreted by the receiver in terms of referential content, or a signal that appears to arise purely from affect may have contextual components that are not derived from affect but from social learning. Hence, even so-called affective/motivational calls may be coded according to the social context or inflected by status and individual identity. Many monkey species can distinguish individuals one from another by their calls alone (e.g., the trill calls of spider monkeys, *Ateles geoffroyi*), and many, for example, Japanese macaques (*Macaca fuscata*), vary their calls according to social situation (May et al., 1989). The ability to recognize individuals is not reserved for primates. Slobodchikoff and colleagues (1991) tested the ability of prairie dogs to recognize individual human "predators" by stalking them in the field, wearing different clothes in different tests. They found that the prairie dogs were able to distinguish one individual human from another and that they incorporated information about the physical features of individual predators into their alarm calls. This is a remarkable result and it tells us that animals are not only perceiving much more detail than we might have considered previously but also that they are able to signal this information to each other. Further, we seem to be dealing with mammalian characteristics here that have been well conserved over evolutionary time.

Smith (1981, 1990) rightly pointed out that a vocal signal can be multiply referent. The identity of a caller may not only be that of an individual but also that of a group or a line, as in the matrilineal signatures in recruitment screams of pigtail macaques (Gouzoules and Gouzoules, 1990), or the recognition of individuals and kin in rhesus macaques (Rendall et al., 1996). There is some evidence of this occurring in other mammals as well (Hanggi and Schusterman, 1994; Hare, 1998; Johnston and Jernigan, 1994; Tooze et al., 1990). The fact that such evidence is relatively scanty is not evidence of its absence but of the infrequency of such studies. We thus know at least that, in some species, the receiver can identify the caller from the vocal signal alone and, in certain instances, such recognition may work in *conjunction with* and *in addition to* the actual message. In these cases, the message to be communicated comes encased with its own envelope, equipped with sender details and possibly even signaling family membership, dialect (Green, 1975), and even subspecific identification (Hodun et al., 1981). Locational cues have often been noted, not just as a consequence of sound attenuation but also via variations or specificity of the call (Harrington and Mech, 1983). In other words, current views of the relative importance of vocal signals in primates have shifted away from simple motivational assessment (affect) to understanding subtle variations as degrees of "referentiality." Moreover, communication is often expressed not just in one modality but in several, requiring visual, auditory or other inputs (Partan and Marler, 1999), and thus the dimensions of communication are even more complicated.

The next set of questions concerns the ability of animals to distinguish individuals not only by their vocal signals but also whether receivers listen for further details and information in the message. This can occur in intra-species and inter-species communication. For instance, it has been shown that one monkey species may respond to the alarm calls of another species, as is the case in ring-tailed lemurs responding to calls by *Verreaux sifakas* (Oda and Masataka, 1996). More extreme, a mammalian species may respond to alarm calling by an avian species. For instance, vervet monkeys know the meaning of the predator calls made by the starlings living alongside them as well as their own species-specific calls (Seyfarth and Cheney, 1990; Hauser, 1988). When the starling's "eagle" alarm call was played over a loud speaker, the monkeys looked up and, when the starling's ground-predator alarm call was played, most of the monkeys ran to the trees. No response was given when the starling's song was played back, and that was a control for the experiment because the song does not indicate the presence of any predator. It would seem that the vervet monkeys have learnt to interpret the alarm signals of the starling. In this way, different species living in the same area may make use of each other's communication signals.

Vervet monkeys are not the only species to interpret the alarm call of another species. Doing so is quite common among bird species and may result in some kind of symbiosis. For instance, when Australian magpies (*Gymnorhina tibicen*)—an extremely vigilant territorial avian species—alarm call, they elicit a response not only from their conspecifics but also from noisy miners (*Manorina melanophrys*), which join them and help fend off intruders (Kaplan, 2004). The reason for the high incidence of inter-species attention to each other's alarm call lies partly in the sound structure of calls, as we have now known for some time (Marler, 1955, 1981). In most instances, cross-species vocal communication does not imply that one species intentionally sets out to warn the other but merely that the other (the receiver) has interpreted the call in a correct way.

Attributing Meaning in Alarm and Food Calling

Some alarm and other short calls are said to contain specific information about the predator. These are referential signals. For instance, different alarm calls may be used for different predators leading to different responses by the receivers. This presupposes the presence of concepts. These alarm calls may not be given intentionally, however. Nevertheless, in some cases some calls are addressed to a receiver who acts on the message *intended* by the signaler, as shown by the fact that the alarm call is given only when at least one conspecific is present and not when the caller is alone. Intentionality and the presence of an audience are thus vital ingredients in communication that is both intentional and "referential."

Research has shown that vervet monkeys (Cheney and Seyfarth, 1990; Seyfarth et al., 1980; Owren, 1990) and Diana monkeys (Zuberbühler, 2000; Zuberbühler et al., 1999) have referential signals for alarm calls and use them with clear distinction between predator types (e.g., snake, eagle, or leopard; ground or aerial predator). Ring-tailed lemurs (*Lemur catta*) also produce different alarm calls to alert their group members to an aerial or a ground predator but they have a general call, a relatively soft "glup" sound as a general alert signal to the group. If an aerial predator has been detected, they follow the "glup" by loud calls, first rasps and then shrieks when the raptor is within attacking range. If the predator is a carnivorous ground predator, the "glup" is followed by "clicks" and "yaps" (Macedonia, 1990). Therefore, they signal about aerial versus ground predators and also the proximity of the predator. So it is possible to establish external referents by different vocal means: in vervet and Diana monkeys this occurs by producing different sounds, in ring-tail lemurs, by different composites, using some identical components but in various combinations with other sounds (Macedonia, 1990; Macedonia and Evans, 1993).

However, this form of communication is not unique to primates. When we look at the growing literature of referential alarm calling, we find that a number of mammals outside the primate line have developed such signals. Examples known so far are marmots (Blumstein and Armitage, 1997), prairie dogs (Slobodchikoff et al., 1991), meerkats (Manser, 2001), and domestic chickens (Evans, 1997). The ground squirrels of California (*Spermophilus beecheyi*) also direct referential signals at their conspecifics. They produce "chatter" calls when they see a predator on the ground and "whistles" when an eagle or hawk flies overhead, but their calls are not as specific as the alarm calls of chickens (discussed later) or the "eagle" and "snake" alarm calls of vervet monkeys (Hare, 1998). Thus, increasingly, more species of different orders are being found that seem to be able to do this, including birds.

These examples make it possible to argue that signals not only have consequences in terms of changing the behavior of others in a generalized way, but also they do so in very specific ways. A generalized response by the receivers of a message would be to simply run when an alarm call is heard, and a specific response would be indicated when the signal predictably elicits the direction of flight (into a scrub, up a tree etc.).

Signals are referential when they name a specific object, although that "object" may still represent a category of stimuli. For instance, an "eagle" alarm call does not have to refer to a specific eagle species, but may also refer to a goshawk, a falcon, and indeed any bird of prey, but may not, as was discovered in research of free-ranging vervet monkeys, include a stork or a vulture (Seyfarth et al., 1980b; Seyfarth and Cheney, 1986; Hauser, 1989). Both of these latter bird species are of no consequence to the monkeys but they are of large size and have wing shapes similar to the dangerous raptors, especially the splayed primary feathers. The reasons why we know that learning is involved and discrimination honed to *predatory* species of specific danger to vervet monkeys, rather than to birds of similar shape, is that young vervet monkeys make mistakes and often alarm call when they see storks and vultures. By the time the same vervet monkeys are adults, they are silent when a stork or vulture flies overhead (Hauser, 1988).

Evidence of referential signaling is not confined to mammals. Domestic chickens show a similar vocal discrimination in alarm calls for ground and aerial predators. Importantly, the research on the domestic chicken has shown that these calls are intentional since they are audience directed (Evans, 1997; Evans and Marler, 1994; Karakashian et al., 1988). A rooster will not call, but go through the physical motions appropriate for a response to aerial or ground danger, if there are no chickens nearby.

If we add food calls to the list of referential calls, informing either about the presence of food, or about the distribution and amount, or even about the quality, as has been found in primates (Hauser and Marler, 1993; Hauser et al., 1993), canids (Well and Bekoff, 1981), and in avian species (Evans and Marler, 1994), we begin to see a pattern of wide-ranging abilities of individual species, not just primates, to communicate aspects of their immediate environment.

The question remains whether some non-human species may be capable of communication that approximates aspects of language. As mentioned before, many heated debates, particularly about primate vocal abilities, took place on precisely these questions in the 1970s, with its critics quick to follow (Terrace et al., 1979) and leading to further research (Zimmerman et al., 1994). In some modern linguistic theories, the interest is more sharply focused on theories of grammar, generative morphology, and syntax. In this sense, all animals are at least pre-linguistic. However, if we focus attention not on the vehicle of communication but on function, that is, social communication, we can then establish which forms of communication might have embedded in them some key elements of "language" which could eventually lead to linguistic abilities.

Phonetically and semantically, there is at least some research evidence, although in its infancy, that not all processes associated with the acquisition of human language (jumping the threshold from pre-linguistic to linguistic) are unique to humans. Even dolphins have been found to have representational and conceptual skills (Herman et al., 1993). Evidence is thus mounting that many other mammals, not connected to the primate order, possess referent signals in their vocal repertoires and may thus show at least rudiments of higher cognitive abilities (Baron-Cohen et al., 2000; Hauser, 2000; Hoage and Goldman, 1986; Rogers, 1997).

Deception in Vocal Signaling

Most of the referent signals (alarm calls, food calls) that have been discussed above are so-called "honest" signals, that is, they mean what they say. There are also "dishonest" signals and some of these have to be intentional. Deception by use of vocal signals is a case in point. Some coyotes have been shown to make a great deal of noise at sites of burrows and scratch the ground as if searching for rodents or prairie dogs for a little while, and then suddenly stop, quietly move away, and hide. The hapless prey, not hearing the sounds anymore, eventually peers from its underground domicile and is then ambushed by the waiting coyote (Fox, 1971). Deception is widespread in domestic and wild dogs (Mitchell and Thompson, 1986) and has been described in some detail in primates (Whiten and Byrne, 1988) and avian species (Kilner et al., 1999).

Tactics for deception, involving signals, also include acoustic concealment, as Whiten and Byrne (1988) established some time ago for great apes and other primates. In each of the anecdotal cases reported, including captive gelada and chimpanzees (de Waal, 1982), the intention was sexual and related to mating and grooming with ineligible males: the examples involved hiding and pretending to do something else and then quietly engaging in sexual activity out of sight of the eligible males. Among primates, acoustic concealment also occurs in contexts of finding a food source that the finder may not wish to share. Instead of announcing the food, the finder may be silent. Among wolves, such tactical concealment is developed to a very high degree. Howling in wolf packs indicates presence and strength of a pack and is usually a distance increasing signal, signaling that any other wolves should stay away. The other pack may reply from a distance and then move on, but, on occasions, they may also fall silent and stealthily approach and then attack the first howling group (Harrington, 1987;

Harrington and Mech, 1983; Schassburger, 1993). The "ominous silence" is a mark of communication: when wolves fall silent, the other pack starts being nervous and listens carefully (Rogers and Kaplan, 2003).

A common reason for employing dishonest signals is also to cause a distraction and by doing so remove a competitor from a favored food source. The Arctic fox has been observed to use warning calls in this manner (Rogers and Kaplan, 2003), and so have domestic dogs and certain species of birds (summarized in Kaplan and Rogers, 2001). Similarly, primates have been known to employ deceptive methods, associated with food and predators. One, described by Whiten and Byrne (1988), involved the pretence that a predator was nearby: a baboon being chased by another baboon was observed to stop and look around as if there were a predator in the near distance. When the pursued individual did so, his pursuer stopped and looked around too, giving the pursued baboon time for escape. This is perhaps the most complex form of communication. Such deception can be expected to occur only when a communication system is firmly in place, and it is likely to be effective only when used very rarely. While, in the short term, it may be a rewarding way of dealing with danger or desire, it is a risky form of communication. An individual using too many false alarm calls may well be ignored and, depending on species, the false signaler may actually be punished.

Deception by issuing food calls is another tactic but this needs to be accompanied by an advantage for the sender. One advantage would be that the food finder signals the incorrect information about the food quality and food amount (Hauser et al., 1993). Another is to lure unsuspecting conspecifics into close proximity of the caller. Evans and Marler (1994) have established a link between food calling and courtship in the chicken. A cockerel may invite a hen to partake of food that he has found and this is often followed by mating. They observed, however, that a cockerel sometimes makes food calls when no food is present. He appears to do this only when the hen is far enough away to be unable to see whether food is actually present where the cockerel is located. On hearing the call, the hen approaches the cockerel, presumably to search for food in that vicinity, and thus the cockerel deceives the hen into approaching. He does so usually in order to achieve a mating. Leslie (1985) described a case of deception based on inter-species communication between two birds, a parakeet and a blue jay. The visiting parakeet, perched on the outside of the jay's cage, seemingly hungry, indicated by eye position and other cues that it wanted the chopped spinach in the cage. The blue jay moved the chopped spinach close to the edge of its cage, but on the inside, and when the parakeet reached for the spinach the blue jay delivered a sharp attack on the parakeet's head. Domestic dogs are also known to use deception in their interaction with people (Mitchell and Thompson, 1986).

The question is whether the existence of even relatively few documented, and sometimes anecdotal, examples suggests intentionality of communication in species as different as chimpanzees, monkeys, dogs, jays, and chickens. Obviously, if deception has been deliberate (as the term implies), the caller has to be able to predict the response of the receiver. In terms of escape, as in the example of the baboon, it would also be a very costly mistake if the desired outcome was not achieved. Deception is difficult to observe because, as said before, such instances work on surprise and should therefore remain rare. Whether or not we do accept these examples as intentional deception, we would have to concede that such forms of communication are not exclusive to primates. They appear to be quite widespread in vertebrates.

Nonvocal Communication

All animals, but to varying degrees of complexity, have a repertoire of nonvocal means of communication (Bradbury and Vehrencamp, 1998). Research on gestural communication has been a relative latecomer in primate research (Locke, 1978; Povinelli and O'Neill, 2000; Tomasello et al., 1989) and is still comparatively scanty. It tends to be of more general interest when such nonverbal behaviors can be shown to be related to questions of intentional communication and, thus, to higher cognition. Eye gazing and eye gaze following is one area of interest (Emery et al., 1997; Gomez, 1996; Kaplan and Rogers, 2002; Tomasello et al., 1998; see also Chapter 9 by Call) because the latter implies an awareness of the state of another; pointing is yet another. Eye gaze following may not necessarily have anything to do with communication. If one individual gazes in a specific direction and another follows its eye gaze, there is no indication that the first individual intended the gazing to communicate anything. However, in some specific situations, eye gazing is used specifically and intentionally to alert another individual to a specific context or object. For instance, it has been shown that orangutan infants use eye gaze as a begging gesture (Bard, 1992; Miles, 1990). When a food item is present but not within their reach, their visual attention keeps shifting back and forth from the mother's face to the food until the mother either turns away with the food to safeguard it from the infant or she passes it to the infant. Judging by her response, and as far as we can ascertain from observation, the mother has understood the infant's request for the food by its eye gaze alone. Pointing, by contrast, is always an intended communicative act and, apparently, widespread amongst human primates and has been observed in some apes (Krause, 1997). The one doing the pointing is trying to alert another to the presence of an object or situation of interest. The other, in response, will follow the direction of the pointing and so acquire information about the reason for pointing. Except perhaps for studies on canid responding to pointing (goal-directed behaviors of domesticated dogs) and more recently by dolphins (Herman et al., 1999), there has been a dearth of information on this topic in non-primate mammals. For reasons of anatomy, there are also some methodological difficulties of comparing different orders. In the sense of presuming certain anatomical features, avian species, for example, suffer *ab ovo* from a genetic disadvantage. Monkeys and apes have frontally placed eyes, but in the majority of avian species the eyes are largely laterally placed at each side of the head and so it would be difficult to determine direction of gaze, let alone gaze following. However, if the idea of "pointing" is translated into species appropriate options, then such pointing has been observed in birds (Farabaugh et al., 1992). When the territory of an Australian magpie is invaded by a bird of prey that may be spotted hiding in a tree, the sentinel magpie will take up a position nearest to the intruder but outside striking range, sound the alarm and use the beak as a guide in the direction of the bird of prey. Other magpies soon fly in and they then look in the direction of the pointing. The newcomers will then also adopt a posture and a beak position that clearly points in the direction of the intruder and they do so, not by landing next to the first magpie but by taking up positions in a semicircle. The "pointing" of the beak is thus not in the same direction as that of the first magpie (i.e., not imitation), but is directed also at the intruder. From the positioning alone, it is possible as a naïve observer to spot the bird of prey (Kaplan, 2004).

Subtle gestures of this kind (intentional pointing and eye gazing as a signal leading to eye gaze following in the receiver) have been identified so far especially in species that tend to live in groups, such as primates, but also dogs and dolphins and possibly even group-living birds. In orangutans, mother–infant pairs are very close bonds and pointing and eye gaze following have been observed in their interactions (Bard, 1992). Other clear patterns of eye gazing have also been observed in juvenile and adult orangutans (Kaplan and Rogers, 2002).

Whether "pointing" is an important form of communication in dolphins in their natural habitat is not fully established, but we know that they too understand the referential character of human pointing gestures as do dogs (Herman et al., 1999). They too live in close-knit social groups, as do dogs, chimpanzees, and gorillas.

Clearly, vocal and nonvocal behavior can be complex and meaningful in the sense that a specific message is intended by a sender and reliably acted upon in an appropriate manner by a receiver. In this domain, there is no basis for arguing that such meaningful communication is exclusive to the primate line. On the contrary, there is increasingly evidence that similar forms of communication are manifested in widely divergent clusters of species—at least in Class Aves and Class Mammalia.

HUMAN LANGUAGE AND ANIMAL STUDIES

Over many decades, researchers have trained great apes to use American Sign Language in order to be able to communicate with them on human terms, and sometimes about matters of abstract thought. They have done so because observable behavior does not necessarily reveal, or even hint at, internal mental events, such as a sense of future or a memory of the past. We have few, if any, means of assessing internal psychological events and thoughts other than by direct question and answer. Hence, a crucially singular and innovative way was to access communication with great apes via human language or human sign language.

Language carries linguistically descriptive and abstract elements of grammatically equal weight. We can speak of a banana (the concrete) and of the memory of a deceased friend (the abstract) in the same grammatical terms and thus with great ease. It has been surmised that the ability to abstract implies the underlying ability to express such abstraction. Expressed in the negative, since a similar system of abstract thought is not evident in animal communication, it must therefore mean that animals have no mental abilities for abstract concepts and thus lack "mind." Language, in this model, is but a reflection of preexisting mental processes and if language is "missing," it logically implies an absence of concepts. Well-known contemporary linguists such as Steven Pinker (1994) and Derek Bickerton (1990) have made a strong case for the species-specificity of human language. They argue so on the grounds that human languages are qualitatively different in structure from systems of animal communication. Defining human language as "species-specific," that is, to argue that human language has attributes found exclusively in humans, is logically valid because all systems of communication are largely species specific, excepting some, albeit important, overlaps in sounds and gestures (see Chapter 5 by Newman). However, it is quite something else to imply or assert that human language is not only unique but also that the processes required to achieve language are a mark of the ability to form concepts and of intelligence, or to assume an intrinsic interrelationship between language and concepts. According to this argument, one should conclude that the absence of language (in our understanding of it) also means the absence of the ability to form concepts.

So why would anyone have wanted to teach great apes to speak? The very idea of training animals, even great apes, in human language or human sign language can mean only that the researchers, in defiance of the dominant view, believed that abstract thought could exist without language. Presumably, they would also have been unable to subscribe to the view, as Pinker and many others have done, that there is an intrinsic interrelationship between concepts and language, unless it might have been assumed that, by teaching human language, concepts would follow. Hence concepts and language were taught simultaneously. Alternatively, more radically, a training program of apes in human sign language might assume that they have effective communication by means other than human language (Locke, 1978) and that such communication may not be restricted entirely to the realm of stimulus–response. In fact, the idea of training apes in human language assumed that apes not only have their own communication but also that such communicative abilities are maintained and may be revealed by translating them into another communication system. For apes, learning a language (symbolic or sign) involves no less than that, and therefore may be far more complex than merely learning a second language. A second language, as the term states, is an exchange of one language for another, not one system for another. Ape communication may largely exist in pre- or non-linguistic form and yet the processes involved in training apes to acquire skills in human language are based on an expectation that prelinguistic forms can be distilled into linguistic expression (Premack, 1972).

The enormity of the task and the difficulties in experimental design are obvious, particularly with hindsight. A number of researchers made long-term commitments to raising great apes (Trixie and Alan Gardner, Roger Fouts, Lyn Miles, Sue Savage-Rumbaugh, Francine Patterson, Herb Terrace, and a few others) thereby hoping to gain insight into their minds. After many earlier futile attempts to teach apes spoken words, it was found that apes cannot learn to speak because of the construction of their vocal apparatus and, even though they were then taught American sign language or use of symbols, some felt that the entire concept of meaningful communication with great apes might have been flawed (Seidenberg and Petitto, 1979; Terrace, 1979). Still, apes did learn sign language or the use of symbols. Washoe, the chimpanzee, Koko, the gorilla, Kanzi the bonobo, and Chantek, the orangutan, were among the most famous but certainly not the only ones (Fouts et al., 1989; Gardner et al., 1989; Miles, 1983, 1990; Savage-Rumbaugh and Lewin, 1994; Shapiro and Galdikas, 1995). By now, we know that all of the great apes, to varying individual and species degrees, can acquire and use sign language and symbols.

Investigations of linguistic and vocal abilities in primates derived from different theoretical standpoints. One was fuelled by the wish to understand the origin of human language rather than animal communication. As recently as 1991, Lewin argued that chimpanzees may hold the only key to the origin of the human language. In terms of genetics, that is, the high degree of DNA match between chimpanzees and humans, the assumption seemed reasonable. Indeed, comparative studies of the ontogeny of vocal development in nonhuman primates have partly continued in the hope of finding a model for studying language development in human children (Newman, 1995). The other was to understand something about abstract thought and concepts in apes, thereby contributing to the theory of mind debate. The latter has clear theoretical relevance to evolution but has also received unlikely impetus from the new fields of "artificial" intelligence and new theorizing on the brain/mind divide (Carruthers and Smith, 1996).

However, whatever the standpoints were, from an evolutionary point of view, the assumption of closeness of communication systems between apes and humans surmised a great deal. First, it assumed that DNA closeness is equivalent to DNA expression. However, these two are different (see Marks, 2002). Second, species live in specific environments and the latter play a role in the expression of anatomical, physiological, and behavioral traits. Third, the material and efficient causes of selection are subject to considerable vicissitudes, and consequently to heated and ongoing debate (see Singh et al. [2001], the second volume of the Festschrift for Lewontin). Fourth, the evolution of language and of brain size are often seen as co-evolutionary events (Deacon, 1992), although this cannot be proven, and hence human language is related to the belief that human language is a sign of species intelligence.

Research concerned with the evolution of (human) language has looked closely into animal signals for clues as to whether animals show signs of some rudimentary elements of what we know constitutes human language. Again, the limelight has largely been on primates, and especially on apes (Byrne,

1995). They are capable, so research has shown, of developing basic grammatical rules and thereby exhibiting a form of proto-grammar (Greenfield and Savage-Rumbaugh, 1990, 1991). Unfortunately, those seeking to explain the origins of intelligence based on the primate line (e.g., Parker and McKinney, 1999) have one insoluble problem. Similar rudiments of language have been found in unrelated taxa as will be explained below.

Co-evolutionary Events

Development of vocalization and vocal learning is particularly pronounced in birds. In fact, less is known about vocal development in primates than in birds, despite two decades of pioneering studies of monkey vocalizations and of the vocal communication systems of great apes.

Notably, while early attempts to teach apes to speak failed, there have been very successful attempts to teach birds to speak because their vocal apparatus can produce human speech sounds with ease. Not just parrots, but many songbirds can mimic human speech (Kaplan, 2000), yet these abilities have largely been dismissed as mindless copying. However, the work by Irene Pepperberg (1999) with a Gray parrot called Alex shows that English words are not uttered by mindless association of certain words with certain events. Unlike apes, Alex can communicate with the researcher actually using words and these are not only comprehensible (Patterson and Pepperberg, 1996), but also appear to be the result of thinking (or awareness) rather than of automatic responses (Pepperberg, 1990, 2001).

Apart from laboratory studies with Alex, there are also some field studies of untrained birds. Many other avian species, foremost among them corvids and parrots, may have abilities similar to those of Alex (the ability to speak or "mimic," the ability to respond to human commands and requests) or, if not in human language, then in a range of behaviors thought to require similar higher cognitive abilities. The now extensive corpus of studies on ravens (Fritz and Kotrschal, 1999; Heinrich, 1995, 1999, 2000; Lorenz, 1931, 1952), and Kaplan's recent studies (2000, 2003) on the Australian magpie, also belonging to corvidae, show similar complexity of communication, such as referential signals (alarm calls), individual recognition and vocal expression of social hierarchies. Tool using of crows in the wild (Hunt, 1996, 2000) and a range of experiments about avian problem-solving abilities, of a kind similar to those undertaken for apes (e.g., Heinrich, 1999), further underscore the emerging view that some species of birds may have higher cognitive abilities that are at least at the same level of complexity as those of apes.

Rudiments of language in avian communication (Epstein et al., 1980; Evans and Marler, 1994; Kluender et al., 1987) are not easily dismissible. They raise questions on how this is achieved without the benefit of a neocortex and, thus, without the benefit of the one remarkable difference between mammals and birds. For a long time it was thought that the development of the neocortex held the key to higher cognition (on the Neocortex and Brain Volume, see Chapter 9). However, functional equivalencies between a part of the avian brain (the neostriatum) and the mammalian prefrontal cortex have become apparent. So it has become difficult to claim uniqueness for the order of primates, or, specifically for the uniqueness of great apes in this context.

COMPLEX COMMUNICATION, SOCIAL ORGANIZATION, AND THE HUNT

The activity of hunting may be a very specific subset of possible group activities and it tends to be a group activity only in cases of either focusing on prey species to afford substantial tactical or other advantages over the predator. One reason for group hunting may be a matter of size. Wild dogs, such as wolves, dholes, and African wild dogs, form packs to hunt prey much larger than themselves. The other reason for group hunting may be a matter of speed. Chimpanzees at Gombe hunt live prey much faster than themselves. Group hunting helps in cornering fast prey. That they hunted at all was a major shock for the primatological community (Goodall, 1986). At first believed to be the gentle ancestors of humans, it was then discovered that chimpanzees had a "dark side" to their natures (Wrangham and Peterson, 1997). Chimpanzees feed on a large variety of plant matter and also like the occasional treat of ants or termites, honey and eggs, but they were not really considered omnivores, let alone carnivores. At first, the shock had probably more to do with the association that was made to prehistoric hominid lifestyle, than with chimpanzees. In the natural environment one can only acquire red meat either by feeding on carcasses, that is, as scavengers, or by killing live animals. The scavenger theory of meat eating by chimpanzees (and early humans) has always been unattractive because it does little to generate pride in one's hominid past. Killing live animals fits the hunter theory of the more recent human past, but here were chimpanzees capturing other primates with their bare hands and ripping them apart while they were still alive. Watching chimpanzees consume another primate, still screaming in pain and thus very much alive, is excruciating and presents a gory scene. Orangutans had very occasionally been observed to pick up a carcass of a loris and consume it but, so far, there is no evidence that orangutans habitually or even irregularly eat, let alone kill for, meat (Sugardjito and Nurhuda, 1981). Females, after giving birth, have been known to consume the placenta, but this may have other reasons than "meat-eating." Many females of many different species eat the placenta after parturition and this is thought to occur to prevent attracting predators to the fresh smell of blood. The absence or presence of meat eating in orangutans has never generated much excitement either because so very few observations of this kind have been made to date or, possibly, because orangutans are not considered direct human ancestors (see discussion in Kaplan and Rogers, 2000).

Carnivorous behavior and habits in chimpanzees do not appear to be genetically pre-programmed: chimpanzees cannot and do not know instinctively how to kill their prey. Their teeth and jaws are the wrong shape. True carnivores, such as the large cats, have pointed, piercing, not flat, teeth that can hold and crush, and are long enough to break the spine of a victim very swiftly. The body shape of chimpanzees is not really that of a hunter either. They are heavy, too heavy to swing elegantly from branch to branch and they have to rely on sturdier branches to support their weight. Also, they are not runners on the ground, at least not for longer distances. By comparison, canids, another family of carnivores, have tremendous stamina and can walk long distances in a day without ill effect. They are excellent runners and can stay with their prey for chases as long as 20 km (this being so in wolves particularly) and some species can perform sprints of up to 70 km/hr. However, it can also be said that, in some respects, canids are not the best-equipped hunters either. They do not have good vision for seeing shapes but have compensated for this by being able to see movement well. More importantly, their teeth are not designed to kill but only to hold. They cannot even kill a rodent by biting alone but need to shake it vigorously to break its neck (Rogers and Kaplan, 2003). So they too, like the chimpanzees, set about devouring their larger prey while it is still alive. Thus, one cannot entirely rely on anatomical features alone in order to decide who might make a good hunter or, even, whether a species hunts for live prey at all.

True carnivores, furthermore, can live and hunt alone. Each felid, of whatever species, is a very efficient killing machine, as are birds of prey. Every design feature about them indicates that they are at the top of the food chain and predatory. Design features in chimpanzees clearly do not point in the direction of killing larger live prey. It appears, though, that chimpanzees, and wild dogs, have compensated for these deficiencies by other means. Both species live, and hunt, in groups. In other words, in order to make sense of any hunting lifestyle in a species that does not seem to be well equipped anatomically, we might need to seek explanations for such feeding habits in their social life and in higher cognition.

Advantages of Living Together

Groups of animals are usually subdivided into four different functional types: breeding, feeding (utilizing the same food source together), foraging (banding together while searching for food), and population (sharing the same home range). The first group type is the most common among all mammals and the latter the least common. African great apes and many other primate species form groups that have all four elements in common. Groups of carnivores usually do not fall into all four of these categories, and often only into one of them. Felids, for instance, such as tigers, can live and hunt alone and mostly do. Canid assemblies, however, reflect different kinds of prey: those seeking small prey may hunt alone and form groups only for breeding. Those species seeking larger prey may, on the other hand, share all four functional types, as do most great apes.

Living permanently in groups and sharing all four elements of group life has a number of serious disadvantages and it is not always clear whether these are offset by the advantages. Here, group size is a particular issue with respect to hunting and foraging. A group is usually not as mobile as a single individual. It may be more visible and this can be more dangerous than traveling alone or in small groups. Living in permanent groups may result in getting less food than do solitary foragers. In groups, the incidence of disease and the risk of parasite transfer are substantially increased. There is a greater chance of squabbling within a group, with prolonged displays of aggression and even injury and death. Finally, it is more difficult to coordinate activities in a larger group than in a pair or small family group. These disadvantages seem to be at the forefront of many of the endangered species (certainly amongst group living carnivores) while the more asocial carnivores (such as foxes and coyotes) are able to thrive even under very adverse human influence.

In the literature on orangutans, there have been suggestions that ecological reasons may account for group cohesiveness and more solitary units. When it was discovered that orangutans in Sumatra tended to live in groups as against the more dispersed orangutans in Borneo, it was thought that survival pressures may have forced the Bornean orangutans apart. Hence, it seems that, in some instances, the disadvantages of group living outweigh the benefits (van Schaik et al., 1983).

For hunters of prey, there are clear advantages of being in groups and most of these have to do with being able to successfully catch and keep food. Wild dog species can succeed in taking down prey as large as, or larger than, themselves only if they make a team effort. Within groups, they are more likely to locate a food source and more likely to keep the kill. In the plains of Africa and even in some regions of Asia, any kill is at once heavily contested and, unless the smaller predators have the numbers to keep the hungry competitors at bay, they will not get a full feed. For great apes, living in groups may confer some similar advantages even in frugivorous foraging, because there will always be a host of competing species for the same kind of food. The ability to feed uncontested or at least long enough to get a meal is more likely in groups than for individuals.

Why great apes grew so large, compared to monkeys, is a question that has been debated at great length in other disciplines and here is not the place to indulge in this debate. It may be important to argue, however, that their size and a group lifestyle should predispose them more than others to becoming hunters (see also Stanford, 1995a). Although not quite at the top of the food chain as true carnivores, they are strong and large enough, and they have the group for support, relying on them as much as individual dogs do in a pack. At the very least, large body size and group cohesion reduces inter-species competition. They have grown larger than the typical size of prey animals without having lost the ability to take themselves into the trees, should ground risks prove too great.

Countless studies show that those species hunting in groups do not simply run/fly or skip along with each other in an aggregated group. As far as we know from any species with characteristics of cooperation, hunting is a well-coordinated activity. Wolves, being much closer to apes in their social characteristics than perhaps any other group of species in this regard, live in close-knit groups. Even among some birds of prey that hunt socially, group cohesion is strong. While only few hunt in groups, as distinct from pairs (as do some falcons, see Hector, 1986), those who do show regular patterns of group hunting have also developed distinct strategies that involve different tasks for the members of the team. This has been described in some detail for Harris' hawks (Bednarz, 1988).

Insurmountable evidence confirms that the behavior of a pack of wild dogs is intentional and tactical. When they hunt, there is plenty of communication as shown by the example of encircling the prey and whistling by dholes (Fox, 1984), or the "testing" techniques used by wolves (testing the quality of their live quarry by tactics of wearing them down). There is no confusion as to who encircles from the left and who does so from the right. It can be argued that wolf packs that stalk another wolf pack have no such coordinated plan: the dominant dog merely has to give the signal and the others may follow. Hunting cannot work in this way because group members have different tasks and often have to take up different positions during the hunt. A simple comparison can be made to domestic dogs that have become feral and form loose marauding troops. These dogs, as studies in Italy have found (Boitani, 1983; Boitani and Ciucci, 1995), ultimately have been shown to have low survival chances because the packs lack a firm group structure and the training and learning necessary to coordinate their hunting activities.

Hunting group expeditions carried out by chimpanzees to kill and then eat a favorite monkey species (white or red colobus monkeys) function very much like those of wolf packs. Colobine monkeys can outrun chimpanzees in trees and thus would be able to escape easily if they were merely chased. They are lighter, more nimble and can move much faster through the canopy than chimpanzees. They have also developed strategies against attacks (Stanford, 1995b). In order for chimpanzees to catch them, an ambush must be set for the colobine monkeys. In one reported case, one or two chimpanzees herd and confuse the monkeys from the rear and they may do so provocatively and openly so that the monkeys look back at their pursuers. At the same time other chimpanzees take up their positions in front, preventing possible escape routes of their quarry and, when the monkey leaps forward, seemingly away from the pursuers, it is caught by the waiting front-guard chimpanzees. Another scenario may involve encircling the victim and cutting off the escape route by slowly advancing on all sides. This is every bit the same range of strategies that wild dogs use.

These hunting activities in chimpanzees have been hailed as being a clear sign of higher cognition and as being unique amongst animals (Stanford,

1995a). The argument to promote cooperative hunting as evidence for "intelligence" in chimpanzees is largely based on the assumption that hunting in chimpanzees is regarded as a deliberate, planned, and coordinated act (Wrangham and Peterson, 1997).

There are at least two reservations one may express in relation to the assumption that hunting by chimpanzees is an indication of demonstrating something special and superior about their intelligence, as compared to hunting by other non-primate species. Tomasello and Call (1997), for instance, are not convinced that each individual in the hunting party actually knows what its role is (as do dogs and socially hunting birds of prey) and the tactics that are being used derive from trial and error, not from insight. There have also been critics who have argued that the success rate of these hunting parties by chimpanzees is relatively low (although that is true of many hunts even among true carnivores, such as the felids).

Second, most pack hunting dogs would fulfill the same characteristics as chimpanzees because they too have alternative foraging activities available to them. In order to capture food several times their own size and strength, their techniques need to be very well coordinated indeed. Chimpanzees are, therefore, not unique either in performing the activity of hunting or in the degree to which they coordinate their activity with other group members. On the contrary, failing comparative research studies in this area, one can only speculate that probably the dogs and Harris' hawks, with a higher percentage of successful kills, are superior in their coordination of the hunt, confirming the views by Tomasello and Call (1997).

The other argument that is made in relation to chimpanzees is that they do not need to hunt for their food by killing another primate. Carnivores, such as many wild dogs, by contrast, are seen to be driven by that need. In other words, hunting in dogs may be unlearned and innate, whereas it is a product of social learning in chimpanzees. However, research into carnivorous behavior has shown that hunger and killing are not necessarily related. Many canids, except the larger species, have many alternative food sources and may equally not be "driven" by that need. It may well be true that chimpanzees may not be driven by hunger to hunt and kill because they have other sources of food. Acts of killing by chimpanzees thus do not seem to be derived from necessity but from a kind of willful malice or, rather, some socially acquired preference.

However, not all foods are consumed for reasons of general hunger but many are sought and consumed as specific "hunger." Some animals ingest stones to aid digestion, some parrots fly long distances to feed on stony outcrops in order to obtain important minerals. Some monkeys feed on charcoal to cope with toxins in their food and settle their stomachs. Chimpanzees may require occasional high doses of protein and their acts may follow some physiological craving rather similar to that of the parrots seeking minerals. Meat provides proteins in an efficient and concentrated form and it is possible that its ingredients are at times essential or desirable additives for the chimpanzee diet.

Alternatively, one can argue that, if hunting is "intelligent" behavior, other species therefore also show "intelligence." Coordination of the social group during hunting requires excellent communication. For individuals to know what to do in situations that require split-second decisions and close coordination with other group members, they need to be well in tune with each other. They bring to the hunt their skills that only close-knit group living bestows. We can thus either argue that the communicative skills required for a successful hunting expedition are a reflection of prior learning of social skills, or we can argue that "intelligence" leads to good communicative skills and hence to coordination. Feral dogs may not be inferior to wolf packs in cognitive abilities but yet they do not succeed as well and are less well-coordinated precisely because they lack the social learning in this regard (Boitani and Ciucci, 1995).

Another point that is made in this line of arguing is that chimpanzees furnish living proof for the origins of human violence (Wrangham and Peterson, 1997). Violence, if this is meant as a demonstration of aggression, is however not clearly associated with hunting. Hunting is not in itself an "aggressive" act in the sense that killing occurs as a result of emotions. There is no "anger" against the victim, there is no "hatred," there are no foolish acts of running out just for the sake of hurting another. Sometimes neighboring groups (be this of chimpanzees or wolves) may kill each other in acts of aggression but this is usually over *territorial* claims, not directly about *hunting*. Indeed, if the emotions played a role in hunting, the cautious timing, observation, and sheer concentration required to successfully capture a most reluctant living participant and the ability to succeed would be greatly diminished. In fact, studies have found no correlation between the level of social aggression and killing success.

However, in chimpanzee hunting parties we do find extremes of excitement and aggression, atypical behavior for carnivores. This element would suggest that the hunt is something qualitatively different from the hunting of carnivores, be these birds or mammals.

If we want to look for aggression, we need to look into the social life of group-living animals. It appears that aggression of canids and primates, be this at feeding sites, in daily life, around dens or among rivals, is related to group complexity and, specifically, to vertical hierarchies, as will be discussed later. Such hierarchies exist equally amongst chimpanzees (de Waal, 1982) and gorillas (Maple and Hoff, 1982; Yamagiwa, 1992), as well as in stable packs of wild canids (Fox, 1971; Rogers and Kaplan, 2003).

Hierarchy, Group Complexity, and Feeding

Social rank is highly important in many primates and in some species of canids. In some species, vertical hierarchy is observed strictly. Being an alpha male or alpha female (holding the highest status) is a matter of power, of breeding rights and privilege for their offspring. There are some performance criteria, such as appearing and looking confident and insisting on the performance of social rituals by subordinates. All these fall into the precincts of power and such hierarchies need to be constantly reinforced by a series of ritualized and stereotyped acts. Only in complex groups do we find extensive forms of communication that prescribe a series of socially ritualized acts and a complex gradation of signals in order to maintain stability. Single signals usually cannot suffice in these contexts. It is not just a matter of expressing and maintaining emotional bonds but acknowledging and reinforcing relative social positions. Making a connection between signal and social complexity, and ultimately intelligence, was first suggested by Alison Jolly (1966a, 1966b) and has been embraced as the social intelligence hypothesis (Byrne and Whiten, 1988).

Stability of complex groups also comes at a price and this is one of the intractable contradictions of vertical hierarchies. While they bestow stability by rules of conduct once the hierarchy is established, there are also opportunities to break the rules or, expressed negatively, failure to observe the rules. A vertical hierarchy invites harmony at one level (same rank alliances) but when conflict occurs even very few acts of aggression and tensions within a group can lead to serious outcomes. Vertical hierarchies can, at times, also be semitransparent and allow for a small trickle of "upward mobility." That is, the alpha individual may not always remain in this position. In chimpanzees, the alpha male may be deposed by a plot of a group of eligible young males who may even expel the alpha male or kill him (de Waal, 1982). In wolf society, a group of young males may equally turn on an alpha male (Mech, 1970).

A good place to test this assumption about the effects of a complex hierarchy, as in the case of wolves and chimpanzees, is to observe the behavior of a group at the site of a favorite food source. In chimpanzee society, an alpha animal will feed and will decide who else is allowed to take part in feeding (usually much younger members of the group who are not the alpha animal's direct competitors) as is the case in wolf society. Chimpanzees, gorillas, and wolves do not have many enemies (other than humans!) and their social system reflects this (Schassburger, 1993). They can afford to turn their attention inwards, to the group, and spend more time socializing an activity that, in turn, might foster complex communication.

Complex communication does not only translate into more gradations of nonverbal behaviors, such as grooming and a detailed gestural repertoire, but also into the vocal repertoire, offering a range of vocalizations that are graded and varied. Noise and graded sounds, transitional elements and slides from one sound type to another may be an indication of this, as Marler and Hamilton (1966) reasoned nearly 40 years ago. They observed that many primates, and especially the great apes, show a poorly defined structure of most sound signals. They concluded that this was related to the lifestyle of the primates, as well as their evolutionary position. One key characteristic that they observed to be common to all of the primates with noisy call structures was that they are socially close-knit, terrestrial or semi-terrestrial, and that they live in relative isolation from species about their same size and structure. Having identified these ecological facts they then argued that these conditions actually favored a loosening of strict species-specific types of calls. Instead, since there is no one near of similar acoustic qualities, they may exploit or risk more variations. These variations may then evolve into more elaborate ways for transmitting information than could be carried in stereotyped sounds (see also Marler and Tenaza, 1977). By analogy, the same principles may apply to the wolf, and to other species with low numbers of predators their own size, such as elephants with a very complex vocal repertoire (McComb et al., 2000). And their social system is also matched by a system of expression that is perhaps amongst the most advanced amongst mammals (Schassburger, 1993). Similar arguments may well also hold for some species of sea mammals (Rendell and Whitehead, 2001; Tyack, 1986).

Hunting and group living thus afford one the view that a similar convergence of a set of specific characteristics can be observed in completely different orders of animals. The view that apes, specifically chimpanzees, are unique in this regard would have to be rejected along with an implied notion of a unidirectional evolutionary trajectory of cognitive development. Some valuable insight can be gained from making comparisons across orders when the focus is on communication and higher cognition.

CONCLUSION

The picture that is emerging from the disparate studies of communication in vertebrates does not generally support the notion that specifically complex communication is unique to primates (although the combination of many traits may be). Rather, there is a fruitful connection to be made between signal complexity and ecological factors, as well as between signal complexity and some adaptive necessity to live in groups. Ecological factors refer to niche equivalents rather than species relatedness. Alternatively, in attempts to understand how behavioral repertoires were shaped, it may be less important, at times, to note relatedness by order than to note similar climatic and ecological conditions (see also Chapter 4 by Box and Russon).

Complexity of group life is usually identifiable by the presence of either reproductive or social vertical hierarchies, or both, and it is these factors that are reflected in complex communication. Ungulates, for instance, have not developed sophisticated communication systems and yet they live in groups for mutual protection against predators. Their group living strategies tend to be contained in the strategy that there is safety in numbers. Group cohesion is generally loose and hierarchy is limited to a ruling male and equally flat or indifferent loose group structure. This may explain why ungulates have not developed similar intelligent social behavior.

Notably, groups with a complex social life largely fall into two very disparate, even opposite, ecological categories. One cluster consists largely of prey animals (including ground-living mammals and birds), while another cluster of species is distinguished by the size of the species (they are large), their exceptional fighting skills, and the absence of a range of serious enemies as well as little inter-species competition. Amongst prey animals, in particular, plenty of examples can be found of referential signaling. The fact that chickens, squirrels, prairie dogs, or meerkats show abilities similar to some monkey species does not necessarily support the view of a gradual evolution of cognitive abilities. More likely, signal complexity is a strategy that comes to the fore when predatory pressures are consistently high and, surprisingly, also when these pressures might be particularly low. Prey species have a need to communicate effectively and swiftly about matters of life and death. Species with few predatory pressures, by contrast, may be able to indulge in vocal and gestural experimentation without risk to life and limb and without acoustic competition from a competitive near relative. One might then be forced to ask to what extent referential signaling, if taken on its own, is a valid measure of higher cognition, a question that this chapter has tried to repudiate.

Amongst those species that have few enemies, one finds very diverse groups of sea mammals, canids, avian species such as corvids, and some species of parrots, elephants, and the great apes. Here too, the known examples involve a range of species from very different orders with few evolutionary links. Nevertheless, a range of experiments that have been conducted on higher cognition have at times made a link between signal complexity and higher cognition. The great apes, as well as wolves, some dolphins, corvids, and parrots certainly appear to show signal complexity and higher cognition in conjunction with one another.

REFERENCES

Bard, K. A., 1992, Intentional behaviour and intentional communication in young free-ranging orangutans, *Child Dev.* **63**:1186–1197.

Baron-Cohen, S., Tager-Flusberg, H., and Cohen, D. J., eds., 2000, *Understanding Other Minds*, Second edition, Oxford University Press, Oxford.

Bednarz, J. C., 1988, Cooperative hunting in the Harris' hawks (*Parabuteo unicinctus*), *Science* **239**:1525–1527.

Bickerton, D., 1990, *Languages and Species*, University of Chicago Press, Chicago.

Blumstein, D. T. and Armitage, K. B., 1997, Alarm calling in yellow-bellied marmots: I. The meaning of situationally variable alarm calls, *Anim. Behav.* **53**:143–171.

Boitani, L., 1983, Wolf and dog competition in Italy, *Act. Zool. Fennica*, **174**:259–264.

Boitani, L. and Ciucci, P., 1995, Comparative social ecology of feral dogs and wolves, *Ethol. Ecol. Evol.* 7:49–72.

Boran, J. R. and Heimlich, S., 1999, Social learning in cetaceans: Hunting, hearing and hierarchies, in: *Mammalian Social Learning: Comparative and Ecological Perspectives*, H. O. Box and K. R. Gibson, eds., Cambridge University Press, Cambridge, UK, pp. 282–307.

Bradbury, J. W. and Vehrencamp, S. L., 1998, *Principles of Animal Communication*, Sinauer, Sutherland, MA.

Brown, C., Gomez, R., and Waser, P. M., 1995, Old World monkey vocalizations: Adaptation to the local habitat? *Anim. Behav.* **50**(4):945–961.

Byrne, R., 1995, *The Thinking Ape: The Evolutionary Origins of Intelligence*, Oxford University Press, Oxford.

Byrne, R. W. and Whiten, A., eds., 1988, *Machiavellian Intelligence. Social Expertise and the Evolution of Intellect in Monkeys, Apes, and Humans*, (Second edition, Cambridge University Press, 1997), Clarendon Press, Oxford.

Carruthers, P. and Smith, P. K., 1996, *Theories of Theories of Mind*, Cambridge University Press, Cambridge, UK.

Cheney, D. L. and Seyfarth, R. M., 1990, *How Monkeys See the World: Inside the Mind of Another Species*, University of Chicago Press, Chicago.

De Waal, F., 1982, *Chimpanzee Politics. Power and Sex among Apes*, Unwin Paperbacks, London.

Deacon, T. W., 1992, Brain-language coevolution, in: *The Evolution of Human Languages*, Vol. 10. J. A. Hawkins and M. Gel-Man, eds., Addison-Wesley, Redwood City, CA, pp. 49–83.

Emery, N. J., 2000, The eyes have it: The neuroethology, evolution and function of social gaze, *Neurosci. Biobehav. Rev.* **24**:581–604.

Emery, N. J., Lorincz, E. N., Perrett, D. I., Oram, M. W., and Baker, C. I., 1997, Gaze following and joint attention in rhesus monkeys (*Macaca mulatta*), *J. Comp. Psychol.* **111**:286–293.

Epstein, R., Lanza, R., and Skinner, B. F., 1980, Symbolic communication between two pigeons (*Columbia livia domestica*), *Science* **207**:543–545.

Evans, C. S., 1997, Referential signals, *Persp. Ethol.* **12**:99–143.

Evans, C. S. and Marler, P., 1994, Food calling and audience effects in males chickens, *Gallus gallus*: Their relationships to food availability, courtship and social facilitation, *Anim. Behav.* **47**:1159–1170.

Farabaugh, S. M., Brown, E. D., and Hughes, J. M., 1992, Cooperative territorial defence in the Australian magpie, *Gymnorhina tibicen* (Passeriformes, Cracticidae), a group-living songbird, *Ethology* **92**:283–292.

Fouts, R. S., Fouts, D. H., and van Cantfort, T. E., 1989, The infant Loulis learns signs from cross-fostered chimpanzees, in: R. A. Gardner, B. T. Gardner, and T. E. van Cantfort, eds., *Teaching Sign Language to Chimpanzees*, State University of New York Press, Albany, NY, pp. 280–292.

Fox, M. W., 1984, *The Whistling Hunters. Field Studies of the Asiatic Wild Dog (Cuon Alpinus)*, State University of New York Press, Albany, NY.

Fox, M. W., 1971, *Behaviour of Wolves, Dogs and Related Canids*, Jonathan Cape, London.

Fritz, J. and Kotrschal, K., 1999, Social learning in common ravens, *Corvus corvax*, *Anim. Behav.* **57**:785–793.

Gardner, R. A. and Gardner, B. T., 1969, Teaching sign language to a chimpanzee, *Science* **165**:664–672.

Gardner, R. A., Gardner, B. T., and Cantfort, T. E., eds., 1989, *Teaching Sign Language to Chimpanzees*, State University of New York Press, Albany, NY.

Gibson, K. R. and Ingold, T., eds., 1993, *Tools, Language, and Cognition in Human Evolution*, Cambridge University Press, Cambridge, UK.

Gómez, J. C., 1996, Ostensive behavior in great apes: The role of eye contact, in: A. E. Russon, K. A. Bard, and S. T. Parker, eds., *Reaching into Thought. The Minds of the Great Apes*, Cambridge University Press, Cambridge, UK, pp. 131–151.

Goodall, J. (1986). *The chimpanzees of Gombe. Patterns of behavior*, Harvard University Press, Cambridge, MA.

Gouzoules, H. and Gouzoules, S., 1990, Matrilineal signatures in the recruitment screams of pigtail macaques, *Macaca nemestrina*, *Behaviour* **115**:327–347.

Green, S., 1975, Dialects in Japanese monkeys: Vocal learning and cultural transmission of locale-specific vocal behaviour, *Z. Tierpsych.* **38**:304–314.

Greenfield, P. M. and Savage-Rumbaugh, E. S., 1990, Grammatical combination in *Pan paniscus*: Processes of learning and invention in the evolution and development of language, in: *Language and Intelligence in Monkeys and Apes: Comparative Developmental Perspectives*, S. T. Parker and K. R. Gibson, eds., Cambridge University Press, New York, pp. 540–578.

Greenfield, P. M. and Savage-Rumbaugh, E. S., 1991, Imitation, grammatical development, and the invention of protogrammar by an ape, in: *Biological and Behavioral Determinants of Language Development*, N. A. Krasnegor, D. M. Rumbaugh, R. L. Schiefelbusch, and M. Studdert-Kennedy, eds., Erlbaum, Hillsdale, NJ, pp. 235–238.

Hanggi, E. B. and Schusterman, R. J., 1994, Underwater acoustic displays and individual variation in male harbour seals, *Phoca vitulina*, *Anim. Behav.* **48**(6):1275–1283.

Harcourt, C., ed., 2000, *New Perspectives in Primate Evolution and Behaviour*, Westbury Publishing, Otley, UK, pp. 91–110.

Hare, J. F., 1998, Juvenile Richardson's ground squirrels, *Spermophilus richardsonii*, discriminate among individual alarm callers, *Anim. Behav.* **55**:451–460.

Harrington, F. H. 1987, Aggressive howling in wolves, *Anim. Behav.* **35**:7–12.

Harrington, F. H., and Mech, L. D., 1983, Wolf pack spacing: Howling as a territory-independent spacing mechanism in a territorial population, *Behav. Ecol. Sociobiol.* **12**:161–168.

Hauser, M. D., 1988, How infant vervet monkeys learn to recognise starling alarm calls: The role of experience, *Behaviour* **105**:187–201.

Hauser, M. D., 1989, Ontogenetic changes in the comprehension and production of vervet monkey (*Cercopithecus aethiops*) vocalizations, *J. Comp. Psych.* **2**:149–158.

Hauser, M. D., 1996, *The Evolution of Communication*, MIT Press, Cambridge, MA.

Hauser, M. D., 2000, *Wild Minds: What Animals Really Think*, Henry Holt & Co., New York.

Hauser, M. D., and Marler, P., 1993, Food-associated calls in rhesus macaques (*Macaca mulatta*), I. Socioecological factors, *Behav. Ecol.* **4**:194–205.

Hauser, M. D., Teixidor, P., Fields, L., and Flaherty, R., 1993, Food-elicited calls in chimpanzees: Effects of food quantity and divisibility, *Anim. Behav.* **45**(4):817–819.

Hector, D. P., 1986, Cooperative hunding and its relationship to foraging success and prey size in an avian predator (*Falco femoralis*), *Ethology*, **73**:247–257.

Heinrich, B., 1995, An experimental investigation of insight in common ravens (*Corvus corax*), *Auk* **112**:994–1003.

Heinrich, B., 1999, *Mind of the Raven. Investigations and Adventures with Wolf-birds*, Cliff Street Books, New York.

Heinrich, B., 2000, Testing insight in ravens, in: *The Evolution of Cognition*, C. Heyes and L. Huber, eds., MIT Press, Cambridge, MA, pp. 289–305.

Herman, L. M., Abichandani, S. L., Elhajj, A. N., Herman, E. Y. K., Sanchez, J. L. et al., 1999, Dolphins (*Tursiops truncatus*) comprehend the referential character of the human pointing gesture, *J. Comp. Psych.* **113**:347–364.

Herman, L. M., Pack, A. A., and Palmer, M.-S., 1993, Representational and conceptual skills of dolphins, in: *Language and Communication: Comparative Perspectives*, H. L. Roitblat, L. M. Herman, and P. E. Nachtigall, eds., Lawrence Erlbaum, NJ, pp. 403–442.

Heyes, C. M., 1998, Theory of mind in nonhuman primates, *Behav. Brain Sci.* **21**:101–148.

Heyes, C. and Huber, L., eds., 2000, *The Evolution of Cognition*, MIT Press, Cambridge, MA.

Hoage, R. J. and Goldman, L., eds., 1986, *Animal Intelligence: Insights into the Animal Mind*, Smithsonian Institution Press, Washington, DC.

Hodun, A., Snowdon, C. T., and Soini, P., 1981, Subspecific variation in the long calls of the tamarin, *Saguinis fuscicollis, Z. Tierpsych.* **57**:97–110.

Hunt, G. R., 1996, Manufacture and use of hook-tools by New Caledonian crows. *Nature* **379**:249–251.

Hunt, G. R., 2000, Human-like, population-level specialization in the manufacture of pandanus tools by the New Caledonian crow (*Corvus moneduloides*), *Proc. Roy. Soc. Lond, B.* **267**:403–413.

Johnston, R. E. and Jernigan, P., 1994, Golden hamsters recognize individuals, not just individual scents, *Anim. Behav.* **48**(1):129–136.

Jolly, A., 1966a, *Lemur Behaviour*, University of Chicago Press, Chicago.

Jolly, A., 1966b, Lemur social behavior and primate intelligence, *Science* **153**:501–507.

Kaplan, G., 2000, Song structure and function of mimicry in the Australian magpie (*Gymnorhina tibicen*) compared to the lyrebird (*Menura* ssp.), *Int. J. Comp. Psych.* **12**(4):219–241.

Kaplan, G., 2003, Alarm calls, communication and cognition in Australian magpiess (*Gymnorhina tibicen*), *Acta Sinica* (IOC symposium paper, Beijing 2002).

Kaplan, G., 2004, *The Australian Magpie* (*Gymnorhina tibicen*), Natural History Series, University of New South Wales Press, Sydney.

Kaplan, G. and Rogers, L. J., 2000, *The Orang-Utans. Their Evolution, Behavior, and Future*, Perseus Publishing, Cambridge, MA.

Kaplan, G. and Rogers, L. J., 2001, *Birds. Their Habits and Skills*, Allen & Unwin, Sydney.

Kaplan, G. and Rogers, L. J., 2002, Patterns of eye gazing in orangutans, *Int. J. Primatol.* **23**:501–526.

Karakashian, S. J., Gyger, M., and Marler, P., 1988, Audience effects on alarm calling in chickens (*Gallus gallus*), *J. Comp. Psych.* **102**:129–135.

Kilner, R. M., Noble, D. G., and Davies, N. B., 1999, Signals of need in parent–offspring communication and their exploitation by the common cuckoo, *Nature* **397**:667–672.

Kluender, K. R., Diehl, R. L., and Killeen, P. R., 1987, Japanese quail can learn phonetic categories. *Science* (Sept.):1195–1197.

Krause, M. A., 1997, Comparative perspectives on pointing and joint attention in children and apes, *Int. J. Comp. Psychol.* **10**:137–157.

Leslie, R. F., 1985, *Lorenzo the Magnificent. The Story of an Orphaned Blue Jay*, John Curley & Associates, South Yarmouth, MA.

Lewin, R., 1991, Look who's talking now. Do apes hold the key to the origin of human language? *New Sci.* **27**:39–42.

Locke, A. J., ed., 1978, *Action, Gesture and Symbol: The Emergence of Language*, Academic Press, New York.

Lorenz, K., 1931, Beiträge zur Ethologie sozialer Corviden, *J. Ornith.* **79**:67–127.

Lorenz, K. Z., 1952, *King Solomon's Ring*, Crowell, NY

Macedonia, J. M. and Evans, C. S., 1993, Variation among mammalian alarm call systems and the problem of meaning in animal signals, *Ethology* **93**:177–197.

Macedonia, J. M., 1990, What is communicated in the antipredator calls of lemurs: Evidence from playback experiments with ringtailed and ruffed lemurs, *Ethology* **93**:177–190.

Manser, M. B., 2001, The acoustic structure of suricates' alarm calls varies depending on predator type and the level of urgency, *Proc. R. Soc. Lond. B* **268**:2315–2324.

Maple, T. L. and Hoff, M. P., 1982, *Gorilla Behaviour*, Van Nostrand Reinhold, New York.

Marks, J., 2002, *What it Means to be 98% Chimpanzee*, University of California Press, Berkeley, CA.

Marler, P., 1955, Characteristics of some animal calls, *Nature* **176**:6–8.

Marler, P., 1981, Bird song: The acquisition of a learned motor skill, *TINS* **4**:88–94.

Marler, P., and Hamilton, W. J. I., 1966, *Mechanisms of Animal Behavior*, John Wiley & Sons, New York.

Marler, P. and Tenaza, R., 1977, Signaling behavior of apes with special reference to vocalization, in: *How Animals Communicate*, T. A. Sebeok, ed., Indiana University Press, Bloomington, IN

Marler, P., Evans, C., and Hauser, M., 1992, *Animal Signals: Motivational, Referential, or Both? Nonverbal Vocal Communication: Comparative and Developmental Approaches*, Cambridge University Press, Cambridge, MA.

May, B., Moody, D. B., and Stebbins, W. C., 1989, Categorical perception of conspecific communication sounds by Japanese macaques, *Macaca fuscata, J. Acoust. Soc. Am.* **85**:837–847.

McComb, K., Moss, C., Sayialel, S., and Baker, L., 2000, Unusually extensive networks of vocal recognition in African elephants, *Anim. Behav.* **59**:1103–1109.

Mech, D. L., 1970, *The Wolf: The Ecology and Behaviour of an Endangered Species*, The Natural History Press, Garden City, NY.

Miles, H. L. W., 1983, Apes and language: The search for communicative competence, in: *Language in Primates: Implications for Linguistics, Anthropology, Psychology and Philosophy*, J. de Luce and H. T. Wilder, eds., Springer-Verlag, New York, pp. 43–61.

Miles, H. L. W., 1990, The cognitive foundations for reference in a signing orangutan, in: *"Language" and Intelligence in Monkeys and Apes*, S. T. Parker and K. R. Gibson, eds., Cambridge University Press, Cambridge, pp. 511–539.

Mitchell, R. W. and Thompson, N. S., 1986, Deception in play between dogs and people, in: *Deception. Perspectives on Human and Nonhuman Deceit.* R. W. Mitchell and N. S. Thompson, eds., SUNY, New York, pp. 193–204.

Newman, J. D., 1995, Vocal ontogeny in macaques and marmosets: Convergent and divergent lines of development, in: *Current Topics in Primate Vocal Communication*, Zimmermann et al., eds., Plenum Press, New York, pp. 73–97.

Oda, R. and Masataka, N., 1996, Interspecies responses of ringtailed lemurs to playback of antipredator alarm calls given by Verreauxs sifakas, *Ethology* **102**:441–453.

Owren, M. J., 1990, Acoustic classification of alarm calls by vervet monkeys (*Cercopithecus aethiops*) and humans (*Homo sapiens*): I. Natural calls, *J. Comp. Psych.* **104**:20–28.

Parker, S. T. and McKinney, M. L., 1999, *Origins of Intelligence: The Evolution of Cognitive Development in Monkeys, Apes, and Humans*, Johns Hopkins University Press, Baltimore, MD.

Parker, S. T. and Gibson, K. R., eds., 1990, *"Language" and Intelligence in Monkeys and Apes: Comparative Developmental Perspectives*, Cambridge University Press, New York.

Partan, S. and Marler, P., 1999, Communication goes multimodal, *Science* **283**:1272–1273.

Patterson, D. K. and Pepperberg, I. M., 1996, A comparative study of human and parrot phonation: Acoustic and articulatory correlates of vowels, *J. Acoust. Soc. Am.* **96**:634–648.

Pepperberg, I., 1990, Cognition in an African grey parrot (*Psittacus erithacus*): Further evidence for comprehension of categories and labels, *J. Comp. Psych.* **104**:41–52.

Pepperberg, I. M., 1999, *The Alex Studies: Communication and Cognitive Capacities of an African Grey Parrot*, Harvard University Press, Cambridge, MA.

Pepperberg, I. M., 2001, Evolution of avian intelligence, in: *The Evolution of Intelligence*, R. Sternberg and J. Kaufman, eds., Erlbaum Press, Mahwah, NJ, pp. 315–337.

Pinker, S., 1994, *The Language Instinct*, Penguin Books, London.

Povinelli, D. J. and O'Neill, D. K., 2000, Do chimpanzees use their gestures to instruct each other? in: *Understanding Other Minds*, Second edition, S. Baron-Cohen, H. Tager-Flusberg, and D. J. Cohen, eds., Oxford University Press, Oxford, pp. 459–487.

Premack, D., 1972, Language in chimpanzees, *Science* **172**:808–822.

Rendall, D., Rodman, P. S., and Edmond, R. E., 1996, Vocal recognition of individuals and kin in free-ranging rhesus monkeys, *Anim. Behav.* **51**(5):1007–1015.

Rendell, L. and Whitehead, H., 2001, Culture in whales and dolphins, *Behav. Brain Sci.* **24**:309–382.

Rogers, L. J., 1997, *Minds of their Own*, Allen & Unwin, Sydney

Rogers, L. J. and Kaplan, G., 2000, *Songs, Roars and Rituals. Communication in Birds, Mammals and other Animals*, Harvard University Press, Cambridge, MA.

Rogers, L. J. and Kaplan, G., 2003, *Spirit of the Wild Dog*, Allen & Unwin, Sydney.

Rumbaugh, D., 1995, Primate language and cognition: Common ground, *Soc. Res.* **62**:711–730.

Russon, A. E., Bard, K. A., and Parker, S. T., eds., 1996, *Reaching into Thought: The Minds of the Great Apes*, Cambridge University Pres, Cambridge, UK.

Savage-Rumbaugh, E.S. and Lewin, R., 1994, *Kanzi, the Ape at the Brink of the Human Mind*, John Wiley & Sons, New Yok.

van Schaik, C. P., van Noordwijk, M. A., Warsono, B., and Sutriono, E., 1983, Party size and early detection of predators in Sumatran forest primates, *Primates* **4**:211–221.

Schassburger, R., 1993, *Vocal Communication in the Timber Wolf, Canis lupus, Linnaeus. Structure, Motivation, and Ontogeny*, Paul Parey Scientific Publishers, Berlin.

Seidenberg, M. S. and Petitto, L. A., 1979, Signing behaviour in apes: A critical review, *Cognition* 7:177–215.

Seyfarth, R. M. and Cheney, D. L., 1986, Vocal development in vervet monkeys, *Anim. Behav.* **34**:1640–1658.

Seyfarth, R. M. and Cheney, D. L., 1990, The assessment of vervet monkeys of their own and another species' alarm calls, *Anim. Behav.* **40**:754–764.

Seyfarth, R. M., Cheney, D. L., and Marler, P., 1980a, Vervet monkey alarm calls: Semantic communication in a free-ranging primate, *Anim. Behav.* **28**:1070–1094.

Seyfarth, R. M., Cheney, D. L., and Marler, P., 1980b, Monkey responses to three different alarm calls: Evidence of predator classification and semantic communication, *Science* **210**:801–803.

Shapiro, G. and Galdikas, B., 1995, Attentiveness in orangutans within the sign learning context, in: *The Neglected Ape*, R. D. Nadler, B. F. M. Galdikas, L. K. Sheeran, and N. Rosen, ed., Plenum Press, New York, pp. 199–212.

Singh, R. S., Krimbas, K., Paul, D. B., and Beatty, J., eds., 2001, *Thinking about Evolution: Historical, Philosophical and Political Perspectives* (Festschrift for Lewontin), Cambridge University Press, Cambridge.

Slobodchikoff, C. N., Kiriazis, J., Fischer, C., and Creef, E., 1991, Semantic information distinguishing individual predators in the alarm calls of Gunnison's prairie dogs, *Anim. Behav.* **42**:713–719.

Smith, W. J., 1981, Referents of animal communication, *Anim. Behav.* **29**:1272–1275.

Smith, W. J., 1990, Animal communication and the study of cognition, in: *Cognitive Ethology: The Minds of Other Animals* (Essays in honor of Donald R. Griffin), C. A. Ristau, ed., Lawrence Erlbaum Associates, Hillsdale, NJ, pp. 209–230.

Snowdon, C. T., 1993, Linguistic phenomena in the natural communication of animals. Language and communication: Comparative perspectives, in: *Language and Communication: Comparative Perspectives*, H. L. Roitblat, L. M. Herman, and P. E. Nachtigall, eds., Lawrence Erlbaum Associates, Hillsdale, NJ, pp. 175–194.

Stanford, C. B., 1995a, *The Hunting Apes. Meat Eating and the Origins of Human Behaviour*, Princeton University Press, Princeton, NJ.

Stanford, C. B., 1995b, The influence of chimpanzee predation on group size and anti-predator behaviour in red colobus monkeys, *Anim. Behav.* **49**:577–587.

Sternberg, R. J. and Kaufman, J. C., eds., 2002, *The Evolution of Intelligence*, Lawrence Erlbaum Associates, Mahwah, NJ.

Sugardjito, J. and Nurhuda, N., 1981, Meat-eating behaviour in wild orang-utans, *Pongo pygmaeus, Primates* **22**:414–416.

Terrace, H. S., 1979, *Nim: A Chimpanzee Who Learned Sign Language*, Knopf, New York.

Terrace, H. S., Pettito, L., Sanders, R., and Bever, T., 1979, Can an ape create a sentence? *Science* **206**:809–902.

Tinbergen, N., 1953, *Social Behaviour in Animals*, Methuen, London.

Tinbergen, N., 1963, On aims and methods of ethology, *Z. Tierpsychol.* **20**:410–433.

Todt, D., Goedeking, P., and Symmes, D., eds., 1988, *Primate Vocal Communication*, Springer-Verlag, Berlin.

Tomasello, M. and Call, J., 1997, *Primate Cognition*, Oxford University Press, New York.

Tomasello, M., Call, J., and Hare, B., 1998, Five primate species follow the visual gaze of conspecifics, *Anim. Behav.* **55**:1063–1069.

Tomasello, M., Gust, D., and Frost, G. T., 1989, A longitudinal investigation of gestural communication in young chimpanzees, *Primates* **30**:35–50.

Tooze, Z. J., Harrington, F. H., and Fentress, J. C., 1990, Individually distinct vocalizations in timber wolves, *Canis lupus, Anim. Behav.* **40**:723–730.

Tyack, P., 1986, Population biology, social behavior and communication in whales and dolphins, *TREE* **1**:175–188.

Well, M. C., and Bekoff, M., 1981, An observational study of scent-marking in coyotes (*Canis latrans*), *Anim. Behav.* **29**:332–350.

Whiten, A. and Byrne, R. W., 1988, Tactical deception in primates, *Behav. Brain Sci.* **11**:233–273.

Wrangham, R. and Peterson, D., 1997, *Demonic Males. Apes and the Origins of Human Violence*, Bloomsbury, London.

Yamagiwa, J., 1992, Functional analysis of social staring behavior in an all-male group of mountain gorillas, *Primates* **33**:523–544.

Zahavi, A., 1979, Ritualisation and the evolution of movement signals, *Behaviour* **72**:77–81.

Zentall, T. R., 2000, Animal intelligence, in: *Handbook of Intelligence*, R. J. Sternberg, ed., Cambridge University Press, Cambridge.

Zimmermen, E., Newman, J. D., and Jürgens, U., eds., 1994, *Current Topics in Primate Vocal Communication*, Plenum Press, New York.

Zuberbühler, K., 2000, Referential labeling in Diana monkeys, *Anim. Behav.* **59**:917–927.

Zuberbühler, K., Cheney, D. L., and Seyfarth, R. M., 1999, Conceptual semantics in a nonhuman primate, *J. Comp. Psych.* **113**:33–42.

PART FOUR

Theory of Mind

CHAPTER SEVEN

Theory of Mind and Insight in Chimpanzees, Elephants, and Other Animals?

Moti Nissani

This chapter attempts to provide an accessible review of a fundamental scientific and philosophical question: Are animals conscious? Instead of trying to address the issue of consciousness as a whole, this partial review will for the most part touch upon just two facets of consciousness—theory of mind and insight.

This review takes it for granted that we need to place a question mark on the understandable but as yet unfounded assumption that our evolutionary next of kin—the great apes—are also our closest cognitive relatives. At the moment, we cannot rule out the view that "a classification of the animal kingdom based on intelligence would probably cut right across the classifications based on structure" (Hobhouse, 1915). As a matter of fact, at the moment we cannot even refute the counterintuitive claim that there are no differences in intelligence between one non-human vertebrate and another (Macphail, 1982; Thomas, 1986).

Moti Nissani • Department of Interdisciplinary Studies, Wayne State University, Detroit, MI, USA
Comparative Vertebrate Cognition, edited by Lesley J. Rogers and Gisela Kaplan. Kluwer Academic/Plenum Publishers, 2004.

On first sight, the case for consciousness in animals seems overwhelming. Given the evolutionary notion of continuity, and given, moreover, the remarkable physiological and genetic similarity between apes and humans, it seems scarcely credible to argue that apes are devoid of any trace of consciousness. Furthermore, it seems reasonable to suppose that consciousness confers an enormous evolutionary advantage, for it allows animals to try out possible actions in their head without actually performing them through costly trial and error (Griffin, 2001). Common intuition seems to point in the same direction, as the following passage suggests: "When I looked into Washoe's eyes she caught my gaze and regarded me thoughtfully, just like my own son did. There was a person inside that ape 'costume'. And in those moments of steady eye contact I knew that Washoe was a child" (Fouts, 1997).

Regretfully, the question of animal consciousness *cannot* be resolved by either intuition or theory. Although intuition often serves as valuable breeding ground for research ideas, it is notoriously fallible. And as far as theory is concerned, it could be just as well supposed that consciousness exerts its own evolutionary price (e.g., hesitation when swift action is called for), or that it is not readily achieved, even when favored by natural selection.

ELEPHANT COGNITION

Much of the original work reported in this chapter deals with elephants, so it may be worthwhile to briefly review their cognition. Because we know so little about the minds of elephants, and because the three known species of elephants are similar, this review will focus on the cognition of elephants in general terms, often leaving aside the *terra incognita* of possible cognitive and behavioral differences among the three species.

Although elephants have been in close association with humankind for thousands of years, and although anecdotes about their wisdom or witlessness are many, their mentality has only been subject to a mere handful of controlled studies. This curious gap in the research literature provides fertile grounds for speculations and controversies. Indeed, views on the subject range from the assertion that "elephants are exceedingly intelligent; that they have a form of intelligence which manifests itself in many ways that are very like our own; and that, in these respects, they stand as far apart as we do from all other living things—the great apes not excluded" (I. T. Sanderson, 1962), to the assertion that the "elephant is a stupid animal; and I can assert with confidence that all the stories I have heard of it, except those relating to feats of strength or docility performed under its keeper's direction, are beyond its intellectual power, and are mere pleasant fictions" (G. P. Sanderson, 1912; see also Carrington, 1958).

At 6kg, the elephant brain is the largest brain of all land mammals (Rensch, 1957). Sheer size however means little, for a good part of the brain must support the bulk of an animal and some of its special, noncognitive, functions. Seen this way, the elephant is a comparative lightweight: an elephant weighing 75 times as much as a human has a brain weighing only 2–3 times as much (Spinage, 1994). Likewise, the elephant's cerebral hemispheres do not cover the cerebellum as they do in humans and apes (Spinage, 1994). But, in common with the brain of dolphins, great apes, and humans, the brain of elephants is highly convoluted—a possible indicator of high intelligence and of an advanced capacity for learning (Alexander, 2000). Elephants also have a large neopallium (the area associated with memory). Above all, most mammals are born with brains that are around 90 percent of their weight as adults, but humans are born with about 27 percent, elephants with 35 percent, and chimpanzees with about 54 percent (Eltringham, 1982; Poole, 1997, p. 38). There is thus considerable postnatal brain growth in these three species, a fact that is believed to be linked to the exceptional learning ability of their young.

The *only* complete study of elephant cognition was conducted on a 5-year-old Asian female at the Munster Zoo. She faced two small wooden boxes whose lids were painted with two different patterns, for example, circle and cross, and had to remove the correct lid to get a food reward. It took her about 330 trials, over a period of several days, to consistently choose the reinforced pattern, for example, the cross (Rensch, 1956, 1957). By the fourth single-pair discrimination task, she reached criterion on the 10th trial. After learning to discriminate between 20 symbol pairs, she performed superbly on a test that combined all 20. The test lasted several hours, yet her performance actually improved toward the end. A year later, her scores ranged from 63 to 100 percent—a scientific demonstration, Rensch says, of the adage that elephants never forget.

In a partial study, Leslie Squier (reported in Stevens, 1978) confirmed the elephant's ability to learn a simple discrimination task and to remember it 8 years later, "a remarkable performance and one that many *Homo sapiens* might have difficulty emulating" (Markowitz, 1982). In another study, Povinelli (1989) reported that elephants do not recognize themselves in the mirror (an observation that we have confirmed with our two Asian elephants—Nissani and Hoefler-Nissani, 2002, unpublished observations). Thus, the few controlled experiments on record seem to lend credence to Carrington's conclusion that "elephants are not sufficiently intelligent to grasp an idea easily or quickly in the early stages, but once it has penetrated to their somewhat slow brains it is virtually ineradicable."

In contrast to the picture which emerges from these zoo studies, field observers of elephants in the wild are often more impressed by their minds (e.g., "the more we learn about elephants, the fainter the line we have drawn between man and other animals will become"—Poole, 1997). In one field study, McComb et al. (2000) found that an average female can recognize the contact calls of about 100 other females. Equally remarkable, the "legend" of coming to the aid of a wounded comrade is true. In one confirmed instance, when a bull was shot, two younger elephants came to his rescue and tried to lead him away (Denis, 1963; cf. Carrington, 1958). Romanes' (1882) seemingly tall tale that, during an operation, "elephants behave like human beings, as if conscious that the operation was for their good, and the pain unavoidable," is likewise probably true (Blashford-Snell and Lenska, 1997; Groning and Saller, 1999; Shand, 1995). Aelian's fantastic 3rd century recount of an Ethiopian tale that "if one elephant sees another lying dead, it will not pass by without drawing up some earth with its trunk and casting it upon the corpse, as though it were performing some sacred and mysterious rite on behalf of their common nature" has been likewise confirmed by contemporary researchers. During drought, the African elephant is the only animal known to dig for water, and there is moreover an account of an elephant chewing bark to fashion a plug, using it to plug a water hole, and hiding the hole from other animals by covering it with sand (Gordon, 1966). In Burma (=Myanmar), according to one observer (Williams, 1950), raiding young elephants often stuff their bell with mud in order to avoid detection.

Elephants are proficient tool users (Chevalier-Skolnikoff and Liska, 1993; Hart and Hart, 1994). They often scratch themselves with sticks and break fences by piling logs—or even young elephants—over them. There are likewise many reliable accounts of elephants hurling objects (Chevalier-Skolnikoff and Liska, 1993; Wickler and Seibt, 1997). A 27-year-old bull, for instance, routinely concealed a rock in his trunk and hurled it at his keeper (Linden, 1999). Once the training of elephants in playing cricket or soccer is over, they "play the game with the enthusiasm of boys having a knock-up on the village green" (Carrington, 1958; I. T. Sanderson, 1962).

I have tried to confine my very partial account of the elephant's mind to the more trustworthy, or to independently corroborated, observations, omitting the more anecdotal, and even more fantastic, tales. Even so, the picture that emerges is of an animal whose mentality deserves closer scientific attention than it has so far received.

DO ELEPHANTS AND CHIMPANZEES KNOW THAT PEOPLE SEE?

An organism can be said to possess a *theory of mind* when it is capable of attributing mental states to others—when it understands that others see, feel, and know. An organism's theory of mind might be inextricably linked to its capacity for consciousness, insight, concept of self, and such emotions as pride, shame, empathy, and compassion. The knowing component of this complex concept can perhaps be illustrated by imagining the following sequence (following Zentall, 2000). In the first scene, little Susan observes Wong Tsu hiding a candy under a pillow and then going outside to play. As the next scene unfolds, Susan sees an adult entering the room, removing the candy without disturbing the pillow, and hiding the candy in a desk drawer. Next, Susan sees Wong Tsu coming back. If Susan expects him to look for the candy under the pillow and not in the drawer (i.e., if she implicitly *theorizes* that Wong Tsu has a *mind* and that in his mind the candy is still under the pillow), we can say that Susan possesses the knowing component of a theory of mind.

As we shall see, attempts to prove the existence of a theory of mind are often exceedingly complex when we divert our gaze from young children to animals. According to Lorenz (cited in Thorpe, 1956), if a raven is observed hiding food, it often retrieves the food, scolds the observer, and hides the food out of human reach. On first sight, this behavior appears to involve the same conceptual understanding as Susan's, but here the cause is slippery—it could be traceable to understanding the perspective of another, *or* to genetic programming, *or* to prior trial-and-error learning.

Probably owing to such conceptual roadblocks, the question of whether any animal besides ourselves possesses a theory of mind has never been scientifically resolved. Hunter-gatherers, ancient civilizations, and Buddhism seem to have taken it for granted that animals are conscious. But with the advent of Christianity and, later, Cartesian philosophy, most educated westerners came to view animals—including the ones closest to us in appearance and behavior—as mindless, unfeeling, automatons. A couple of centuries later came yet another paradigmatic shift, for evolutionary theory seemed to require a continuity of physical and mental characteristics among humans and their closest relatives.

Darwin's views prevailed until the mid-1990s or so. On the experimental side, numerous studies and field observations seemed to independently converge on the belief that some animals, at least, possess a theory of mind. Chimpanzees, for example, were said to be capable of insightful problem solving, acquiring a rudimentary form of language, self-referential behavior in front of a mirror, and deception (cf. Russon et al., 1996). Moreover, chimpanzees split from the human line only a few million years ago, share close to 99 percent of the human genome, and show remarkable behavioral similarities to human beings. All these observations seemed to imply that chimpanzees, and perhaps other species, possess a theory of mind.

Many researchers, to be sure, rejected this seemingly irresistible avalanche, but theirs appeared to be the kind of rearguard action that is often seen in science. In the 1990s, however, the pendulum began swinging again in a neo-Cartesian direction, and here one set of elegant experiments stands out, for it appeared to provide the clearest experimental evidence to date that chimpanzees—and perhaps all other non-human animals—possess "clever brains but blank minds" (N. Humphrey, cited in Gallup, 1998).

In these experiments, Daniel J. Povinelli and his colleagues capitalized on captive chimpanzees' tendency to beg from their keepers to ask: Do chimpanzees understand the seemingly elementary fact that people see? (Povinelli and Eddy, 1996; Povinelli et al., 2000; Reaux et al., 1999). Six or seven young chimpanzees took part in a complex series of longitudinal studies. In a typical pre-training phase, a youngster entered a room where it faced, across a clear Plexiglas barrier, a familiar experimenter. The experimenter sat on either the right or left side of the barrier, in each case directly across a hole in the Plexiglas. In an average of 48 consecutive sessions consisting of 10 trials each, the chimpanzee learned to insert a hand through the hole which faced the experimenter, as opposed to the hole which did not. Once this skill was acquired, the chimpanzee had to choose between one experimenter who had appealing food in her hand and one who held a neutral block of wood. In this case, the chimpanzee received a food reward only when it inserted its hand through the hole facing the food-carrying experimenter.

In the experiment itself, the chimpanzees faced variations along a single theme: in all variations they had to spontaneously decide whether to insert their hand toward an experimenter who could see them or toward an experimenter who could not. Only a few of these variations need to be described here (cf. Povinelli and Eddy, 1996). In all these variations, one of the experimenters faced the right hole of the barrier while the other experimenter faced the left hole (cf. Figure 1). Thus, in a typical blindfold trial, both experimenters stood motionlessly in front of the right or left hole, one with a blindfold covering her mouth and the other with an identical blindfold covering her eyes. Thus, the former could see the chimpanzee, while the latter could not. In the buckets condition, the experimenters stood or sat in the same place, but one held a large bucket on her shoulder and could see the animal, while the other experimenter covered her head with a bucket and could not see the animal. In the similar screen condition, one experimenter held a cardboard screen on her shoulder while the other held it so that it completely obscured her face from the chimpanzee. In the Back/Front condition, one experimenter faced the Plexiglas partition and the chimpanzee, while the other had her back turned to both. In the looking-over-shoulder condition, both experimenters assumed the same position and both had their backs turned to the chimpanzee, but one experimenter had her face turned toward the animal, while the other experimenter turned her face away.

Startlingly, in the first few trials, in all but one variation, the chimpanzees consistently performed at chance level. That is, and in contrast to the performance of young children (3 years of age or older) in a similar task, Povinelli's chimpanzees were just as likely to beg from a person who could see the begging

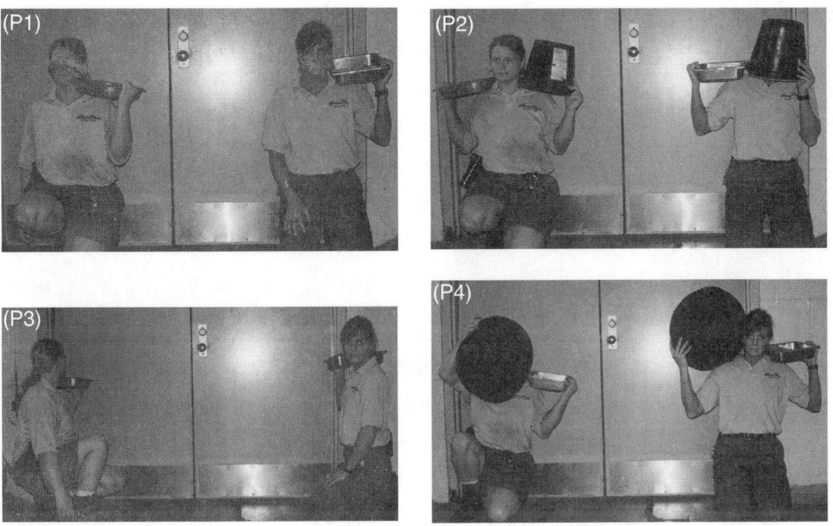

Figure 1. Four probe conditions in chimpanzee seeing experiments. (P1) blindfolds, (P2) buckets, (P3) sideways, and (P4) screens.

gesture as from a person who could not. The one exception in the original set of experiments involved the back/front combination (Figure 2, frame B3), but further experiments suggested to the authors that the chimpanzees' proficiency in this task was merely "a consequence of their reinforcement history in the training phase" (Povinelli and Eddy, 1996). Thus, chimpanzees "do not seem to appreciate that others 'see' things" (Povinelli and Giambrone, 2001).

These results, along with similarly discouraging studies of pointing and gaze-following, led Povinelli to the neo-Cartesian view that only humans can represent explicitly their "own psychological states and those of others" (Povinelli, 1998). Later, Carruthers (1998) based his view that sentience is exclusively human in part on Povinelli's results.

The question of animal consciousness in turn has profound implications for our very conception of the world around us, for evolutionary biology, for comparative psychology, and perhaps also for the treatment of human autism (Baron-Cohen et al., 2000; Carruthers and Smith, 1996). It has also practical and controversial implications, ranging from a call for putting into place "more rigorous standards for conducting and reporting empirical research" in comparative psychology (Povinelli and Giambone, 2001), to endorsing intrusive research on apes (Wynne, 1999).

Thus, if the neo-Cartesians are right, only one animal possesses a measure of understanding of its own and others' actions and mental states—and that animal is us! It follows that the ongoing, extensive search for culture, imitation, deception, or genuine self-awareness in animals is misguided in principle, somewhat akin to the quest for the philosopher's stone. Looking closely at any great ape, we cannot, indeed, help feeling that it thinks and feels, at least in some rudimentary fashion, as we do. But this, the neo-Cartesians insist, is an illusion.

Povinelli's work is consistent with some well-known observations. For one thing, his work throws light on our failure to come up with a single unequivocal demonstration of a theory of mind in animals; we may have failed to come up with a clear-cut proof, neo-Cartesians may say, for the very good reason that such proof does not exist. Computer models likewise raise the possibility that complex behaviors, of the kind that are often shown by chimpanzees, can emerge mechanically, through mere learning and the pressure of natural selection, from the interaction of simple behavioral building blocks (cf. van Hooff, 2000; see also Morgan, 1920).

Povinelli's experiments, and the neo-Cartesian hypothesis they led to, have however been questioned (Byrne, 2001; Byrne and Whiten, 1988; Call, cited in Tuma, 2000; Gallup, 1998; Hare et al., 2000; Heyes, 1998; Russon et al., 1996). Before describing collaborative research efforts with elephants and chimpanzees at the Detroit Zoological Institute, and in an effort to provide a rationale for their design features, I wish to highlight a few additional uncertainties in Povinelli's experiments.

Povinelli tacitly assumes that, cognitively speaking, the chimpanzee is our closest relative. This assumption is not necessarily true, for reasons that emerge elsewhere in this chapter. Thus, the neo-Cartesian hypothesis might be wrong, even if it can be shown that chimpanzees lack a theory of mind.

Although inconclusive, there is a body of circumstantial and experimental evidence suggesting that apes do have a theory of mind (e.g., Povinelli et al., 1990; Premack, 1988; Uller and Nichols, 2000). There is, in particular, experimental evidence that "chimpanzees are capable of modeling the visual perspectives of others" (Povinelli et al., 1990) and that, at least in some situations, they know "what conspecifics do and do not see" (Hare et al., 2000; but cf. Povinelli and Giambrone [2001] for a dismissal of this latter claim). The tentative confirmatory evidence extends to other species as well; for instance, marmosets are claimed to be capable of true imitation (Voelkl and Huber, 2000), bottlenose dolphins seem capable of mirror self-recognition (Reiss and Marino, 2001), and even the green bee-eaters of India seem capable of distinguishing a person who can see them from a person who cannot (Smitha et al. 1999, cited in Griffin 2001). Given these observations, more than one set of experiments, one basic design, and six or seven subjects of one species (Povinelli and Eddy, 1996) are needed to resolve the theory of mind controversy.

Chimpanzees are not only physiologically our closest kin (de Waal, 2001), but they often behave similarly to human beings, for example, like us, they too follow the gaze of another, use tools, and engage in tribal warfare. Occam's Razor would seem to imply similarity of the cognitive processes which underlie such analogous behavioral patterns. But, if cognitive similarity is rejected, one must resort to an auxiliary evolutionary hypothesis. Thus, Povinelli and Giambrone (2001) are forced to put forward a *reinterpretation hypothesis*, which counterintuitively suggests, for example, that what appears to us as deception in animals is in fact traceable to the evolution of "brain systems dedicated to processing information about the regularities of the behaviors of others."

One unresolved issue of Povinelli's work involves the age of his chimpanzees. Since the initial work was carried out with young chimpanzees roughly between the ages of 5 and 6 years, the experiments left open the possibility that theory of mind arises later in chimpanzee development, as Povinelli and Eddy (1996) noted. To meet this objection, Povinelli and colleagues re-tested the *same* chimpanzees twice, up to the age of 9 years. In the first retest, a year or so after the original work was completed, the chimpanzees forgot the little they had learned, and again seemed unable to tell the difference between a human being who could see them and a human being who could not. But in the second test, when the same chimpanzees approached 9 years of age, they performed extremely well *from the start* in the buckets, screens, and looking-over-shoulder tests (see Figure 1), and only did poorly in tests that involved a direct understanding of the role of human eyes in visual perception (Reaux et al., 1999). These striking improvements at close to age 9 could then be interpreted in two radically different ways. One could assume, with Povinelli et al., "that there was no development between 5 and 9 years of age in the animals' understanding of visual perception as an internal state of attention," and that throughout this longitudinal study, they merely learned stimulus-based rule structures. Alternatively, one could assume that the older chimpanzees of this third series of experiments have by now matured enough to understand the attentional aspects of seeing (or at least, have acquired on their own the broad rule of begging from the person whose face is visible), and, in turn, explain their random performance in subtle tests involving covering the eyes by assuming that chimpanzees, like some of Povinelli's children, may understand that there is indeed someone behind those eyes—without understanding the working of eyes (Hare et al., 2000). That is, the first hypothesis is congruent with the radical assertion that chimpanzees lack a theory of mind; the second, with the trivial conclusion that chimpanzees do not understand how human eyes work. Needless to say, to decide which of these two hypotheses is more nearly correct, one must test adults who have *not* taken part in such tests before.

It is probable that the seeing experiments of Povinelli and colleagues lack *ecological validity*, a related set of problems that could only have been dealt with through a radical alteration of some features of their experiments.

To begin with, the experiments are claimed to have relied on the species-typical begging gesture. In point of fact, however, the chimpanzees had to learn the unnatural gesture of inserting a hand through one of two holes of a transparent plastic barrier. As might be expected, it took them an average of *479* trials, and dozens of sessions, to learn this allegedly natural response. Human observers are tempted to interpret such insertions as begging, but to the chimpanzees themselves they may have had no meaning whatsoever, thus accounting for their chance performance in the critical tests. In a comparable context, Gomez (1998) argues that the prolonged training of subjects in artificial tasks is a mistaken approach.

It is not clear either what the natural begging response of captive chimpanzees is. Our chimpanzees (but not our elephants) seemed to employ a variety of idiosyncratic begging postures, instead of just one. Thus, when begging services, toys, or food from their keepers, one of our captive chimpanzees often nodded her head while another whimpered, banged on the wire mesh, or clapped his hands. Indeed, often throughout our own experiments, the chimpanzees employed one of these more natural responses first, and only then resorted to the required learned gesture of inserting their fingers through or under the wire mesh.

Moreover, Povinelli's chimpanzees begged from people who often assumed unnatural, seemingly indifferent, postures. For instance, throughout any buckets trial, even the seeing experimenter remained motionless and made no eye contact with the animal.

Povinelli's overworked subjects have been experimented with virtually all their lives. They were all born in captivity. Four were reared in a nursery from birth and two were reared by their mother for 1 year and then transferred to the nursery. This background raises serious questions about their natural curiosity and their emotional and cognitive development (Harlow and Harlow 1965; Thompson and Melzack, 1956). Needless to say, few if any of the children in Povinelli and Eddy's initial experiment—which provided the backdrop, controls, and confirmation of the results—suffered the traumas of lifelong experimentation and maternal deprivation.

A related problem involves the over-reliance in most trials on the food versus block of wood configuration. This "baseline probe" was frequently used during extensive pretraining sessions, and it often outnumbered the "treatment probe" in experimental sessions by a ratio of 2 : 1. This might have led the chimpanzees to believe that their task in some of the critical trials involved *guessing* who had the food, not begging from the person who could see them. Indeed, Povinelli and Eddy's data lend a measure of support to this interpretation. Thus, in their Experiment 7, when the majority of trials involved begging from a single person only, and only comparatively few trials involved food versus block of wood before and during critical sessions, the chimps performed significantly above chance in the screens test, leading the authors to suggest that some learning had occurred in this case. But, in their Experiment 8, when only food versus block of wood were used as spacer trials (and no singles), the chimpanzees' performances in the screen test returned to chance level. In this context, Povinelli and Eddy are puzzled: "It is difficult to know why the subjects performed so poorly on the screen-over-face probe trials in this experiment as compared to Experiment 7." Such unaccountable results can in turn be readily explained by the assumption that the chimpanzees were confused about the nature of the task they were to perform in these experiments. And again, when working with children, the equivalent of food versus wood was used in training, but only once during the trials themselves, thus raising doubts about the validity of the key comparison between the young chimpanzees and the younger children.

Another possible design flaw involves inadequate payoffs. In a typical experiment (for instance, their Experiment 1), 80 percent of the trials presented no choice at all, 8.9 percent presented the easy, immediate choice between a block of wood and a favorite food item, 4.4 percent the easy choice between an experimenter's back and front, and only 6.7 percent presented the more difficult, experimental choices (e.g., looking over the shoulder). Thus, the subjects were assured of being rewarded in 168 of 180 trials (93.3 percent) by acting mechanically, and were likely to be rewarded in 174 of every 180 trials (96.7 percent). This high reward ratio seems to beg mechanical, behavioristic action, rather than thought and contemplation—if you wish to encourage thinking, you must create situations where the payoffs for thinking before acting involve a bit more than a 3.3 percent increase in the reward schedule. Thus, we may conclude that the choice between a seeing and non-seeing person was not as automatic as the choice between back and front, or food versus wood, but not that it could not be made at all. This difficulty is compounded by the fact that the relevant fractions are entirely different for the human subjects, who had to make the more difficult choice in 30 percent of the trials. Acting mechanically, for the toddlers, would have involved a 15 percent loss of rewards versus a 3.3 percent for the chimpanzees.

To the best of our knowledge, Povinelli's ingenious experiments, despite their procedural simplicity, serious methodological and conceptual flaws, counterintuitive character, wide acclaim, and profound implications, have not been replicated with chimpanzees (although other researchers addressed similar questions by relying on different approaches, e.g., Hare et al., 2000), nor have they been extended to other species. It seemed worthwhile therefore to reproduce Povinelli's essential procedure while safeguarding against some of its design flaws. Thus, to avoid reliance on just one species before rejecting Darwin's cognitive continuity hypothesis, the experiments described here were carried out with seven chimpanzees *and* two elephants. To meet the age objection, six of the seven chimpanzees we worked with, and both elephants, were adults. To avoid extensive pretraining of the subjects, we tested the animals in their own compounds and, instead of requiring them to insert their fingers or trunks through one or another hole of an artificial barrier, we monitored the almost as reliable criteria of insertion and direction of hand (and trunk) gesture (see below for further details). To bypass the problem of excessive experimentation, our work was carried out in a zoo setting, with subjects that were thought to have never taken part in psychological experiments. To partially meet the objection of emotional deprivation, at least some of the subjects we chose were raised by their mothers. To maximize the chance that the chimpanzees understood that their task was not guessing who had the food, we drastically curtailed the use of the food versus block of wood condition, relying instead on the back/front and on the new lying-down (Figure 2, B4) conditions. To avoid the inadequate payoffs difficulty, the fraction of critical trials in each session was raised, this fraction was varied from one session to another, and the number of trials in each session was varied as well.

Thus, if we too observe chance performance, despite deliberate attempts to alter Povinelli's design, his work and its far-reaching conclusions would receive much-needed confirmation. On the other hand, a partial failure to replicate his results would require additional work to tease apart the variable(s) that account for the difference, and would require new experiments to decide whether some

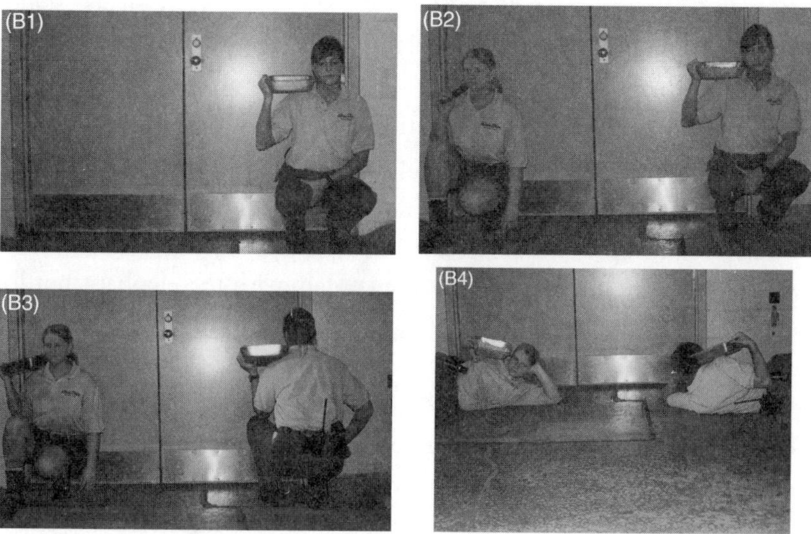

Figure 2. The four background conditions in chimpanzee seeing experiments. (B1) One experimenter with food tray. (B2) Food versus flashlight. (B3) Back versus front. (B4) Lying toward or away from subject.

chimpanzees "understand the attentional aspect of gaze" (Povinelli and O'Neill, 2000). A partial failure, in other words, will return the pivotal question of theory of mind in chimpanzees and other animals to the status quo ante, prior to Povinelli and Eddy's (1996) experiments.

Experiment 1: Do Elephants know that People See?

Two Asian elephants (*Elephas maximus*) of the Detroit Zoological Institute served as subjects. Winky, the older of the two, was wild-born in Cambodia, probably in 1952, acquired by the Sacramento Zoo in 1955, and moved to Detroit in 1991. Wanda was wild-born in India in 1958, brought to the United States around 1960, and moved to Detroit in 1994. During their tenure in Detroit, these elephants have never served as experimental subjects. Little is known about their history prior to arrival in Detroit.

In the experiments described here, which took place from May to December of 2001, a group of 3–5 researchers and keepers entered an enclosed courtyard inside the elephants' one-acre compound, typically around 12:45 p.m., three times a week. If not already by the cables separating the courtyard from their open-air enclosure, both elephants typically approached the cables and stayed nearby for the duration of the experiment.

In each experimental trial, the elephant was first brought in line by one keeper (the commander of that trial), so that she faced the center of the courtyard. In each trial, one or two other keepers served as givers, standing approximately 4 m away from the elephant with food in their pockets or on a stand behind them, and with their backs to the elephants. The givers then turned, assumed the experimental position, stood motionless in that position for a few seconds, and then approached the elephant in that position (e.g., one walking forward toward the elephant while the other waking backward at the same pace), coming to a stop some 2 m apart from each other and about 1 m away from the elephant, avoiding eye contact with the elephant but looking instead about 2 m below the eyes, at the center of the elephant's trunk.

In over 90 percent of the trials, the elephant waited for this scene to unfold before begging. If her trunk crossed the cables before the givers came to a standstill, she was given the familiar verbal command "trunk down" by the keeper/commander (but not by the givers). Following this command, she readily withdrew her trunk beyond the cables and waited for all movements to cease before begging again. The roles of commander and givers were changed in consecutive trials, with the keeper who last gave the food either not taking part or serving as the commander. In each trial, positions and roles were predetermined in a quasi-random fashion (i.e., never assigning a plus value to the same side in more than two consecutive trials).

When the elephant withdrew from the experiment (especially at the beginning, where the novelty of the situation seemed to distress them), the experiment continued on the following session.

At least two observers independently recorded their observations of each session, while a third person photographed each session on a digital video camera. Interobserver reliability was calculated from the independent records of the two observers and from the computer log of a third person who independently observed the video record.

The single condition, back/front, buckets, and screen variations were almost identical to those of Povinelli and Eddy's (1996). The rock/food condition was comparable to their block of wood/food condition. The sideways posture (Figure 3, Frame P1), which is comparable to their looking-over-the-shoulder posture, was chosen to provide sharper visual definition. We have, as well, introduced a lying-down posture (Figure 4, B4), which involved one keeper lying down on her side and facing the elephant, and the other lying on her side and looking away from the elephant.

Figure 3. Three experimental conditions in the elephant seeing experiments. (P1) Sideways, right choice. (P2) Bucket over head versus bucket over shoulder, wrong choice. (P3) Screen over face versus screen over shoulder, right choice.

Figure 4. Four background (B1–B4) conditions in the elephant seeing experiment. (B1) A captive elephant's natural begging pose. (B2) Food versus rock, wrong choice. (B3) Back versus front, correct choice. (B4) Lying down, wrong choice.

At the start, we had to conduct preliminary vision tests. Wild elephants are often active at night, they are more dependent on olfaction than vision, and there are even reports of a blind matriarch competently leading her troop (Groning and Saller, 1999). Moreover, their vision often deteriorates with age (Stevens, 1979). When food was placed inside a hollow PVC tube (15 cm in diameter), both our elephants seemed to grope for the opening with their trunk, not by relying on vision. Fortunately, other lines of evidence suggested that our two elephants had adequate vision. In routine veterinary eye examinations, Wanda's eyes were judged normal while Winky appeared to have limited vision in the right eye and normal vision in the left. Similarly, the two elephants showed little interest in the back side of a $86 \times 193\,\text{cm}^2$ mirror, but became agitated, manifesting behavior never seen before by their keepers, when the mirror was turned toward them. When a turkey's feather was taped to Wanda's forehead on three different occasions, Winky deftly removed it, suggesting that she saw it clearly enough. Both elephants immediately retrieved distant food items attached to nearby ropes and chains with diameters as small as 1 cm. When 1.4 cm^3 (4 g) white sugar cubes were tossed near them, the elephants appeared to notice the movement, looked in the direction of the rolling cubes, and retrieved them with ease. Finally, although Wanda and Winky daily respond to a variety of commands, each consisting of both arm movements and words, in 10 trials with each (5 in one session and 5 in another about 1 month later) they obeyed immediately 19 of 20 commands when only the gestural component of each was used, again suggesting that their vision sufficed to detect the experimental conditions depicted in Figure 3. To be on the safe side, however, our elephant design avoided fine discrimination tasks (e.g., involving blindfolds or closed eyes).

Another critical assumption of the seeing experiment with the elephants (but not with the chimpanzees) involved the question of imitation or social learning, for the elephants could not be physically separated from each other during the experiment. While we cannot go into this question here, a variety of control procedures, and our observations of the two elephants in many other tasks, suggest absence of imitation. Our results can therefore be safely viewed as an indication of individual, not collective, responses.

For reasons given earlier, and because the elephants performed poorly on the rock/food begging task (Table 1), we chose, as spacer trials, the more reliable single (Figure 4, B1) and back/front (Figure 4, B3) variations. On any given session, 6–8 such trials were interspersed with 2–4 experimental conditions.

We have also attempted to rule out the possibility that the elephants were given unconscious cues by their keepers, experimenters, and onlookers. To begin with, the elephants performed comparatively poorly on the rock/food task (Table 1), and thus failed not only to distinguish the presence of food, but also to receive any subtle cues from the experimenters. Moreover, at some tasks (e.g., screens; Figure 3, P3), the two elephants performed at chance level, again suggesting that they were not responding to subtle, unintentional cues. Finally, in preliminary experiments aimed in part at inquiring into this question of unintentional cues, we presented each elephant with a choice between garlic- and vanilla-soaked identical blocks of wood, and rewarded them only when they tapped the vanilla block. In over 100 trials, although the experimenters knew what the correct choice was, each elephant performed at chance level. This again suggests that our elephants were not responding to unintentional cues.

Table 1. Do elephants know that people see?

	Rock/ food	Back/ front	Lying down	Sideways[a]	Buckets	Screens	Total (3 conditions)
Winky	5/8	8/8	6/8	6/8	3/8	5/8	20/32 0.63
Wanda	5/8	6/8	7/8	7/8	7/8	5/8	26/32 0.81
Total	10/16 0.63	14/16 0.88	13/16 0.81	13/16 0.81	10/16 0.63	10/16 0.63	46/64 0.69
Six chimps[b]	0.98	0.84	NA	12/24 0.50	14/24 0.58	26/48 0.54	52/96 0.54

Note: NA: Not available.
[a] In chimps, the actual position was the very similar one of looking over the shoulder.
[b] Povinelli and Eddy (1996).

In the experiment itself, when an elephant begged correctly, holding her trunk steady for at least 2 s directly in front of the appropriate keeper, she was praised and given a portion of a choice morsel. When she begged from the wrong person, or when her choice was not clear, the keepers withdrew.

Probably because the experimental setup mimicked so closely the elephants' normal routine and natural begging gesture, the elephants reached criterion of begging correctly (9 of 10 times) from a single person by the *first* pretraining session.

The results are given in Table 1. While the small sample, and the limited number of trials (to minimize learning effects), do not allow us to carry out meaningful statistical tests, the elephants seemed to have performed in the back/front condition as well as Povinelli and Eddy's (1996) chimpanzees, and to outperform them in the three experimental conditions. On average, in the three conditions on the right, our elephants were correct in 69 percent of the trials (as opposed to Povinelli and Eddy's 54 percent). Moreover, in the lying-down condition—a configuration they have most likely never experienced before in their life—they again did exceptionally well.

To throw additional light on our data, in subsequent natural trials we capitalized on Winky's (but not Wanda's) proclivity to beg from people some distance away. In 10 consecutive trials spaced 1–5 min apart, a keeper stood just outside her trunk's reach, holding a 1-L bottle of carbonated soda drink. In odd-numbered trials, the keeper faced Winky; in even-numbered, the keeper had her back turned to Winky. When the keeper faced her, Winky begged silently by extending her trunk toward that person. When the keeper had her back turned to Winky, in four trials Winky forcefully blew air on the keeper; in the last trial she begged and then blew. In another set of similar trials, the keeper was just within trunk's reach. In the five front trials, Winky begged and retained the begging posture without being told to, for at least 5 s, never touching the keeper. In all five back trials, Winky never begged; instead, she immediately touched the keeper.

Our preliminary results could of course be ascribable to chance—we had only two elephants, and we could only test them a few times (to minimize trial-and-error learning). It is likewise possible that elephants are better at this task than chimpanzees. But, we felt, the difference could also stem from the comparatively older age of our two elephants and from our modified design. It seemed worthwhile therefore to apply our design to adult chimpanzees.

Experiment 2: Do Chimpanzees know that People See?

Six adult (range 15–31 years; mean age 23.8 years) and one subadult (age 11) chimpanzees, all members of the single social group of ten of the Detroit Zoological Institute, took part in this study (Table 2). The study started on November 2001 and ended on May 2002. All sessions were conducted while subjects were in their ordinary $3 \times 3 \times 3\,m^3$ feeding and sleeping morning cages.

At any given trial, some non-participating chimpanzees could hear their fellows and the experimenters, see the experimenters and follow their preparations for a coming session, but could see neither chimpanzee nor the experimenters at the critical point when the subject inserted her fingers through the mesh.

Table 2. Name, age (at time of experiment), sex, birthplace, and extant rearing history of chimpanzee subjects

Name	Age (years)	Sex	Birthplace	Extant rearing history	Known family relationship to other subjects
Bubbles	31	F	Wild	Unknown	
Joe Joe	30	M	Wild	Unknown	Father of Tanya
Beauty	29	F	Wild	Unknown	
Barbara	19	F	Captivity	Unknown	
Abby	19	F	Captivity	Mother raised	
Akati	15	F	Captivity	Hand raised	
Tanya	11	F	Captivity	Mother raised	Daughter of Joe Joe

Table 3. Do chimpanzees know that people see?

Name	Food/ flashlight	Back/ front	Lying	Sideways	Screens	Blindfolds	Buckets	Total (4 probe trials)
Bubbles	9/10	8/8	4/4	3/4	3/4	3/4	2/4	11/16 (0.69)
Joe Joe	15/15	9/11	3/4	4/4	2/4	2/4	2/4	10/16 (0.63)
Beauty	9/10	6/7	4/4	4/4	3/4	4/4	2/4	13/16 (0.81)
Barbara	15/16	7/8	4/4	3/4	3/4	3/4	3/4	12/16 (0.75)
Abby	11/13	9/11	3/4	4/4	3/4	2/4	2/4	11/16 (0.69)
Akati	6/6	8/8	4/4	2/4	4/4	3/4	1/4	10/16 (0.63)
Tanya[a]	3/4	4/4	2/4	2/4	3/4	2/4	1/4	8/16 (0.50)
Total	68/74 (0.92)*	51/57 (0.89)*	24/28 (0.86)*	22/28 (0.79)*	21/28 (0.75)*	19/28 (0.68)**	13/28 (0.46)	75/112 (0.67)
Young chimps[b]	0.98	0.84	NA	0.50[c]	0.54	0.49	0.58	
33-month-old humans[b]		0.76			0.52			
42-month-old humans[b]		0.94			0.83			

Notes: * $p < 0.01$; ** $p < 0.05$ (t-test, one-tailed, hypothetical mean: 0.5).
[a] Subadult.
[b] Povinelli and Eddy (1996).
[c] Data for the similar looking-over-the-shoulder condition.

For the most part, we adhered to the design features described earlier in connection with the elephants, so we only need to highlight some differences between the two. For the chimpanzees, begging was defined as the insertion of at least a portion of one finger through or under the mesh toward the correct experimenter, on that experimenter's side of the cage, beyond the imaginary vertical line which separated the right and left sides of the cage.

Pretraining Phase I. Finger Insertion to Beg Food: This phase began with the head keeper approaching the right or left side (preselected at random) of the wire mesh with a trayful of tidbits, kneeling down (to be on the subject's eye level), facing the subject about 0.5 m away from the wire mesh, modeling for her the desired finger-insertion behavior, and only rewarding her (by placing the food in her outstretched fingers, not through the zoo's customary procedure of mouth feeding) when she correctly inserted her fingers. At this phase of the experiment, the head keeper used her customary training procedures, which included direct eye contact and soothing, encouraging words. Criterion was defined as correct insertion in four out of five successive approaches.

Pretraining Phase II. Finger Insertion Toward Experimenter Carrying Food: The second phase of pretraining involved a choice between two givers. At the beginning of each trial, the givers consulted that day's written plan. They then stood 2 m apart from each other, with their backs turned to the subject, about 3 m away from the wire mesh. Following a predetermined schedule, one experimenter held a tray of desired food items on her left or right shoulder, while the other held a 0.4 m long, red flashlight on her same (right or left) shoulder. Both experimenters simultaneously turned to face the subject and moved toward her in tandem while fixing their gaze on the center of the wire mesh. Both experimenters halted at the same time, when they were within 0.5 m of the mesh, and knelt. They now waited for the subject to move towards the food side if necessary, and to insert one or more fingers through the mesh on that side. If the subject failed to respond within 30 s, responded with her own earlier idiosyncratic begging gesture, or inserted her fingers toward the experimenter carrying a flashlight, the two experimenters withdrew. Such failures were followed by pretraining Phase I (single experimenter). When the subjects now begged correctly from the single experimenter, Phase II was resumed. Following the completion of each trial, each giver withdrew and independently recorded the subject's behavior in her individual scoresheet. Criterion again was defined as correct insertion toward the food-carrying giver in four of five successive trials.

On any given session, a third experimenter operated a video camera. The roles of givers and photographer were kept constant in a single session but rotated on successive sessions among the six keepers. Within sessions, experimenters' position (right or left) and their task (e.g., carrying a food tray or a flashlight; having a bucket over their shoulder or head), and tray-carrying shoulder (right or left) were preselected and written down in that day's experimental plan. All background and probe trials involving two experimenters followed the general procedure described earlier in relation to Pretraining Phase 2.

As with the elephants, a typical session consisted of 2–4 probe trials, interspersed among 6–8 background trials, with inter-trial intervals lasting 30 s or more.

Background trials consisted of the single person variation of Pretraining Phase 1 (Figure 2, B1), choice between food and flashlight (B2), and two distinct choices (B3, B4) between one giver who could see them and one who could not.

Following Povinelli and Eddy (1996), four probe conditions were used (Figure 1). In all four, the chimpanzees had to choose between begging from a person who could see them and a person who could not.

As was the case with the elephants, the chimpanzees were thoroughly familiar with the givers and photographer, but not with the two main experimenters who stood at all times some 3 m away and choreographed the actions, addressed problems as they arose, and independently recorded key events. The chimpanzees were likewise familiar by sight with most of the objects used in this experiment, including the flashlight, trays of food, buckets, and screens; they had not, however, previously handled these objects. The blindfolds were unfamiliar.

By the end of the third pretraining session (learning to insert one more finger through the wire mesh toward a single human), all our chimpanzees achieved the predetermined criterion of inserting one or more fingers through or under the wire mesh in four of five consecutive trials. By the end of four additional sessions, all subjects achieved the 80 percent criterion in the food tray versus flashlight choice.

To our surprise, in some of the experimental tasks, our six adults (but not the one subadult) performed significantly above chance (Table 3). Our results cannot therefore rule out the notion that both chimpanzees and elephants partially understand that it makes more sense to request food from a human being whose face was directed toward them than from a human being whose face was not. The results are likewise compatible with the alternative notion that the animals acquired this partial knowledge mechanically, through long association with human beings. One thing does, however, emerge unequivocally from these data: *Povinelli's key, striking, implication—that adult chimpanzees perform at chance level in such tasks—is mistaken.* Thus, much more empirical work with chimpanzees, elephants, and other animals will have to be carried out before we can answer the age-old question: Cognitively speaking, is *Homo sapiens* one species among many, a species apart, or somewhere in-between?

INSIGHT IN ANIMALS?

Insight may perhaps be defined as either an understanding, in one's mind, of the key elements of a situation, or the ability to mentally restructure such elements to solve a problem.

To qualify as insight, the observed behavior must not be directly attributable to genetic programming, trial and error, or a combination of both. Given the confusion that surrounds this subject, it may be worthwhile to illustrate here both sources of error.

One example of genetically determined behavior comes from *Drosophila melanogaster*. Even when raised in total isolation, upon first encountering a conspecific female, most mature males manifest the species-specific courtship behavior. Moreover, these flies court even when the behavior is dysfunctional, for instance, they futilely and tirelessly court sex mosaics with male genitalia, provided only that these mosaics emit the species-specific sex-pheromone (Nissani, 1975). Obviously, insight can only be attributed to behavioral sequences which sharply differ from such innate, inflexible courtship.

The task of ruling out trial-and-error learning is far more complicated. One classical example of such learning is Morgan's dog (1920). The iron gate leading to Morgan's house was held shut by a latch, but swung open by its own weight when the latch was lifted. Morgan's fox terrier could smoothly raise the latch with the back of his head, release the gate, and escape. But because the origins of this deceptively insightful sequence were carefully observed, it could be ascribed to "continued trial and failure, until a happy effect is reached," not to "methodical planning."

Captive hamadryas baboon can similarly learn to use tools through trial and error. When a pan with favorite treats was placed out of their reach, and an L-shaped rod was placed within reach, the baboons tried to reach the food directly but eventually gave up; occasionally directing their attention to the rod. After about 11 hr, one baboon chanced to have flipped the rod over the pan; upon retrieving the rod, the pan was brought within reach. Gradually, and without an understanding of the task, he learned to use the rod to retrieve the pan (Beck, 1986).

Both dog and baboon examples forcefully recall Hobhouse's (1915) admonition that "what appears when complete to be an act of great sagacity, may, when its genesis is carefully traced, reveal itself as the outcome of accident, or perhaps of training."

It may help to clarify the meaning of insight by relying on a fanciful exemplar. In Aesop's days, many people took it for granted that a crow could indeed solve this puzzle: dying of thirst, a crow *figured out* how to get water from the bottom of a pitcher by sequentially dropping pebbles into it.

What then are the key elements of this performance?

- The solution has neither been genetically wired nor acquired by trial and error.
- Either this kind of problem is new to the animal, or the animal came up with a new solution to an old problem.
- The correct behavioral sequence has been arrived at suddenly and completely, not in a fumbling, piecemeal fashion.

- The behavior gives every indication of having been arrived at through a representation of its constituent elements in one's mind.
- The action is carried out relatively smoothly, with all its constituent elements purposefully subserving a single goal.
- The solution is retained after a single performance (Koestler, 1964).
- The behavior is flexible, the animal able to transfer its abstract rules across a range of physical stimuli (Mackintosh et al., 1985; e.g., the crow should be able to solve the problem with different pitchers, pebbles, or liquids).

Anecdotal accounts of insight are many. Romanes (1882), for instance, relates the story of a rogue elephant chasing a few hapless people up a tree. Unable to topple the tree, and seeing a freshly-cut pile of timber nearby, the elephant removed all 36 pieces one at a time to the root of the tree, and piled them up in a regular, business-like manner. He then placed his hind feet on the pile, raised the fore part of his body, and lifted his trunk (fortunately still failing to reach the tree sitters).

Similarly, Gary Alt describes many instances of tracking a black bear in fresh snow and inexplicably coming to the end of its trail. Apparently, to give the slip to its pursuer, the bear backtracked precisely its own footsteps a short distance before leaping off the trail (cited in Shedd, 2000).

Although such stories raise the possibility of animal insight, appear to satisfy the seven criteria above, and may serve as the starting point for careful scientific studies, they cannot by themselves prove the existence of insight in animals—Morgan's dog and Thorndike's cats tell us that much. Yet, we have little more to go on than these anecdotes.

In chimpanzees, one of the earliest insight experiments was carried out by Köhler (1925). For instance, at least one of his chimpanzees could stack up three or four crates and climbed on them to obtain a banana that could not otherwise be reached. But such behaviors do not meet the smoothness criterion: years later, the conduct of his star performers still appeared "laughably inept" (Gould and Gould, 1994). It also appears that chimpanzees stack and climb crates even when no food is in sight, suggesting that, in the experimental conditions too, the behavior has been acquired unintentionally, not through a representational scheme. Moreover, when a single crate was out of sight in a readily accessible corridor, the chimpanzees found it almost impossible to rely on memory, retrieve the crate, and get the fruit (Köhler, 1925), suggesting absence of mental representation of the key elements of the problem. Yet another question mark comes from the work of Epstein and Medalie (1983) who observed similar behavior in a pigeon, again raising the possibility that Köhler's chimpanzees stumbled upon the correct solution by unthinkingly trying everything, not by grasping the basic features of the situation.

Surprisingly, the most carefully documented example of insight in animals comes from the common raven, not from the great apes. When food is tied to a string which is then suspended from a branch, many species of birds reach down, grasp the string with their bill, pull it up, hold it with their foot, release it from their bill, reach down again to grasp the string again, and repeat this sequence until they reach the food. Ravens have been shown to do this too, but in this case this behavior could involve insight (Heinrich, 1999, 2000). Ravens, it should be noted, do not draw strings in the wild, thus suggesting the absence of genetic programming (Hertz, cited in Duncker, 1945). Similarly, Heinrich raised ravens in his own aviary, and could thus rule out prior trial-and-error learning. Only some ravens solved this problem, they took sometimes hours to reach a solution, and successful ravens resorted to two different techniques of string drawing, thus again arguing against prior learning or direct genetic causation. These experiments led Heinrich to the conclusion that string drawing "was a new behavior that was acquired without any learning trials. They acted as though they had already done the trials. The simplest hypothesis is that they had—in their heads" (Heinrich, 1999, p. 319). But even in this carefully documented case, insight is just one possibility among many. While it is true, for example, that the entire sequence is extremely unlikely to have arisen by chance, its first two key elements—pulling a string and stepping on it—could very well be. Once this two-step sequence is accidentally stumbled upon, the raven is partially rewarded by the closer proximity of the food source, and, like Morgan's dog, is likely to repeat the behavior that brought about such a fortuitous outcome.

Retractable Cord-Pulling in Elephants

This and the next section describe the application of two new experimental procedures to elephants. In both procedures, the two elephants and the setting were identical to the ones described earlier in this chapter.

To apply the string-drawing paradigm to our elephants, we first had to determine whether elephants can spontaneously pull a rope when that rope is attached to a desirable object. We have seen that ravens and many other birds can do it, and we know that monkeys (Harlow and Settlage, 1934) and chimpanzees (Finch, 1941; Köhler, 1925) can pull rope too. On the other hand, rats (Tolman, 1937) and probably cats and dogs, must be specially trained to do so (Hobhouse, 1915). We subjected our two elephants to hundreds of string-pulling trials. From the very start, and regardless of the kind of string, reward, and rope, the elephants invariably pulled in the reward immediately and unhesitatingly.

We next had to adapt the string-drawing task to elephants. This was achieved by means of a stout retractable ("bungee") cord securely tied to a post on one side and to a heavy rope on the other, with the retractable cord now spanning 2 m, the stout rope spanning additional 5.5 m, and with the nonattached side of the rope placed within reach of the elephant. Along the rope, about 0.5 meter from the point of attachment to the retractable cord and 5 m from the edge of the rope on the elephant's side, a bagel or a whole corn cob was securely tied. To prevent tearing the retractable cord (an easy task for an elephant!), parallel to this cord and attached to the same post and rope, a second, longer rope was used which only allowed the retractable rope to grow in length by approximately 2 m.

In this setup, the elephants could obtain the treat by pulling the rope with their trunk, anchoring it with another part of their body, pulling it again, and repeating this sequence, if necessary. Once they reached the food, they had to go on holding the rope with mouth or foot until they could comfortably disentangle the food from the rope.

In all, 10 daily sessions, ranging from 1 to 8 min, were conducted with Winky, and 9 sessions, ranging from 2 to 15 min, with Wanda. In both cases, the sessions spanned a period of 6 weeks.

Both elephants mastered the pulling-anchoring drawing sequence, with Winky settling on her mouth as an anchor and Wanda on her feet. As in the case of Heinrich's ravens, Winky tended to remain in one place while drawing the rope, while Wanda either walked on the rope or remained in one place while using her trunk and left foot.

To our disappointment, our elephants' performance fell short of the ravens'. To begin with, preliminary and extensive rope-pulling observations showed that Wanda tended to pull on any rope (and not only retractable ones) by using both her trunk and feet, so using both trunk and foot with the retractable cord task would fail to satisfy the novelty criterion (Figure 5). As we expected, she mastered the trunk-foot sequence on the first session. Winky, on the other hand, only used her trunk to pull the rope in preliminary trials. She thus

Figure 5. Wanda drawing a bagel attached to a heavy rope which in turn is attached to a retractable rope (not seen in photo), through a coordinated use of her trunk and feet.

satisfied the novelty criterion (barring the unlikely possibility that she faced such tasks in the past) and only obtained the treat on the fourth session.

Given Winky's initial failures, we can surmise that this behavior is not genetically wired. But in both cases, the solution was not sudden, complete, or smooth, nor was it retained after just one performance. Instead, it was fumblingly acquired over time. For example, on the session immediately following the session at which she successfully used the trunk/mouth sequence, Winky, while clearly motivated, failed to obtain the food and resorted instead to a variety of unproductive strategies. Likewise, the elephants' retrieving strategy improved from session to session. Moreover, both elephants failed a variety of transferability tests. For instance, when a second rope, not tied to any treat, was attached to a post alongside the first rope in a more readily accessible location, Wanda spent at least 10 min futilely stepping on that rope and trying to draw it in. Neither elephant demonstrated flexibility—for example, when Wanda was faced with a situation where a mouth anchor would have been superior, she continued to rely on her feet.

It is worth noting perhaps that the ten or so participants in this experiment were convinced that the elephants genuinely grasped the basic elements of this problem. It was only later, while analyzing the video record, that I was led to reject their unanimous view. At the moment, retractable cord pulling in our two elephants seems to have arisen through trial-and-error learning, not through insight.

Do Elephants know when to Suck or Blow?

The next experiment is partially based on the premise that each of our two captive elephants always maximizes her food intake at the expense of her companion. Reports of altruism among wild elephants make a brief defense of this premise necessary. Yet, months of observations categorically convinced us that, when it comes to food, their behavior is self-seeking. For instance, before they enter the indoor compound for the night, their keepers distribute favorite treats such as coconuts and carrots throughout their four-room indoor enclosure. In such cases, the dominant Winky successfully takes hold of all the treats. Likewise, we have never seen them willingly offer food to each other.

The experiment itself capitalized on our two elephants' ability to obtain objects from an inflexible tube by expertly placing their trunk tightly over one of the tube's openings and either sucking the object toward them or blowing it out and retrieving it when it falls to the ground from the other side of the tube.

During the first session, a short, 5 cm in diameter, PVC tube (in Figure 6, the short tube below the elephant's left ear) was horizontally clamped to the cable which separated the elephants' outdoor quarters from the keepers' courtyard. A small piece of bagel, or a 1.4 cm^3 (4 g) white sugar cube, was placed on either side of the tube, at least 2 cm away from the edge so that it could not be directly retrieved by the trunk. One elephant and then the other served as a subject, while a keeper enticed the other away. The results for the first 10 trials are given in Table 4, suggesting a random retrieving strategy.

A few days later, both elephants were allowed to remain near the tube, with the dominant Winky standing directly in front of it, while Wanda stood some 2 m away, along the cables. In this setup, Winky could invariably obtain the treat by sucking it from either side of the tube. She could also get it by blowing it toward the side farther away from Wanda, but she could lose it by blowing it toward the other side. In 12 out of 13 trials, and in contrast to her behavior on the preceding session, she sucked the morsel from one side of the tube or the

Table 4. Initial results of tube experiments

Elephant	Sucked to obtain morsel	Blew to obtain morsel	Total
Wanda	5	5	10
Winky	4	6	10

Figure 6. A blowing contraption. A treat is placed at the open edge of the short, horizontal lower tube. If sucked, the treat is blocked by the researcher's fingers and does not reach the elephant, causing the trial to end unsuccessfully. On the other hand, blowing readily sends the morsel toward the elephant.

other. Only in the seventh trial did she blow, in Wanda's direction, lightly; the morsel thus fell close to Winky and she was able to obtain it too. This test could not be carried out with the subordinate Wanda.

Next, a researcher held a long tube horizontally, with one side directly facing the elephant and the other facing the courtyard (to which the elephant had no access). In this setup, only sucking availed them of the morsel. Both elephants adjusted immediately to this new situation, grabbing the open side available to them and tightening their grip. Winky sucked in six out of six trials and Wanda in seven out of eight.

A similar performance was repeated on the next session, with similar results: Winky sucked in all five trials and Wanda sucked in four out of five.

To overcome the dominance effect and to make competition possible, we now clamped a 1.5 m tube horizontally to the cables. The elephants were directed to stand on either side of the tube, facing the courtyard. In this case, blowing would give the treat to one's companion while sucking would land it in one's trunk. Wanda was more adept at this tusk; not once in her 24 attempts, nor Winky in her 6, did either elephant resort to blowing. In each and every case, both elephants sucked the morsel.

The elephants were now separated, and placed successively near the short horizontal tube, alone. Now, Wanda blew in two out of five trials and sucked in three, while Winky blew in three and sucked in two, again suggesting a random retrieving strategy when alone.

In the next session, the animals again stood on either side of the 1.5 m tube. And again, in over 15 min and 35 trials of rather heated (and delightful) competition, they sucked the treat.

Although it seemed at this point that both elephants understood when to blow the treats and when to suck them, we wanted to reassure ourselves that they did not adopt the correct strategy by trial and error. To that end, in the next session we presented the animals with a contraption which only rewarded blowing behavior (Figure 6).

When first presented with this contraption, both elephants resorted to both sucking and blowing. In her last 10 trials, however, Wanda only resorted to blowing and applied just enough pressure to have the treat land right by her front feet. Winky's record was not as impressive; she used a combination of sucking and blowing for trials 1–6, and only resorted to blowing on trials 7–10. Nor was she as adept as Wanda at controlling the landing location.

Having just faced a situation where only blowing was rewarded, the two subjects were now brought together to compete along the two sides of the 1.5 m tube, where only suction will do. In 15 min of steady, intense, competition, and over 40 trials, neither elephant blew once.

The next session began with the blowing contraption above (Figure 6). Again, this task proved more difficult than all the others. In trials 1–13, Wanda used a combination of sucking and blowing (of which only blowing was effective), while in trials 14–20 she only blew. In 10 trials, Winky sucked on the first and then switched to blowing-only strategy. With each elephant, these trials were immediately followed by exposure to a hand-held long tube (in which sucking yielded the reward while blowing landed it out of reach). In six consecutive trials, Wanda only sucked while Winky sucked in 13 out of 16 trials.

Although the results appear consistent with the supposition "that elephants must be credited with the ability to anticipate what will come of certain actions

(=ideation)" (Rensch, 1957), they cannot rule out prior exposure to similar tasks, sampling errors, and other explanations.

CONCLUSION

All past and present claims to the contrary, science cannot yet resolve the question of animal consciousness. For over a century, scientists believed that evolutionary theory required at least some traces of consciousness in animals. Empirical studies (e.g., on insight, language acquisition, mirror self-recognition, and deception) lent further support to this viewpoint. In retrospect, however, neither theory nor experiments proved decisive.

A similar historical pattern seems to apply to the more recent experiments of Povinelli and his colleagues. For a few years, these experiments provided strong support to the view that animals lacked a theory of mind. But the similar experiments recounted in this chapter seem to have once again restored greater uncertainty to this field. Given these recent negative results, and given moreover the picture which emerges from the field as a whole, we ought perhaps to suspend judgment at the moment and instead find new ways of tackling this old problem.

The search for insight in animals likewise confronts us with more questions than answers. As we have seen, the few controlled experiments in this area lend themselves to alternative interpretations. Likewise, our own search for insight in elephants failed to resolve this question. Like ravens and other avian species, our two elephants mastered the string-drawing task—but in a manner that appeared more consistent with trial-and-error learning than with insight. In contrast, our elephants seemed to have possessed from the start a basic grasp of the tube task; but here too, trial-and-error interpretations cannot be ruled out.

Decades of innovative experimentation with many species of animals besides the great apes may be needed before we know whether or not we are the only conscious beings on earth. One way or the other, such knowledge is sure to alter our conception of what it is like to be an animal—or a human being.

ACKNOWLEDGMENTS

This work was made possible through the generous support of Wayne State University and the Detroit Zoological Institute. I likewise thank Scott Carter, Michelle Seldon-Koch, and Ron Kagan for their interest in this project. For camaraderie, advice, cooperation, and friendship, I am deeply indebted to

Donna Hoefler-Nissani, Bettie McIntire, Maria Manuguerra-Crews, Rick Wendt, Chris O'Donnell, Erin McEntee, Erin Porth, Jennifer Goode, Kelly Wilson, Kim Van Spronsen, Marilynn Crowley, Mary Mutty, Megan Brunelle, Melanie Hiam, Patrick Smyth, and Patti Rowe. For patience and first-class editing, Gisela Kaplan and Lesley J. Rogers have my heartfelt thanks.

REFERENCES

Aelian (3rd century, BC) *On the Characteristics of Animals*, Harvard University Press, Cambridge (English translation by A. F. Scholfield, 1958).

Alexander, S., 2000, *The Astonishing Elephant*, Random House, New York.

Baron-Cohen, S., Tager-Flusberg, H., and Cohen, D. J., eds., 2000, *Understanding Other Minds*, second edition, Oxford University Press, Oxford.

Beck, B. B., 1986, Tradition and social learning in animals, in: *Animal Intelligence: Insights into the Animal Mind*, R. J. Hoage and L. Goldman, eds., Smithsonian Institution Press, Washington, DC.

Blashford-Snell, J. and Lenska, R., 1997, *Mammoth Hunt*, HarperCollins, London.

Byrne, R. W. 2001, Social and technical forms of primate intelligence, in: *Tree of Origin*, F. B. M. de Waal, ed., Harvard University Press, Cambridge, MA.

Byrne, R. W. and Whiten, A., eds., 1988, *Machiavellian Intelligence*, Clarendon Press, Oxford.

Carrington, R., 1958, *Elephants*, Chatto and Windus, London.

Carruthers, P., 1998, Natural theories of consciousness, *Eur. J. Phil.* **6**(2):203–222. <http://psyche.cs.monash.edu.au/v4/psyche-4-03-carruthers.html>.

Carruthers, P. and Smith, P. K., 1996, *Theories of Theories of Mind*, Cambridge University Press, Cambridge.

Chevalier-Skolnikoff, S. and Liska, J., 1993, Tool use by wild and captive elephants, *Anim. Behav.* **46**:209–219.

Denis, A., 1963, *On Safari*, Collins, London.

Duncker, K., 1945, On problem solving, *Psychol. Monog.* **58**:1–113.

Eltringham, K., 1982, *Elephants*, Blandford, Dorset.

Epstein, R. and Medalie, S. D., 1983, The spontaneous use of a tool by a pigeon, *Behav. Anal. Lett.* **3**:241–247.

Finch, G., 1941, The solution of patterned string problems by chimpanzees, *J. Comp. Psychol.* **32**:83–90.

Fouts, Roger., 1997, *Next of Kin*, William Morrow, New York.

Gallup, G. G., Jr., 1998, Can Animals Empathize? Yes, *Sci. Am.* http://www.sciam.com/1998/1198intelligence/1198gallup.html.

Gomez, J. C., 1998, Assessing theory of mind with nonverbal procedures: Problems with training methods and an alternative "key" procedure, *Behav. Brain Sci.* **21**: 119–120.

Gordon, J. A., 1966, Elephants do think, *Af. Wild Life* **20**:75–79.

Gould, J. and Gould, C. G., 1994, *The Animal Mind*, Scientific American Library, New York.

Griffin, D. R., 2001, *Animal Minds*, second edition, University of Chicago Press, Chicago, IL.

Groning, K. and Saller, M., 1999, *Elephants: A Cultural and Natural History*, Konemann, Cologne.

Hare, B., Call, J., Agnetta, B., and Tomasello, M., 2000, Chimpanzees know what conspecifics do and do not see, *Anim. Behav.* **59**:771–785.

Harlow, H. F. and Harlow, M. K., 1965, The affectional systems, in: *Behavior of Non-Human Primates*, A. M. Schrier, H. F. Harlow, and F. Stollnitz, eds., Academic Press, New York.

Harlow, F. and Settlage, P. H., 1934, Comparative behavior of primates. VII. Capacity of monkeys to solve patterned strings tests, *J. Comp. Psychol.* **18**:423–435.

Hart, B. L. and Hart, L. A., 1994, Fly switching by Asian elephants: Tool use to control parasites, *Anim. Behav.* **48**:35–45.

Heinrich, B., 1999,*The Mind of the Raven*, Harper Collins, New York.

Heinrich, B., 2000, Testing insight in ravens, in: *The Evolution of Cognition*, C. Heyes, and L. Huber, eds., MIT Press, Cambridge, MA.

Heyes, C. M., 1998, Theory of mind in nonhuman primates, *Behav. Brain Sci.* **21**:101–148.

Hobhouse, L. T., 1915, *Mind in Evolution*, Macmillan, London.

Hooff, J. A. R. A. M., van., 2000, Primate ethology and socioecology in the Netherlands, in: *Primate Encounters*, S. C. Strum, and L. M. Pedigan, eds., University of Chicago Press, Chicago.

Koestler, A., 1964, *Act of Creation*, Macmillan, New York.

Köhler, W., 1925, *The Mentality of Apes*, Harcourt, New York.

Linden, E., 1999, *The Parrot's Lament*, Dutton, New York.

Mackintosh, N. J., Wilson, B., and Boakes, R. A., 1985, Differences in mechanisms of intelligence among vertebrates, in: *Animal Intelligence*, L. Weiskrantz, ed., Clarendon Press, Oxford.

Macphail, E. M., 1982, *Brain and Intelligence in Vertebrates*, Clarendon Press, Oxford.

Markowitz, H., 1982, *Behavioral Enrichment in the Zoo*, Van Nostrand, New York.

McComb, K., Moss, C., Sayialel, S., and Baker, L., 2000, Unusually extensive networks of vocal recognition in African elephants, *Anim. Behav.* **59**:1103–1109.

Morgan, C. L., 1920, *Ani. Behav.* Second edition, Edward Arnold, London.

Nissani, M., 1975, A new behavioral bioassay for an analysis of sexual attraction and pheromones in insects, *J. Exp. Zool.* **192**:271–275.

Poole, J., 1997, *Elephants*, Voyageur Press, Stillwater, MN.

Povinelli, D. J., 1989, Failure to find self-recognition in Asian elephants (*Elephas maximus*) in contrast to their use of mirror cues to discover hidden food, *J. Comp. Psychol.* **103**:122–131.

Povinelli, D. J., 1998, Can Animals Empathize? Maybe not. *Sci. Am.* http://www.sciam.com/1998/1198intelligence/1198povinelli.html.

Povinelli, D. J. and Eddy, T. J., 1996, What young chimpanzees know about seeing, *Monogr. Soc. Res. Child Dev.* **61**(No. 2, Serial No. 247).

Povinelli, D. J. and O'Neill, D. K., 2000, Do chimpanzees use their gestures to instruct each other? in: *Understanding Other Minds*, Second edition, S. Baron-Cohen, H. Tager-Flusberg, and D. J. Cohen, eds., Oxford University Press, Oxford.

Povinelli, D. J. and Giambrone, S., 2001, Reasoning about beliefs: A human specialization? *Child Dev.* 72:691–695.

Povinelli, D. J., Nelson, K., and Boysen, S., 1990, Inferences about guessing and knowing by chimpanzees, *J. Comp. Psychol.* **104**:203–210.

Povinelli, D. J., Bering, J., and Giambrone, S., 2000, Towards a science of other minds: Escaping the argument by analogy, *Cogn. Sci.* **24**:167–201.

Premack, D., 1988, "Does the chimpanzee have a theory of mind?" revisited, in: *Machiavellian Intelligence*, R. W. Byrne and A. Whiten, eds., Clarendon Press, Oxford.

Reaux, J. E., Theall, L. A., and Povinelli, D. J., 1999, A longitudinal investigation of chimpanzees' understanding of visual perception, *Child Dev.* **70**:275–290.

Reiss, D. and Marino, L. 2001, Mirror self-recognition in the bottlenose dolphin: A case of cognitive convergence, *Proc. Natl. Acad. Sci. USA* **98**:5937–5942.

Rensch, B., 1956, Increase of learning capability with increase of brain-size, *Am. Nat.* **90**:81–95.

Rensch, B., 1957, The intelligence of elephants, *Sci. Am.* **196**:44–49.

Romanes, G.J., 1882, *Animal Intelligence*, K. Paul, London.

Russon, A. E., Bard, K. A., and Parker, S. T., eds., 1996, *Reaching into Thought*, Cambridge University Press, Cambridge.

Sanderson, G. P., 1912, *Thirteen Years among the Wild Beasts of India*, J. Grant, Edinburgh.

Sanderson, I. T., 1962, *The Dynasty of Abu*, Knopf, New York.

Shand, M., 1995, *Queen of the Elephants*, Jonathan Cape, London.

Shedd, W., 2000, *Owls aren't Wise and Bats aren't Blind*, Harmony, New York.

Smitha, B., Thakar, J., and Watve, M., 1999, Theory of mind and appreciation of geometry by the small green bee eater (*Merops orientalis*), *Curr. Sci.* **76**:574–577.

Spinage, C. A., 1994, *Elephants*, Poyser, London.

Stevens, V. J., 1978, Basic operant research in the zoo, in: *Behavior of Captive Wild Animals*, H. Markowitz and V. J. Stevens, eds., Nelson-Hall, Chicago, IL.

Thomas, R. K., 1986, Reasoning and language in chimpanzees, in: *Animal Intelligence: Insights into the Animal Mind*, R. J. Hoage and L. Goldman, eds., Smithsonian Institution Press, Washington, DC.

Thompson, W. R. and Melzack, R., 1956, Early environment, *Sci. Am.* **194**:38–42.

Thorpe, W. H., 1956, *Learning and Instinct in Animals*, Harvard University Press, Cambridge.

Tolman, E. C., 1937, The acquisition of string-pulling by rats—conditioned response or sign-gestalt? *Psychol. Rev.* **44**:195–211.

Tuma, R. S., 2000, Thinking like a chimp, *HMS Beagle*, Issue 90, http://news.bmn.com/hmsbeagle/90/notes/feature2-illustrator.

Uller, C. and Nichols, S., 2000, Goal attribution in chimpanzees, *Cognition* **76**:B26–B34.

Voelkl, B. and Huber, L., 2000, True imitation in marmosets, *Anim. Behav.* **60**:195–202.

Waal, F. B. M. de, ed., 2001, Introduction, *Tree of Origin*, Harvard University Press, Cambridge, MA.

Wickler, W. and Seibt, U., 1997, Aimed object-throwing by a wild African elephant in an interspecific encounter, *Ethology* **103**:365–368.

Williams, J. H., 1950, *Elephant Bill*, Doubleday, Garden City, NY.

Wynne, C., 1999, Do animals think? The case against the animal mind, *Psychol. Today* **32**(6):50–53.

Zentall, T. R., 2000, Animal Intelligence, in: *Handbook of Intelligence*, R. J. Sternberg, ed., Cambridge University Press, Cambridge.

CHAPTER EIGHT

The Use of Social Information in Chimpanzees and Dogs

Josep Call

Thorndike tested large numbers of dogs and cats [...] came to the conclusion that, so far from "reasoning," they do not even associate images with perception, as humans do, but remain limited chiefly to the experiential linking of mere "impulses" with perceptions.

However stupid a dog may seem compared to a chimpanzee, we suggest that in such simple cases as have just been described, a closer investigation would be desirable.

W. Köhler (1925)

When people are asked to rank animals with regard to their intelligence, dogs and chimpanzees are often cited as prime examples of smart animals. A couple of studies even show that not only students of an introductory psychology class but also professionals in animal learning and cognition rank dogs and chimpanzees as particularly intelligent animals (Eddy et al., 1993; Rasmussen et al., 1993). Intelligence, however, is a difficult concept to define because it has different meanings. Sometimes intelligence is used as a synonym of

Josep Call • Max Planck Institute for Evolutionary Anthropology, Leipzig, Germany.

Comparative Vertebrate Cognition, edited by Lesley J. Rogers and Gisela Kaplan. Kluwer Academic/Plenum Publishers, 2004.

creativity and flexibility. Chimpanzees are clearly intelligent by this definition. They use tools, solve problems in various ways, and they can make inferences. Others use intelligence as equivalent to learning ability (Rumbaugh and Pate, 1984). This is extended to trainability in the case of dogs. Dogs clearly excel in this aspect as their many uses and training regimes show that dogs can learn various tasks with accuracy and they can be trained to cooperate with others.

A third use of intelligence emphasizes the ability to solve social problems. This deployment of intelligence has recently received considerable research attention, particularly in non-human primates (see papers in Byrne and Whiten, 1988; Tomasello and Call, 1997; Whiten and Byrne, 1997). Social situations are particularly challenging to individuals who have to compete and cooperate with conspecifics to obtain limited resources such as food or mates. One important skill in both cooperative and competitive situations is the ability to read the behavior of conspecifics (Krebs and Dawkins, 1984). However, reading and anticipating the behavior of others does not necessarily need to be restricted to picking up certain stimuli that have been detected in the past. It can also involve the formation of knowledge that allows individuals to predict others' behavior rather than merely learning to respond to specific cues. Predictive power may be enhanced if animals can not only read certain cues in certain situations but also can anticipate the behavior of conspecifics in novel situations based on the knowledge they have of their group mates. This knowledge about the social field includes things like what others can and cannot see, or what body parts are responsible for having visual access in various species. Another aspect of social knowledge consists of individuals picking up information about the environment from their conspecifics, for instance, by following their conspecifics' gaze to resources that they cannot directly perceive. Finally, individuals can also manipulate the knowledge of conspecifics by directing the attention of their partners toward certain locations.

Until recently, the social intelligence hypothesis has been concentrated in non-human primates, particularly chimpanzees. Numerous cases of deception, social strategies, and political maneuvering have been reported in chimpanzees. However, other non-primate social species can also offer important insights into comparative social cognition. One such species is the domestic dog. A comparison between dogs and chimpanzees on the use and manipulation of social information during social interactions seems especially attractive for two reasons. First, both species share a number of features in their social behavior that makes them good candidates to develop social cognitive skills. These features include a high degree of sociality, living in permanent or semipermanent aggregations, a long life span, and the formation of strong attachments with their social companions (either conspecifics or humans).

The second reason for this comparison is that each species shares important aspects of its history with humans. On the one hand, chimpanzees are our closest living relatives with whom we share a great deal of our phylogenetic history. On the other hand, dogs are possibly our closest associates in the animal kingdom. This closeness is due not only to the process of domestication but also to the fact that dogs are commonly raised among humans. Thus, the dog–chimpanzee comparisons offer us the prospect of a non-primate perspective on social cognition while at the same time highlighting similarities and differences between species that share important aspects of our own evolutionary and developmental history.

This chapter will be structured in three major sections dealing with the use of social information during social interactions: Reading Attention, Following Attention, and Directing Attention. The order of these sections follows a rationale of passive reception of information to more a proactive role in manipulating the behavior of others. Each section will present the evidence available in each topic for each species. In the final section of this chapter, I discuss the implications of the research reviewed in the previous sections paying special attention to similarities and differences between chimpanzees and dogs and their respective ontogenies and phylogenies.

READING ATTENTION

For many animals, the mere presence of conspecifics alters their behavior dramatically. For instance, male domestic chicken preferentially vocalize about food in the presence of females but not other males (Gyger et al., 1986). This phenomenon is known as an audience effect. A more subtle (and cognitively speaking more intriguing phenomenon) is whether animals can read the attentional states of their partners. That is, whether animals respond not just to the mere presence of others, but also their attention states. Reading attention in others is a deceptively complex topic since there are several aspects that fall under this heading. One can ask about what others can or cannot see, what organs are responsible for visual perception, or how would an object look from a different perspective, or whether others are focusing attention on the same aspect (e.g., color as opposed to shape) as oneself. In this section, I will address the first two issues.

What can Others See

As the saying goes, information is power. Needless to say that this principle also applies to social-living animals since knowing what information conspecifics possess is critical both in cooperative as well as competitive situations. For instance, in order to enlist the cooperation of others by means of gestures is important that potential recipients of those gestures have visual access to the signaler of those gestures. Similarly, for subordinate animals whether dominant animals have visual access to monopolizable food sources is an important factor in deciding whether they should attempt to obtain those resources. Currently, there is some data in the literature regarding these questions in chimpanzees and to a lesser extent in dogs.

Young chimpanzees use around 30 gestures to communicate with group-mates in various contexts (Tomasello et al., 1994, 1997b). For instance, they extend their arm to beg for food, slap the ground to call attention to themselves, and put their arm around their mother's back to request being carried to a different location. These gestures fall within three basic sensory modalities: visual, auditory, and tactile. Visual gestures rely solely on visual information, auditory gestures rely mainly on sound production, and tactile gestures depend mainly on establishing physical contact with the recipient (e.g., arm on, see Figure 1). Tomasello et al. (1994, 1997b) found that chimpanzees use gestures from these three sensory modalities differentially depending on the spatial orientation of the recipient. In particular, young chimpanzees use visual gestures preferentially when the recipient is looking at them. In contrast, auditory and especially tactile gestures are used regardless of the attention state of the recipient. One possible conclusion from these data is that chimpanzees know that, if their visually based gestures are to work, others must see them.

Although there are a number of studies on communication in dogs, beginning with the classical work of Konrad Lorenz (1954), there is little information regarding the use of communicative displays in relation to the attention states of potential recipients. Nevertheless, there are some indications that may show that dogs are also sensitive to what others can and cannot see. Hare et al. and Tomasello (1998) found that dogs would communicate when a human was present, but stopped once the human had disappeared. Miklósi et al. (2000) also found that dogs reacted differently when a human was present as opposed to absent. However, these two findings may be explained invoking a very general audience effect based on the presence or absence of the intended target. More convincing are those studies in which the human is present

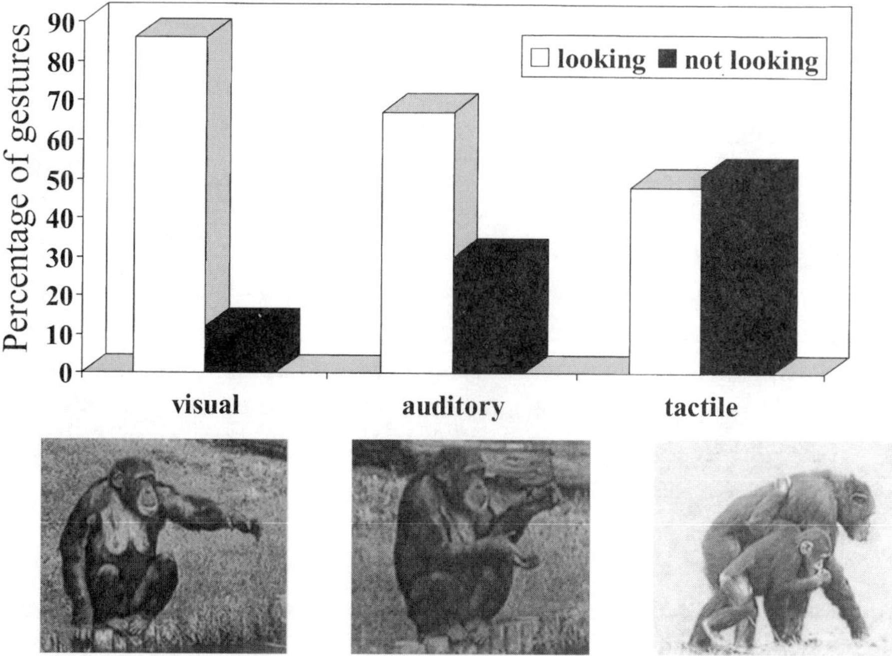

Figure 1. Percentage of visual, auditory, and tactile gestures as a function of the state of the recipient.

throughout the testing period but engages in different attention states. Hare et al. (1998) found that dogs playing a fetch game would retrieve a ball thrown by the owner, bring it back and drop it in front of the owner. In some trials this meant that dogs had to go around the owner to deposit the ball in front of him because the owner had turned his back while the dogs where fetching the ball. Although this may suggest that dogs were also sensitive to the front/back distinction like chimpanzees, another potential explanation is that dogs dropped the ball near the body part that was responsible for ball throwing, in this case, the hands. Alternatively, dogs may find the ventral human area more attractive than the dorsal area, just because the former is a more active area than the latter. As previously said, these results on dogs are still quite fragmentary and future studies will be needed to confirm these results and to reveal the mechanisms behind these findings.

In the competitive arena, chimpanzees also show great sensitivity to their conspecifics' visual perspective. Hare et al. (2000) presented pairs of subordinate-dominant chimpanzees with two pieces of food placed in a cage. A small barrier blocked the dominant's visual access to one of the pieces, whereas the subordinate was able to see both pieces. Results showed that subordinates approached and took the piece that was hidden from the dominant animal in preference to the piece the latter could see. Control tests ruled out the possibility that subordinates were using either behavioral cues from the dominant (e.g., intention movements) or contextual cues (e.g., perhaps the piece in front of the barrier looked as if it was closer). In another experiment, Hare et al. (2000) replaced the opaque barrier with a transparent barrier that did not prevent the dominant's visual access to the food behind the barrier. In this novel situation, the subordinate's preference for the "hidden" piece disappeared, seemingly because the subordinate recognized that the transparent barrier was not serving to block the dominant's visual access to the food. This situation is particularly important because it represented a novel situation for the chimpanzees who had not encountered transparent barriers before when competing with conspecifics.

The experiments previously described dealt with chimpanzees competing over food when all the relevant information was available at the time of choice. This means that subordinate chimpanzees did not have to represent the critical elements of the task because they were able to directly perceive them. For instance, in the critical condition, subordinates could see two pieces of food in two very different perceptual arrangements. One piece of food was in front of a barrier (i.e., blocked from the dominant's view) while the other piece was in the open (i.e., visually accessible to the dominant). In a sense, it could be argued that strictly speaking this task did not require representational cognition (particularly regarding the dominant's mental states) since direct perception may suffice to solve the task. In another series of experiments, Hare et al. (2001) investigated whether chimpanzees were also able to represent and use critical past information such as whether the dominant one had seen the baiting before the competition took place. Compared to the previous study, this represents a substantial qualitative shift because subjects had to remember (rather than merely directly perceive) what others had seen in the past. For these experiments two barriers and one piece of food were used, and the authors manipulated what the dominant saw. In experimental trials, dominants did not see the food hidden, or food they had seen hidden was moved to a different location when they were not watching. In contrast, dominant chimpanzees saw the food being hidden or moved in control trials. Subordinates, on the other hand, always saw the entire baiting procedure and could monitor the visual access of their dominant competitor as well. Subordinates preferentially retrieved and approached the food that dominants had not seen hidden or moved, which suggested that subordinates were sensitive to what dominants had or had not seen during baiting. In an additional experiment, Hare et al. (2001) switched the dominant chimpanzee who had witnessed the baiting for another individual that was also dominant to the subject but who had not witnessed the baiting and compared it with a situation in which the dominant was not switched. Results indicated that subjects retrieved more food when the dominant chimpanzee had been switched than when not, thus demonstrating their ability to keep track of precisely who had witnessed what. This result also ruled out the possibility that subordinates were using just the sequence of door opening and closing to decide which food to take.

What Organ is Responsible for Vision

Given the previous results on visual access in others, one question that comes to mind is whether those species also attach a special value to the eyes as the organ responsible for vision. Although this is an interesting question in its own right, it is important to emphasize that it is not the same question as the one explored in the previous section. One thing is to ask whether other species know what others can or cannot see, and a different question is whether they also know what are the morphological features that enable visual perception. Using a simple analogy, it is one thing to be able to drive a car and another to know how the car's engine works. These are two related yet independent questions. In this section, we will explore how sensitive are individuals to body orientation, or the presence of certain cues, such as the face or the eyes.

Povinelli and Eddy (1996a) tested seven 4- and 5-year-old chimpanzees' understanding of how humans must be bodily oriented for successful communication to take place (see also Reaux et al., 1999). Chimpanzees were trained to extend their hand through a hole toward one of two human experimenters to request food. In test trials, one human was capable of seeing the chimpanzees while the other was not and subjects had to chose which of these two humans to beg from by begging from the corresponding hole. Begging from the human that could not see the chimpanzee resulted in no food being given.

Povinelli and Eddy (1996a) found that chimpanzees consistently gestured toward the human who was facing toward them, as Tomasello et al. (1994) had previously described for the use of gestures. In a number of other experimental conditions, however, chimpanzees did not seem to distinguish between more subtle differences between the two humans. For example, they did not gesture differentially for a human who wore a blindfold over his eyes (as opposed to one who wore a blindfold over his mouth), or for one who wore a bucket over his head (as opposed to one who held a bucket on his shoulder), or for one who had his eyes closed (as opposed to one who had his eyes open), or for one who was looking away (as opposed to looking at the subject), or for one whose back was turned but who looked over his shoulder to the subject (as opposed to one whose back was turned and was looking away). The authors concluded that although these young chimpanzees could learn to exploit one or more cues that signaled conditions conducive to communication with humans (e.g., bodily orientation facing toward versus away from the subject), overall they did not seem to understand precisely how visual perception works, especially the role of the eyes (see also Chapter 7 by Nissani).

These negative results contrast with other studies that suggest that chimpanzees may be sensitive to the role of the eyes during communicative interactions. Gómez et al. (cited in Gómez, 1996) found that six chimpanzees requested the attention of a human by touching her when she had her back turned, her eyes were closed, or was looking in a different direction. A comparison between these three conditions with one in which the experimenter was staring at the subject with her eyes open, indicated that the eyes closed condition was the hardest one and looking in a different direction was the easiest one. Similarly, Krause and Fouts (1997) and C. R. Menzel (1999) reported that three language-trained chimpanzees obtained the attention of their human partners before using visual gestures such as pointing. Call and Tomasello (1994) found that one of two orangutans pointed more often to request food when the experimenter who was facing him had open eyes compared to closed eyes, or had his or her back turned. As in the case of the chimpanzees, eyes closed condition was harder than the back turned condition. Interestingly, those subjects who discriminated better between the attentive and non-attentive conditions had received more human contact during their upbringing.

A recent study has shown that dogs are sensitive to the role of the eyes in visual perception. Call et al. (2003) found that dogs were sensitive to specific cues of attention. Dogs were given a series of trials in which they were always forbidden to take a piece of visible food. In some trials the human continued to look at them throughout the trial (control condition), whereas in others the human either (a) left the room, (b) turned her back, (c) engaged in a activity and paid no attention to the dog, or (d) closed her eyes. Dogs behaved in clearly different ways in all the conditions in which the human did not watch them, as opposed to the control condition in which she did. In particular, when the human looked at them dogs retrieved less food, and there were no significant differences among those conditions in which the human did not have access to the dog. Moreover, in those trials in which dogs took the food, they approached it using slight detours rather than in a straight line. Finally, dogs also sat (as opposed to laid down) more often when the human could see them compared to those conditions in which she was not visually attending to the dog. This shows that dogs are acutely sensitive to the attention states of humans, including sensitivity to the eyes.

In summary, chimpanzees know more about seeing than previously thought. They adjust to gestures and compete appropriately depending on what others can and cannot see. The evidence regarding the precise role of the eyes in visual perception is more mixed. While some studies suggest that chimpanzees do not take into account the role of the eyes, other studies offer more positive results. The abilities of dogs in this topic are much more fragmentary than those found for chimpanzees. It is unclear whether dogs know what others can and cannot see because some obvious alternatives have not been ruled out yet. Nevertheless, there is one study that shows that dogs are sensitive to the various attention states of humans, including sensitivity to the eyes. However, additional studies are needed to assess the cognitive mechanisms responsible for this sensitivity.

FOLLOWING ATTENTION

We have seen that reading attention and, particularly, having an appreciation about what others can and cannot see, can be a useful skill to make appropriate foraging and social decisions. In a sense, reading attention permits individuals to adjust their behavior according to the partner's attentional state in order to obtain a goal. On the other hand, following attention, for instance to certain locations that the individual cannot directly perceive, allows the individual to exploit others' knowledge regarding the location of those resources. For instance, individuals may use others' gaze orientation to find hidden food or to discover predators. Thus, whereas reading attention involves having some information about the location of a resource (i.e., hidden food) and choosing when to get to it depending on the partner's attentional state, attention following involves obtaining information about the precise location of a previously unknown resource. Research on attention following has revolved around two main paradigms: following attention into distant space and following attention in the object choice situation.

Attention Following into Distant Space

Chimpanzees follow the gaze of conspecifics and humans (Call et al., 1998; Itakura, 1996; Povinelli and Eddy, 1996b; Tomasello et al., 1998). Additional studies showed that gaze following could not simply be explained as a low level mechanism of the type "turn in the direction in which others are oriented and then search until you find something interesting" (Povinelli and Eddy, 1996b). In particular, Tomasello et al. (1999) showed that chimpanzees followed gaze around barriers of different types by moving to new locations where they could see what was behind the barriers—seemingly to see what the experimenter was looking at. More importantly, Tomasello et al. (1999) found that chimpanzees who saw a human experimenter looking above and behind them ignored a distractor object that they found when they turned and continued to track the human's gaze to the back of the cage. Also, Call et al. (1998) found that upon following human gaze to the ceiling and not finding any interesting sight (just the cage ceiling), chimpanzees would look back at the experimenter presumably to ascertain if they were looking to the same location. These results are incompatible with a low-level explanation and suggest that chimpanzees may understand that the informants are looking at something specific in a particular location.

There is only one study that investigated gaze following into distant space in dogs. Agnetta et al. (2000) sat in front of the dog and once it was looking at the experimenter, he looked toward one of three different locations (up, left, and right). They also included a control condition in which the experimenter stared at the dog. The authors videotaped and scored the dog's looking responses to each of the three locations. Dogs failed to follow the experimenter's gaze toward the designated locations. The authors speculated that dogs might not have followed the direction of the experimenter's gaze because, in those situations in which they cannot directly perceive a potential target object, they may rely on auditory cues rather than visual ones.

Attention Following in Object Choice

The numerous positive results on attention following into distant space in chimpanzees contrasts with the mixed results to use spontaneously gaze (head) direction from either conspecifics or humans to locate hidden food in the so-called object choice paradigm (Anderson et al., 1995). In a typical situation, subjects are presented with two opaque containers, only one of which contains food. A human experimenter then looks at the baited container and after a variable delay, allows the subject to select one of the containers. With the exception of one chimpanzee (Itakura and Tanaka, 1998), the only positive results in this paradigm have come with several human raised chimpanzees in various studies (Call et al., 2000; Itakura and Tanaka, 1998). Several other studies have shown a very inconsistent ability to spontaneously use the gaze direction of humans to help them to locate hidden food (Call and Tomasello, 1998; Call et al., 1998, 2000). This failure to use this gaze consistently to find hidden food contrasts with the ability to follow gaze to distant locations. This discrepancy seems particularly puzzling because some of the chimpanzees that failed this test were the same ones that were able to follow gaze to distant locations (e.g., Call et al., 1998). Itakura et al. (1999) used a trained chimpanzee conspecific to give the gaze direction cue, but still found mostly negative results. Moreover, failure to use gaze cues extends into other more overt cues such as pointing gestures, touching the stimuli, or placing a marker on top of the baited container (Call et al., 2000; Tomasello et al., 1997a).

Despite this lack of spontaneous use of gaze direction and other cues, several studies have shown that several species can learn to use cues such as head direction and gaze after multiple trials (Itakura and Anderson, 1996; Vick and Anderson, 2000). However, additional tests have shown that it is likely that their performance was based on learning a cue rather than having an appreciation of the visual experience of others. For example, when the experimenter turned his head in the correct direction of the baited container, but looked to the ceiling and not at the baited container, chimpanzees chose the correct container just as often as if the experimenter looked directly at it (Povinelli and Eddy, 1997). In contrast, Soproni et al. (2001) found that dogs, like children, did not take the "looking-at-the-ceiling" as a cue but chose randomly between the two containers.

Other studies have investigated the conditions that help subjects improve their performance. For instance, Call et al. (2000) found that vocalizations (and other noises) and various behavioral cues such as touching, approaching, or lifting and looking under the correct container facilitated the performance of a minority of chimpanzees. Tomasello et al. (1997a) also found that chimpanzees do not transfer from one cue to another. Thus, after chimpanzees learned one cue, for instance, a marker indicating food, they were unable to use another cue such as pointing to solve the same problem.

In contrast to the difficulties shown by chimpanzees in the object choice paradigm, several studies have shown that dogs are capable of spontaneously using a variety of cues in object choice situations (Hare and Tomasello, 1999; Hare et al., 1998, 2002; McKinley and Sambrook, 2000; Miklósi et al., 1998). All these studies presented dogs with two or three opaque containers, only one of which contains food. A human experimenter situated in front of the dog indicates the baited container and after a variable delay, allows the subject to select one of the containers. Results have shown that dogs spontaneously use head direction of a human or a dog (Hare and Tomasello, 1999) to locate food. Furthermore, there is some evidence that dogs can also use glancing (without head direction) spontaneously (McKinley and Sambrook, 2000) or after some training (Miklósi et al., 1998), although some studies did not find an effect (Hare et al., 1998). McKinley and Sambrook (2000) argued that this difference across studies may reflect the different position of the experimenter across studies. Whereas in the McKinley and Sambrook (2000) study the experimenter sat on the floor at eye level with the dog, Hare et al. (1998) administered the cue while the experimenter was standing.

Two other cues have been tested that do not involve the experimenter's head or eyes as sources of information: pointing and marker. Dogs spontaneously use human pointing, that is, an extension of the arm and index finger in the direction of the baited cup, to locate hidden food (Hare et al., 1998, 2002; McKinley and Sambrook, 2000; Miklósi et al., 1998). Clearly, the pointing cue is the most robust of all those that have been tested across studies. Moreover, the use of pointing is not restricted to a specific pointing configuration, on the contrary, dogs can use different types of pointing such as static or distal (Hare et al., 1998; McKinley and Sambrook, 2000) pointing. Static pointing involves the presentation of an extended arm and index finger without any arm or finger movement. In other words, the dog is allowed to observe the cue only after it has reached a certain position. Distal pointing consists of the experimenter moving away from the correct container and then pointing to it so that the hand is closer to the empty container but is directed toward the baited one. McKinley and Sambrook (2000) found positive evidence of distal pointing use, albeit its effectivity was reduced compared to non-distal pointing. Pointing and head direction have also been used to compare different breeds of dogs or canid species. McKinley and Sambrook (2000) found that trained gundogs were better at pointing comprehension than pet gundogs, although this difference did not exist with other cues such as head direction or glancing. Also, there were no differences between pet gundogs and pet non-gundogs. Finally, two studies comparing dogs with wolves reported that dogs clearly outperformed wolves (which failed to get above chance) in point and gaze or in gaze alone (Agnetta et al., 2000; Hare et al., 2002).

Another cue that has received some attention is the use of a marker to indicate the location of food. In this situation, the experimenter hides food in one of two boxes and informs the subject about the correct box by placing a marker in front of it. Agnetta et al. (2000) found that dogs used the marker effectively if they had seen the experimenter place the marker, even when a barrier (at shoulder's height) occluded the human's actual placing of the marker. In contrast, dogs failed to use the marker if they did not see the human placing it, but just found the marker by one of the boxes. This suggests that dogs needed to see the experimenter's action rather than the position of the marker alone to succeed in this test. This interpretation is reinforced with Hare and Tomasello (1999) findings in which dogs were capable of using the approach toward one of the containers as a cue to find the food location. Recently, Hare et al. (2002) also found that dogs outperformed chimpanzees in the use of the marker cue to find food.

In summary, the results on attention following present a double paradox. On the one hand, chimpanzees are good at following gaze into distant space or around barriers, but they have only shown an inconsistent ability to use those same cues to locate hidden food in an object choice situation. On the other hand, dogs present the reverse situation. They fail to follow human gaze into distant space (although this is based on a single study), but they can use a variety of visual cues to locate food in the object choice situation.

DIRECTING ATTENTION

The previous two sections have dealt with gathering and extracting social knowledge about what conspecifics can or cannot see. In both situations, the subject can be perceived as a passive entity who either reads or extracts some knowledge from her social partners. In this section, we will observe the individual taking on a more active role by directing her partner's attention toward particular aspects of the ecological or social environment.

As noted previously, chimpanzees have a rich repertoire of gestures. They poke their group mates to initiate play, place their hand palm up under another's chin to beg for food, put their arm around their mother's back to request being carried to a different location, or slap the ground to call attention to themselves. The vast majority of these gestures (if not all) are intentional requests to obtain something, be it food, play, or reassurance. It is much harder to find gestures in which chimpanzees intentionally direct others' attention to outside entities other than their own bodies. This lack of attention directing gestures such as pointing is particularly striking given that chimpanzees and humans share other gestures such as kissing, begging, or bowing (de Waal, 1988; Tomasello et al., 1985, 1994, 1997b).

Some authors argue that the lack of pointing with an extended arm and finger is actually not surprising at all because chimpanzees point with their body orientation (E. W. Menzel, 1974). In other words, chimpanzees do not display pointing (with an extended arm) because they already have other ways to point. Moreover, de Waal and van Hooff (1981) observed that during recruitment episodes following an aggressive incident, chimpanzees use "side-directed" behavior to indicate enemies to their potential allies. They do this by using an extended arm toward the aggressors. Although this behavior could be construed as pointing at the enemies to inform potential allies, it is also true that chimpanzees use an extended arm toward their potential allies. Therefore, it is not clear whether pointing with the body or side-directed behavior constitute genuine cases of intentionally directing attention as opposed to behaviors that another can read but that the sender does not intend to send as a referential act.

Despite the lack of pointing gestures among chimpanzees, it is clear that chimpanzees spontaneously develop this gesture when they interact with humans. Previously, Call and Tomasello (1996) had argued that only enculturated apes, that is, those raised by humans, in most cases as part of a language-training program, developed this gesture. Indeed, there are numerous examples of apes, not just chimpanzees, acquiring this gesture in the company of humans (Savage-Rumbaugh et al., 1986; Krause and Fouts, 1997; Miles, 1990; Patterson, 1978). For instance, Krause and Fouts (1997) systematically studied the pointing of two language-trained chimpanzees and found that they were capable of pointing very accurately to various targets. In this study, humans could reliably locate the position of rewards on the basis of the information provided by the chimpanzee gestures. In contrast, pointing behavior has not been documented either in the wild or in groups of captive chimpanzees interacting with other chimpanzees that have not received special training (Goodall, 1986; Tomasello and Call, 1997).

Recently, however, Leavens and Hopkins (1998, see also Leavens et al., 1996) have shown that even non-enculturated chimpanzees can develop genuine pointing gestures to direct the attention of humans. These authors found that 40 percent of the chimpanzees they studied pointed with the whole hand (with 5 percent using an index finger extension) to indicate the location of food to humans. These pointing gestures were accompanied with gaze alternation between the human and the food, and they persisted in using this gesture until they received the food. These features suggest that pointing gestures were both intentional and referential. These gestures were observed only when chimpanzees interacted with humans, not with other chimpanzees. Since captive chimpanzees (both mother- and human-reared) interact with humans extensively, and humans are especially attuned to the pointing gesture, it is not surprising that chimpanzees can quickly capitalize on this and develop pointing gestures. Therefore, the current evidence indicates that chimpanzees can develop the pointing gesture to direct the attention of humans to locations or objects that they desire regardless of their rearing history.

The well-documented ability of chimpanzees to engage in intentional communication, including attention direction gestures like pointing when they interact with humans, contrasts with the paucity of data in dog intentional communication. As indicated previously, there are a number of studies that have investigated social communication in dogs and canids more generally (Lorenz, 1954; Mitchell and Thompson, 1986). Yet, there are only two studies that have investigated the ability of dogs to inform naïve humans about the location of hidden food. Hare et al. (1998) hid a food reward in one of three boxes in the presence of the dog. A second experimenter (naïve regarding the exact food location) appeared after a delay and attempted to find the food for the dog. The dog barked and directed the human (mostly by leading him) toward the correct location in 80 percent of the trials. Leading behavior was accompanied by gaze alternation between the human and the hidden reward and by barking.

Miklósi et al. (2000) also found that dogs were capable of informing a naïve human about the location of food. These authors presented dogs with three experimental conditions. In one condition, dogs were placed in a room accompanied by their owners without any food hidden in the room. In a second condition, dogs were left alone in the room with a piece of hidden food. In a third condition, dogs remained in the room with their owner and a hidden piece of food. In those conditions with food present in the room, the owner was naïve regarding the exact food location whereas dogs invariably knew the food location but needed the owner's help to get the food since the food was out of the dog's reach. Miklósi et al. (2000) found that dogs looked more often at the owner when food was present than when there was no food in the room, and looked more at the food when the owner was present than when the owner was absent. Moreover, when both the owner and the food were in the room, dogs alternated between the food and the owner, and vocalized when either looking at the food or the owner.

These authors as well as Hare et al. (1998) interpreted these findings as evidence of intentional communication in the dog. However, it is unclear whether the dog was intentionally communicating or the owner was able to read the behavior of the dog. Partly the difficulty in the interpretation lies in the lack of a ritualized attention directing gesture such as pointing in the great apes. This is not to say that dogs lack intentional communication, even intentional communication aimed at directing others' attention. It is simply that without an overt behavior it is harder to prove. One possibility to resolve this issue would be to study the pointing behavior of some breeds of pointer dogs to assess the degree of intentionality and referentiality involved in this behavior.

In summary, chimpanzees can direct the attention of humans toward hidden locations. Pointing is the most prominent behavior used to achieve this end. A detailed analysis of pointing behavior suggests that apes use this gesture intentionally and referentially to designate distal targets. Dogs can also direct humans toward certain locations. However, the lack of a "pointing" gesture makes it harder to evaluate to what extent dog behavior constitutes a case of intentional and referential communication.

DISCUSSION

Chimpanzees have some knowledge about each of the three aspects of attention that have been explored. They can read the attention states of others so that they can adjust their gestures and compete appropriately depending on what others can and cannot see, even though it is still unclear whether chimpanzees take into account the precise role of the eyes in these situations. Chimpanzees can also spontaneously follow the gaze of others to distant locations but fail to use gaze and other overt communicative cues to find hidden food. Nevertheless, chimpanzees can direct the attention of humans to specific locations by means of pointing gestures that are referential and intentional. The evidence for dogs in this area is more fragmentary. Although there is little information on attention reading (other than general audience effects), one study has shown that dogs take into account the precise role of the eyes when confronted with humans. On the other hand, it is well established that dogs can flexibly use gaze following and other cues to locate hidden food, but fail to follow gaze into distant space. Dogs can also direct humans to certain locations, but it is not totally clear that these behaviors constitute cases of intentional and referential communication.

One thing that becomes apparent is that additional studies along the lines similar to those described in chimpanzees are clearly needed in dogs (and other animals). Despite the recent interest on social cognition in dogs, there are still many gaps in our knowledge. Although dogs show some interesting results in reading and directing attention, they are still insufficient to reach solid conclusions. Tentatively, we can say that dogs may show skills comparable to those found in chimpanzees but additional studies are clearly needed to confirm these initial results. The domain of following attention, particularly following attention in object choice situations, does afford more solid conclusions and a more meaningful (and intriguing) comparison between chimpanzees and dogs. Whereas dogs can use multiple signs to locate hidden food, chimpanzees' performance in this task is weak. Dogs clearly outperform chimpanzees in this domain. This intriguing difference between species highlights the importance to investigate further the mechanisms underlying the performance of dog (and chimpanzee) in each of these domains of attention.

Despite its importance, the mechanisms implicated in behavioral prediction are currently poorly understood. In general, two classes of mechanisms have been proposed. On the one hand, there is the cue-based approach in which individuals learn (through experience) to respond in certain ways to particular situations with little understanding of the phenomena, and a limited ability to solve novel problems without re-learning each problem. On the other hand, there is the knowledge-based approach in which animals not only learn to associate some stimuli with certain responses, but also extract knowledge from those experiences and have some understanding of the phenomena. In the current context, it is conceivable that dogs are merely good at using human cues to detect food, whereas chimpanzees are not. It is possible that dogs' performance could be reduced to a simple mechanism of cue detection. However, there are two arguments against this position. First, dogs respond to a variety of behaviors including pointing, bowing, approaching, looking, glancing, and touching, and placing a marker—not just one. It is unlikely that dogs have been selected to automatically pick up each of these cues. Second, and perhaps more importantly, dogs, unlike chimpanzees, do not respond to cues that do not designate a target. Soproni et al. (2000) found that dogs, like children, used gaze to select a container, when the experimenter was looking at the container, but refrained from selecting it when the experimenter was looking above it. In contrast, chimpanzees selected the container both when the human was looking at it and above it (Povinelli and Eddy, 1997).

On the other hand, this study highlights that chimpanzees are not so good at using any kind of cues in any given situation. One possibility is that chimpanzees only use those cues when there is a causal (logical) connection between the cue and the reward. Call (2003) presented two containers to chimpanzees (and other great apes) one of which is baited. Both containers are then shaken so that the baited one makes a rattling noise. Results showed that apes readily use the noise made by food inside to select the baited container. In a follow-up experiment, only the empty container was shaken whereas the baited one was lifted so there was no noise emanating from either container. Again, apes selected the baited container above chance, even in this case in which there was no auditory cue that could guide their choices (control tests also ruled out the possibility of inadvertent cueing). In contrast, apes failed to use a tapping noise on the baited container to locate the food. A possible interpretation of this finding is that chimpanzees can use those cues that bear a causal (logical) connection with the reward but not those in which the connection between the cue and the reward is arbitrary. Needless to say that with enough trials, chimpanzees, like many other animals, would learn to associate the arbitrary cue with the reward, but it is also true that a different mechanism such as discriminative learning as opposed to inferential reasoning may explain a species performance after some practice. We are currently exploring the possibility that dogs can also use causal cues as well as arbitrary ones.

Whether dogs are just excellent cue readers as opposed to knowledge builders is a question that will have for the moment to remain unanswered. We know that chimpanzees can do more than cue reading (see Call 2001; Suddendorf and Whiten, 2001, for recent reviews). An intriguing possibility is that chimpanzees' ability to build knowledge hinders their ability to quickly pick up cues that do not have a causal relation with its consequences. This may explain why chimpanzees do so poorly in the object choice situation (or simple discrimination problems) when the cues are arbitrary. Having said that chimpanzees generally perform poorly in object choice, we have to note an exception to this poor performance. And these are the apes that have been raised by humans. Call and Tomasello (1996) argued that those effects are seen in social cognition such as social learning and gestural communication. Additional research has shown that enculturated chimpanzees are also good at reading the role of the eyes (Gómez, 1996) and using cues to find hidden food in object choice situations (Call et al., 2000; Itakura and Tanaka, 1998). The role of ontogeny seems, therefore, very important to develop those skills. Yet, the precise changes that occur during ontogeny remain undetermined. One possibility is that chimpanzees learn that humans communicate in certain ways, for instance, eye contact and pointing are important pieces of the human communicative repertoire. Alternatively, the changes may be more general in nature, for instance, enculturated chimpanzees are more attached to humans and therefore pay special attention to them.

The role of ontogeny in dogs is also largely unexplored although it may be less modifiable than that of the apes. Agnetta et al. (2000) found that puppies with minimum contact with humans were already quite proficient at using cues. Also Hare et al. (2002) found that mother-reared and human-reared puppies did not differ in their ability to use such cues. Similarly, although McKinley and Sambrook (2000) found that training improved performance for pointing comprehension in gundogs compared to pets (both gundogs and non-gundogs), other cues remained unaffected and trained gundogs were indistinguishable from other non-trained dogs. Interestingly, two studies found that wolves did poorly in the object choice situation (Agnetta et al., 2000; Hare et al., 2002). So whereas dogs and humans are good at the object choice, chimpanzees and wolves are not so skilful, but chimpanzees improve when they are raised by humans. It is an open question how wolves raised by humans would perform in object choice situations and other tasks probing the sensitivity to attention states in others.

Another aspect of interest is the evolution of reading, following, and directing attention in chimpanzees and dogs. It is very likely that some of the skills described in this chapter correspond to a suite of skills common to many mammals, particularly, those mammals that are social, long-lived, and form relations. Among them we should find the prime candidates for advanced social cognition of the sort discussed in this chapter. In addition, dogs are very skilful at detecting and using cues from humans to find hidden food. It is possible that during domestication, humans have selected dogs that respond appropriately in communicative situations, which would certainly involve being sensitive to the various aspects of the attention in others covered in this chapter. This would explain why dogs are such good readers of human behavior. However, this skill does not constitute a unique canine trait because, as indicated previously, chimpanzees raised with humans are also proficient behavior readers in those situations. Moreover, other species such as dolphins (Herman et al., 1999; Tschudin et al., 2001) and seals (Scheumann and Call, 2002) are also capable of spontaneously (without specific training) using cues such as head direction and pointing in object choice situations. Nevertheless, it should be noted that those marine mammals had received extensive contact with humans and training in other tasks. Consequently, it is very likely that they learned to interpret certain signals during past interactions with humans. Although we have indicated that most dogs lack a ritualized signal such as index pointing, it is true that some breeds do indeed display such ritualized signals consisting of freezing and pointing with the whole body. Again, this is a trait that has been selected for its usefulness in cooperative hunting with humans. Yet, the cognitive mechanisms underlying this behavior remain totally unexplored.

In conclusion, we started this chapter by indicating some peculiarities such as creativity or trainability that are thought to be part of the cognitive repertoire of chimpanzees and dogs, respectively. We also indicated that both were good candidates for social cognition and found both commonalties and specializations in each species. Recent research on dogs' sensitivity and manipulation of attention in others has produced particularly intriguing results. Future studies should be aimed at deepening our knowledge about the mechanisms implicated in the performance that we have described for dogs (and chimpanzees). As Köhler suggested, it is important that no matter how simple a behavior may appear, its mechanisms are closely scrutinized. Quite often results as those reported in this chapter tend to be explained by referring to relatively simple mechanisms such as conditioning, but no research is conducted to ascertain this possibility. For instance, dogs or chimpanzees may have merely learned this or that behavior as opposed to having some understanding about it. Since such a result could be obtained with classical conditioning, it is often assumed that classical conditioning is the most likely explanation. The fact that classical conditioning was initially discovered in dogs does not help the case of dogs as more-than-simple-learning organisms. Although the use of Morgan's cannon is a practice that should be encouraged, it is also important to realize that an excessive reliance on such practice without fresh data as a counterweight may lead to gross oversimplifications of the phenomena under scrutiny. In other words, the systematic invocation of simple explanations, despite their appeal for simplicity, without empirical verification may perpetuate the thinking that animals like dogs are simple whereas others like chimpanzees (or humans) are complex.

REFERENCES

Agnetta, B., Hare, B., and Tomasello, M., 2000, Cues to food location that domestic dogs (*Canis familiaris*) of different ages do and do not use, *Anim. Cogn.* **3**:107–112.

Anderson, J. R., Sallaberry, P., and Barbier, H., 1995, Use of experimenter-given cues during object-choice tasks by capuchin monkeys, *Anim. Behav.* **49**:201–208.

Byrne, R. W. and Whiten, A., 1988, *Machiavellian Intelligence. Social Expertise and the Evolution of Intellect in Monkeys, Apes, and Humans*, Oxford University Press, Oxford.

Call, J., 2001, Chimpanzee social cognition, *Trends Cogn. Sci.* **5**:369–405.

Call, J., 2003, Inferences about the location of food in the great apes, *J. Comp. Psych.* (in press).

Call, J., Agnetta, B., and Tomasello, M., 2000, Social cues that chimpanzees do and do not use to find hidden objects, *Anim. Cogn.* **3**:23–34.

Call, J., Bräuer, J., Kaminski, J., and Tomasello, M., 2003, Domestic dogs are sensitive to the attentional state of humans, *J. Comp. Psych.* **117**:257–263.

Call, J., Hare, B. H., and Tomasello, M., 1998, Chimpanzee gaze following in an object-choice task, *Anim. Cogn.* **1**:89–99.

Call, J. and Tomasello, M., 1994, Production and comprehension of referential pointing by orangutans (*Pongo pygmaeus*), *J. Comp. Psych.* **108**:307–317.

Call, J. and Tomasello, M., 1996, The effect of humans on the cognitive development of apes, in: *Reaching into Thought*, A. E. Russon, K. A. Bard, and S. T. Parker, eds., Cambridge University Press, New York, pp. 371–403.

Call, J. and Tomasello, M., 1998, Distinguishing intentional from accidental actions in orangutans (*Pongo pygmaeus*), chimpanzees (*Pan troglodytes*), and human children (*Homo sapiens*), *J. Comp. Psych.* **112**:192–206.

de Waal, F. B. M., 1988, The communicative repertoire of captive bonobos (*Pan paniscus*), compared to that of chimpanzees, *Behaviour* **106**:3–4.

de Waal, F. B. M., and van Hooff, J. A. R. A. M., 1981, Side-directed communication and agonistic interactions in chimpanzees, *Behaviour* 77:164–198.

Eddy, T. J., Gallup G. G., Jr., and Povinelli, D. J., 1993, Attribution of cognitive states to animals: Anthropomorphism in comparative perspective, *J. Soc. Issues* **49**:87–101.

Gómez, J. C., 1996, Non-human primate theories of (non-human primate) minds: Some issues concerning the origins of mind-reading, in: *Theories of Theories of Mind*, P. Carruthers and P. K. Smith, eds., Cambridge University Press, Cambridge, pp. 330–343.

Goodall, J., 1986, *The Chimpanzees of Gombe. Patterns of Behavior*, Belknap Press of Harvard University Press, Cambridge, MA.

Gyger, M., Karakashian, S. J., and Marler, P., 1986, Avian alarm calling: Is there an audience effect? *Anim. Behav.* **34**:1570–1572.

Hare, B., Brown, M., Williamson, C., and Tomasello, M., 2002, The domestication of social cognition in dogs, *Science* **298**:1634–1636.

Hare, B., Call, J., Agnetta, B., and Tomasello, M., 2000, Chimpanzees know what conspecifics do and do not see, *Anim. Behav.* **59**:771–785.

Hare, B. H., Call, J., and Tomasello, M., 1998, Communication of food location between human and dog (*Canis familiaris*), *Evol. Comm.* **2**:137–159.

Hare, B., Call, J., and Tomasello, M., 2001, Do chimpanzees know what conspecifics know and do not know? *Anim. Behav.* **61**:139–151.

Hare, B. and Tomasello, M., 1999, Domestic dogs (*Canis familiaris*) use human and conspecific social cues to locate hidden food, *J. Comp. Psych.* **113**:173–177.

Herman, L. M., Abichandani, S. L., Elhajj, A. N., Herman, E. Y. K., Sanchez, J. L. et al., 1999, Dolphins (*Tursiops truncatus*) comprehend the referential character of the human pointing gesture, *J. Comp. Psych.* **113**:347–364.

Itakura, S., 1996, An exploratory study of gaze-monitoring in nonhuman primates, *Jpn. Psych. Res.* **38**:174–180.

Itakura, S., Agnetta, B., Hare, B., and Tomasello, M., 1999, Chimpanzees use human and conspecific social cues to locate hidden food, *Dev. Sci.* **2**:448–456.

Itakura, S. and Anderson, J. R., 1996, Learning to use experimenter-given cues during an object-choice task by a capuchin monkey, *Curr. Psych. Cogn.* **15**:103–112.

Itakura, S., and Tanaka, M., 1998, Use of experimenter-given cues during object-choice tasks by chimpanzees (*Pan troglodytes*), an orangutan (*Pongo pygmaeus*), and human infants (*Homo sapiens*), *J. Comp. Psych.* **112**:119–126.

Köhler, W., 1925, *The Mentality of Apes*, Vintage Books, New York.

Krause, M. A. and Fouts, R. S., 1997, Chimpanzee (*Pan troglodytes*) pointing: Hand shapes, accuracy, and the role of eye gaze, *J. Comp. Psych.* **111**:330–336.

Krebs, J. R. and Dawkins, R., 1984, Animal signals: Mindreading and manipulation, in: *Behavioural Ecology: An Evolutionary Approach*, J. R. Krebs, and N. B. Davies, eds., Blackwell, Oxford, pp. 380–402.

Leavens, D. A. and Hopkins, W. D., 1998, Intentional communication by chimpanzees: A cross-sectional study of the use of referential gestures, *Dev. Psych.* **34**:813–822.

Leavens, D. A., Hopkins, W. D., and Bard, K. A., 1996, Indexical and referential pointing in chimpanzees (*Pan troglodytes*), *J. Comp. Psych.* **110**:346–353.

Lorenz, K., 1954, *Man meets Dog*, Penguin Books, New York.

McKinley, J. and Sambrook, T. D., 2000, Use of human-given cues by domestic dogs (*Canis familiaris*) and horses (*Equus caballus*), *Anim. Cogn.* **3**:13–22.

Menzel, C. R., 1999, Unprompted recall and reporting of hidden objects by a chimpanzee (*Pan troglodytes*) after extended delays, *J. Comp. Psych.* **113**:426–434.

Menzel, E. W. Jr., 1974, A group of young chimpanzees in a one-acre field: Leadership and communication, in: *Behavior of Nonhuman Primates*, A. M. Schrier and F. Stollnitz, eds., Academic Press, New York, pp. 83–153.

Miklósi, A., Polgárdi, R., Topál, J., and Csányi, V., 1998, Use of experimenter-given cues in dogs, *Anim. Cogn.* **1**:113–121.

Miklósi, A., Polgárdi, R., Topál, J., and Csányi, V., 2000, Intentional behavior in dog–human communication: An experimental analysis of "showing" behaviour in the dog, *Anim. Cogn.* **3**:159–166.

Miles, H. L. W., 1990, The cognitive foundations for reference in a signing orangutan, in: *"Language" and Intelligence in Monkeys and Apes*, S. T. Parker and K. R. Gibson, eds., Cambridge University Press, Cambridge, pp. 511–539.

Mitchell, R. W. and Thompson, N. S., 1986, Deception in play between dogs and people, in: *Deception. Perspectives on Human and Nonhuman Deceit*, R. W. Mitchell and N. S. Thompson, eds., SUNY, New York, pp. 193–204.

Patterson, F., 1978, Linguistic capabilities of a lowland gorilla, in: *Sign Language and Language Acquisition in Man and Ape*, F. C. C. Peng, ed., Westview Press, Boulder, CO, pp. 161–201.

Povinelli, D. J. and Eddy, T. J., 1996a, What young chimpanzees know about seeing, *Monogr. Soc. Res. Child Dev.* **61**(3):1–152.

Povinelli, D. J. and Eddy, T. J., 1996b, Chimpanzees: Joint visual attention, *Psych. Sci.* 7:129–135.

Povinelli, D. J. and Eddy, T. J., 1997, Specificity of gaze-following in young chimpanzees, *Br. J. Dev. Psych.* **15**:213–222.

Rasmussen, J. L., Rajecki, D. W., and Craft, H. D., 1993, Humans' perceptions of animal mentality: Ascriptions of thinking, *J. Comp. Psych.* **107**:283–290.

Reaux, J. E., Theall, L. A., and Povinelli, D. J., 1999, A longitudinal investigation of chimpanzees' understanding of visual perception, *Child Dev.* **70**:275–290.

Rumbaugh, D. M. and Pate, J. L., 1984, The evolution of cognition in primates: A comparative perspective, in: *Animal Cognition*, H. L. Roitblat, T. G. Bever, and H. S. Terrace, eds., Lawrence Erlbaum Associates, Hillsdale, NJ, pp. 569–587.

Savage-Rumbaugh, E. S., Mcdonald, K., Sevcik, R. A., Hopkins, W. D., and Rubert, E., 1986, Spontaneous symbol acquisition and communicative use by pygmy chimpanzees (*Pan paniscus*), *J. Exp. Psych.: Gen.* **115**:211–235.

Scheumann, M. and Call, J., 2002, The use of experimenter-given cues by South African fur seals (*Arctocephalus pusillus*), *Anim. Cogn.* (manuscript submitted for publication).

Soproni, K., Miklósi, A., Tópal, J., and Csanyi, V., 2001, Comprehension of human communicative signs in pet dogs (*Canis familiaris*), *J. Comp. Psych.* **115**:122–126.

Suddendorf, T. and Whiten, A., 2001, Mental evolution and development: Evidence for secondary representation in children, great apes and other animals, *Psych. Bull.* **127**:629–650.

Tomasello, M. and Call, J., 1997, *Primate Cognition*, Oxford University Press, New York.

Tomasello, M., Call, J., and Gluckman, A., 1997a, Comprehension of novel communicative signs by apes and human children, *Child Dev.* **68**:1067–1080.

Tomasello, M., Call, J., and Hare, B. H., 1998, Five primate species follow the visual gaze of conspecifics, *Anim. Behav.* **55**:1063–1069.

Tomasello, M., Call, J., Nagell, K., Olguin, R., and Carpenter, M., 1994, The learning and use of gestural signals by young chimpanzees: A trans-generational study, *Primates* **35**:137–154.

Tomasello, M., Call, J., Warren, J., Frost, G. T., Carpenter, M., and Nagell, K., 1997b, The ontogeny of chimpanzee gestural signals: A comparison across groups and generations, *Evol. Comm.* **1**:223–259.

Tomasello, M., George, B. L., Kruger, A. C., Farrar, M. J., and Evans, A., 1985, The development of gestural communication in young chimpanzees, *J. Hum. Evol.* **14**:175–186.

Tomasello, M., Hare, B., and Agnetta, B., 1999, Chimpanzees follow gaze direction geometrically, *Anim. Behav.* **58**:769–777.

Tschudin, A., Call, J., Dunbar, R. I. M., Harris, G., and van der Elst, C., 2001, Comprehension of signs by dolphins (*Tursiops truncatus*), *J. Comp. Psych.* **115**:100–105.

Vick, S. J. and Anderson, J. R., 2000, Learning and limits of use of eye gaze by capuchin monkeys (*Cebus apella*) in an object-choice task, *J. Comp. Psych.* **114**:200–207.

Whiten, A. and Byrne, R. W., 1997, *Machiavellian Intelligence II. Extensions and Evaluations*, Cambridge University Press, Cambridge.

PART FIVE

Brain, Evolution, and Hemispheric Specialization

CHAPTER NINE

Increasing the Brain's Capacity: Neocortex, New Neurons, and Hemispheric Specialization

Lesley Rogers

INTRODUCTION

Brain size has been linked to cognitive capacity and intelligence, but this association is largely an intuitive assumption rather than one that has received unequivocal empirical support. This assumption has a long history dating back to the flurry of active measurement of the human brain in the 1800s, when "races" were ranked according to brain size and the assumed superiority of men over women was said to be couched in their larger brain size (Gould, 1981; Kaplan and Rogers, 1994, 2003). A similar line of thinking has been applied to non-human animals. Over many years, species have been ranked according to brain size, and now there is revived interest in doing so, especially in ranking species according to both brain size and cognitive abilities. It is timely, therefore, to see what light these studies, with emphasis on the recent ones, might shed on our topic of debate, "Are primates special in terms of their cognitive abilities?"

Lesley Rogers • Centre for Neuroscience and Animal Behavior School of Biological, Biomedical and Molecular Sciences, University of New England, Armidale, NSW, Australia.

Comparative Vertebrate Cognition, edited by Lesley J. Rogers and Gisela Kaplan. Kluwer Academic/Plenum Publishers, 2004.

From the outset, I would like to say that I have very little affinity with the view that brain volume measures anything that reflects either intelligence, which in itself is a nebulous and ill-defined characteristic, or behavioral flexibility, again not clearly defined or easily measured. Variation in brain size between different species may be associated with some aspects of cognitive ability, but it has to be recognized that the notion that larger brains are more "intelligent" has very little empirical evidence to support it. A larger brain containing more neurons should be able to transmit more information at the same time, and it is also possible that a brain with more neurons might form more memories or more detailed memories than a smaller one, although we do not know exactly how this might occur. Broadly speaking, within mammals, correlations have been found between the size of the cortex and certain measures of behavior (discussed below), but such relationships do not apply across orders. In fact, size is an unreliable criterion by which to assess neural functioning, since a smaller brain with different circuitry may process information more efficiently than a larger one, as shown by studies of the smaller avian brain compared to the larger mammalian brain (discussed below): despite their smaller brain size relative to body weight, the cognitive abilities of birds have been shown to match, or even surpass, those of many primates (see Chapter 1 by Emery and Clayton and Chapter 6 by Kaplan).

BRAIN SIZE RELATIVE TO BODY WEIGHT

Until quite recently, it was standard practice to consider brain weight (or volume) not as an absolute but relative to body weight to take into account the fact that a certain amount of the brain is given over to controlling muscular movement and to maintaining physiological functioning. This correction factor allows for the fact that a bigger body has a larger mass of muscles to control, a larger surface area to monitor and a larger mass of internal organs to control. Hence, as known for several decades now, within a taxonomic category, species with bigger bodies have proportionately larger brains, and there is a direct relationship between brain and body weight across all of the species: the log of brain weight plotted against the log of body weight for the different species within a class (or order) generates a straight-line relationship (Jerison, 1973, 1976).

Although the plotted points for fish and reptiles fall on close to the same line, those of lower mammals and birds are on lines slightly above this (Figure 1), meaning that they have consistently larger brains for a given body weight. The line for anthropoid primates is a little above that of birds and lower mammals.

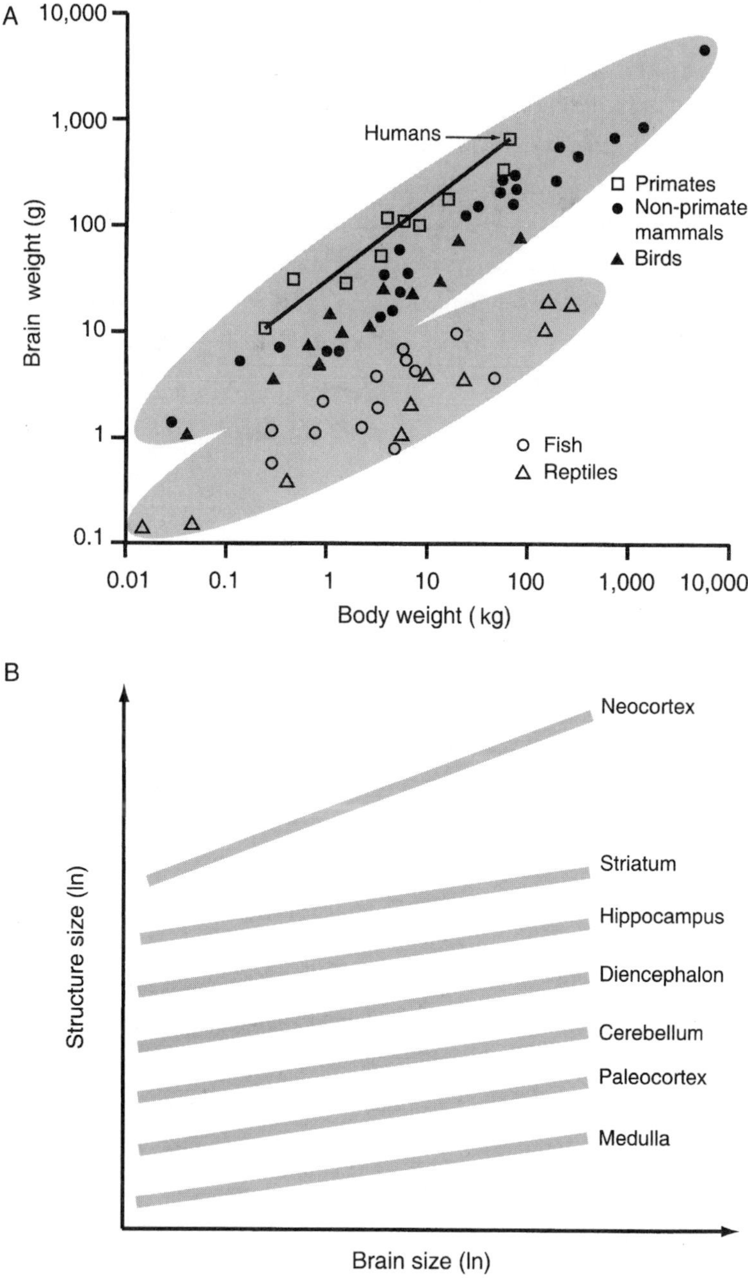

Figure 1. (A) A simplified plot of brain weight against body weight for different groups of species (modified from Jerison, 1973). (B) A simplified representation of the results of Finlay and Darlington (1995) showing the size to weight relationships of four regions of the brain. Note that the neocortex distinguishes itself by increasing in size at a greater rate than the other three regions.

To give an example to indicate the scale of the differences, a hedgehog weighing 860 g has a brain weight of around 3 g, whereas a galago of the same body weight has a brain weight of around 10 g, and a squirrel monkey weighing around only 700 g has a brain weight of over 20 g (Northcutt and Kaas, 1995). In other words, when adjusted for body size, the brain weight of primates is greater than for the other groups, although there is still variation within the groups. The human brain is, as so often stated, the largest in proportion to body weight compared to all other species. The volume of the human cranium, and hence the brain, relative to body weight increased over 2 or 3 million years and may have done so somewhat more sharply around 1.5 million years ago (Noble and Davidson, 1996), but there is no agreement on exactly when and how fast the increase was.

Alternatively, the plotted data points can be fitted within convex polygons (see figures 3 and 5 of Jerison [1994]). This is just a slightly different way of looking at the same data. The polygon encompassing the data points for fish overlaps that for reptiles, but both of these separate from the polygon encompassing the data points for birds and the polygon for mammals (which includes primates). The polygon for birds, however, overlaps that for the lower (smaller) mammals. Only recently have sufficient data been obtained for the order Cetacea (dolphins, whales, and porpoises) and they fit within a polygon that is a continuation of the data for anthropoid primates (Marino, 1998).

The allometric relationship between brain size and body size differs between different groups of vertebrates. The separate regression lines (or polygons) drawn for fish and reptiles and then birds and mammals shift, stepwise, up the Y-axis. Separate regression lines can also be drawn for lower primates (prosimians) and anthropoid primates and these too shift stepwise up the Y-axis. As can be seen, these steps from fish and reptiles to birds and lower mammals and then to lower primates and on to anthropoid primates reflect the order of evolution. There is a further step up from the anthropoid primates and cetaceans to the apes, including humans, of course. The steps are said to represent increasing encephalization, which in turn is said to indicate increasing neural information processing (Jerison, 1994). On the measure of encephalization, some cetaceans are closer to humans than is the chimpanzee (Marino, 1998).

Does this way of measuring brain size indicate superior intelligence in higher primates and cetaceans compared to all other species? Although this question is often answered in the affirmative, and it is a valid starting point for investigations (e.g., Marino, 2002), to find any answer at all requires glossing over the complexities and making an enormous leap from brain size to a general assessment of behavior. On the one hand, brain size is a crude measure of neural tissue: a likely better measure might be number of neurons versus glial cells, number of dendrites per neuron, or the number of synaptic connections per neuron, all of which are known to vary across mammals (Braitenberg and Schüz, 1998), or some other measure reflecting neural mechanisms. On the other hand, in my opinion, there is no reliable way to estimate higher cognitive ability, or intelligence, as a comparison across species. Not surprisingly, different methods of scaling behavior produce different results when they are correlated with the size of brain structures (Deaner et al., 2000).

It is even highly debatable whether valid comparisons about some sort of general intelligence can be made between individuals within one species. Nowhere is this debate more apparent than when it concerns the human species. The controversy surrounding the interpretation of studies of IQ, and the recent claim to be able to measure something known as "general intelligence" in humans (e.g., Posthuma et al., 2001) substantiates this point (discussed by Kamin, 1977; Kaplan and Rogers, 2003). Here it is noted that much the same arguments apply to similar attempts to measure "intelligence" in non-human species: tasks designed to test animals in the laboratory may have little or no bearing on the behavior of individuals or the species in the wild. In fact, to expect that there is any unitary way of measuring cognitive function is contrary to the concepts of evolution and adaptation of species to their particular niches (summarized in Rogers, 1997).

This circumspect approach to tackling the problem of intelligence and brain size does not deny that certain studies have found positive correlations between indices of brain size and behavioral ability provided that they have been confined to one class of animals (usually mammals) and specific behavioral measures (always made in the laboratory). An early example, by Riddell and Corl (1977), found a number of positive correlations between indices of brain size, in mammals, and measurements of behavior on operant learning tasks. Their starting point was Jerison's calculation that the total information capacity of a unit volume (or mass) of the cortex is constant across mammalian species, which leads to the deduction that species differences in information processing must be a function of the size of the cortex (Jerison, 1973). While such associations may be of value in understanding some general aspects of brain function, it remains unknown whether they translate to behavior in species' natural habitats and to animals that have been raised outside of the confines of captivity.

This is a reminder of the point made by Emery and Clayton (Chapter 1) that the focus should now move to consider ecological relevant behavioral measures.

There is another important point to consider about the straight-line plots of brain weight relative to body weight. For each group of animals, the slope of the line plotted is less than one, which means that, although brain weight increases with body weight, it does not quite keep up (Jerison, 1973). This could be interpreted as the heavier species having a greater proportion of their brain capacity devoted to moving and monitoring their large bodies, and so a lesser proportion of their brain would be available for higher cognition.

But is it appropriate to speak of the proportion of the brain to body weight ratio when considering higher cognition? Body weight would have no association to this aspect of brain function unless higher cognition is, in some way, needed to move around a larger body. At least for non-arboreal mammals, it seems that higher cognition is not compromised by having to control a large body. To take an extreme example, the cognitive abilities of elephants do not seem to be curtailed by their large size, as indicated by their learning capacity, long memories and ability to use tools: Chevalier-Skolnikoff and Liska (1993) have reported over 20 different types of tool use in elephants (see also Chapter 7 by Nissani). On the other hand, several researchers have hypothesized that orangutans, the heaviest arboreal species, express their higher cognition in solving the problems of movement and suspension (e.g., Galdikas, 1982; Povinelli and Cant, 1995; summarized in Kaplan and Rogers, 1999). Smaller arboreal primates are able to learn fixed ways of moving about in the forest and, once they have acquired these skills, they have to devote very little cognitive capacity to that activity. According to Povinelli and Cant (1995), the great apes may be too big to perform in this way. They may always have to think carefully about what they are doing, and this applies mainly to the orangutan since it is the most arboreal of the great apes. As the orangutan's limb movements vary with each particular problem of locomotion, they have to be much more resourceful (flexible) than monkeys in the way they go about moving. According to Povinelli and Cant, they have to "produce creative, on-the-spot solutions to immediate problems" and so their body weight is associated with their higher cognitive abilities. If so, it is likely that this association between cognition and moving a heavy body in a complex environment has consequences for other aspects of their higher cognition, and two opposing hypotheses relate to this. Both currently await supporting evidence.

The first is encapsulated in the suggestion by Galdikas (1982) that it is because wild orangutans have to use so much of their cognitive capacity on solving the problems of locomotion that they use fewer tools than orangutans at rehabilitation centers. This argument implies that there is a limit to higher cognitive capacity and that it has to be parceled out according to demands for survival. The second, stated by Povinelli and Cant (1995), suggests that, once a brain has evolved the capacity of planning ahead and solving the complex problems of moving a heavy body through the trees, it might have acquired the capacity to plan other actions, and so this has increased their mental powers for self-awareness (i.e., to have a concept of self).

In the absence of any published empirical evidence testing these two alternative hypotheses, I would like to mention a recent, and as yet unpublished, finding by one of my students, Helga Peters. She has scored the sequence of limb movements (left and right, hand and foot) used by wild orangutans as they move between trees. This is the time when they are at greatest risk of falling and it should, therefore, be the most cognitively demanding. Despite this, she finds a significant degree of stereotypy in their sequences of limb movements, particularly when they use brachiating movements (i.e., the "clambering" posture defined by Cant [1987]) and when they cross from one tree to another by swaying back and forth in the first tree. In other words, even the large bodied, arboreal orangutans may acquire fixed patterns of movements that permit them to apply a lesser amount of cognitive capacity to locomotion and, perhaps, leave more capacity for higher cognition.

This somewhat detailed discussion of the cognitive abilities of orangutans serves mainly to caution against drawing any unitary and reliable conclusion about brain size versus body size, cognitive capacity, and ecological demands. Nevertheless, speculation along these lines can be valuable and may provide a valuable stimulus to research.

Primates are of particular interest in these discussions. They have demanded special attention in research, largely because their capacities pertain to the evolution of humans. Many have proposed, over a long period of time, that there is a discontinuity between humans and other apes both in terms of brain size and behavioral ability, the abilities to use language and tools being the most important of these capacities (Deacon, 1997). More recently, some researchers have conducted comparative studies of primate cognition to see whether there might be a discontinuity between, say apes, and other primates. Rumbaugh (1997), for example, looked at the ability of different species of primates to transfer learning from one task to another (see also Rumbaugh et al., 2000). He gave the primates a series of two-choice discrimination problems and looked at how flexible they were in transferring that learning to tasks on which they had to reverse the associations that they had made previously. When lower primates (also the smaller ones) were asked to learn a little more on one task, they showed a decrease in ability to transfer this learning to another task. In other words, their reversal learning was impeded. By contrast, the larger monkeys (macaques) and the apes tested in the same way showed an increase in learning transfer. Their reversal learning was facilitated. This result suggests that there might be a discontinuity (a qualitative shift) in transfer ability, despite the quantitative and systematic change in cortex (and whole brain) size with respect to body weight across these species (see Rilling and Insel, 1999a). According to these results, increasing encephalization in primates has led to flexibility in learning, and learning has become relational and rule based rather than being bound simply to a stimulus–response paradigm. However, although this general trend may be valid, there are some notable exceptions to the rule: although macaque monkeys show facilitation of reversal learning, the more highly evolved gibbons, which are apes, show no such facilitation. More attention to these exceptions may shed as much light on the evolutionary processes as does generalizing to see the overall trends. In any case, the exceptions to the rule should not simply be ignored, as so far seems to be the case.

NEOCORTEX/ISOCORTEX

Mammals evolved from reptiles over 200 million years ago and with them emerged a new structure of the cerebral hemispheres, known as the neocortex or, more recently, as the isocortex. (Note that, due to its familiarity, I will use the term neocortex.) The neocortex became layered on top of the more primitive paleocortex, also called the allocortex (Figure 2). Although the exact origin of the neocortex is disputed, it appears that it had six different layers even in the earliest mammals, the monotremes (Krubitzer, 1995, 1998). With further evolution of mammals, the neocortex expanded in size relative to the rest of the brain, and it appears to have done so many times over to give rise to different lines of mammals with different cortical organizations. In mammals with large brains, the neocortex is expanded relative to the rest of the brain and the expansion of size of the neocortex has occurred mainly along its surface rather than thickness, thereby causing more fissures or crevices. In fact,

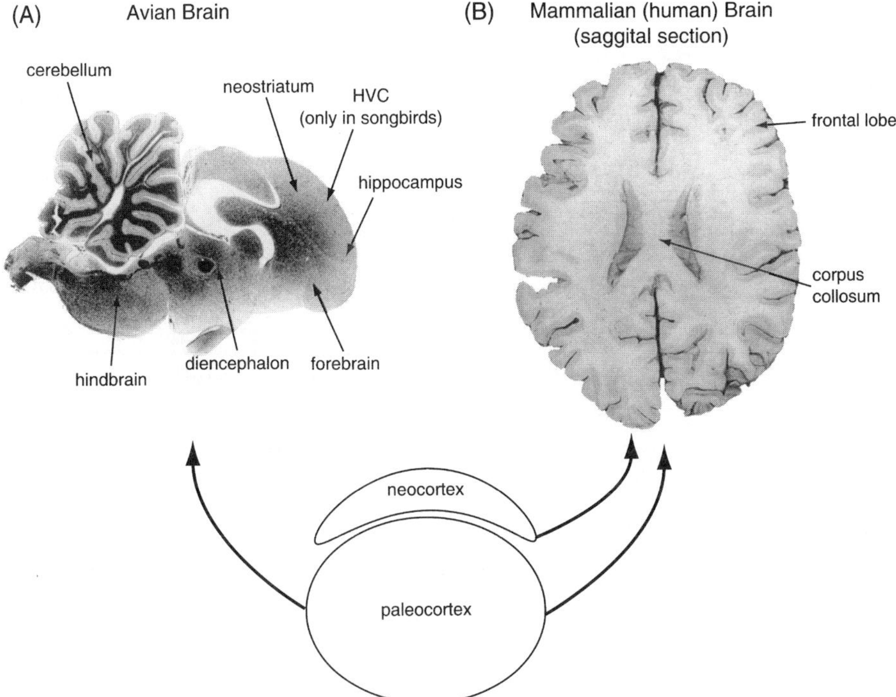

Figure 2. Representations of the (A) avian and (B) mammalian (primate) brain to show some of the structures discussed in the text. Note that the brains are not drawn to scale. The arrows indicate that the modern avian brain has resulted from elaboration of the paleocortex and has no neocortex, whereas the modern mammalian brain has elaborated on both the neocortex and the paleocortex.

over the period of evolution of mammals, the surface area of the neocortex increased more than 1000-fold with no comparable increase in thickness. To give some examples, the surface area of the neocortex of the rhesus macaque is a 100 times greater than that of a mouse, and the human's is a 1000-fold greater than that of the rhesus macaque.

The neocortex and the striatum are referred to collectively as the "executive brain," and both are considered to be involved in all aspects of higher cognition, as distinct from the hindbrain functions of controlling the muscles and other functions of the body. Hence, brains may be scaled on the basis of the volume of neocortex and striatum relative to the volume of the hindbrain, the latter being closely tied to body weight. This relative measurement of brain volume (or weight) has superseded the measurement of brain size relative to body weight (discussed above). However, some recent evidence suggests that, whereas the neocortex expanded with the evolution of mammals, the size of the striatum remained little changed (Clark et al., 2001), which suggests that it would be better to consider it as a separate entity from the neocortex.

It is widely assumed that the evolution of the neocortex was associated with the evolution of higher cognition or intelligence and ultimately, in humans, consciousness (Bayer and Altman, 1991; Innocenti and Kaas, 1995; Kaas, 1995; Krubitzer, 1995). In fact, a recent study (Rilling and Insel, 1999a) using magnetic resonance imaging to compare the brains of different species of primates showed that, compared to expectations for a primate of our size, the human neocortex is larger, our prefrontal cortex is more convoluted, and we have relatively more white matter to gray matter. These structural specializations, however, tell us nothing about the functional outcomes of the brain (i.e., intelligence and consciousness). Nevertheless, there have been a number of attempts to link consciousness to particular structural aspects of the human brain. Eccles (1989) developed a hypothesis about the evolution of consciousness based on the presence in the neocortex of pyramidal cells clustered into bundles, called dendrons. Each dendron bundle has a large number of connections and is also connected to other dendrons. As far as can be deduced from comparisons between various species, the dendrons may have first appeared in the primitive mammalian neocortex and then multiplied increasingly to give the approximate 40 million of these dendrites in the human neocortex. Eccles speculated that electrical activity in the dendrons interacts with the "world of the mind" to produce what he called units of consciousness, called psychons. Thus, he tied consciousness to a particular cell type but this is, so far, not a testable hypothesis. It does, however, lay a starting point for electrophysiological approaches to the study of higher cognition and consciousness.

The hypothesized association between the neocortex and higher cognition is, of course, a human-centered position. Birds have usually been ignored or underestimated by the scientists who have written about the evolution of intelligence and consciousness. Although the neocortex might have provided mammals with the neural substrate required for their higher cognitive abilities, *without* a neocortex birds have complex cognitive abilities that rival those of species *with* a neocortex, as discussed in several previous chapters. However, it should be mentioned that Karten (1991) pointed out that the neocortex may have a functional homologue in birds, this being the neostriatum.

FRONTAL LOBES

Some researchers designate a subregion of the neocortex, known as the prefrontal cortex and located in the frontal lobes (Figure 2), a special role in the evolution of cognition in hominids. The prefrontal cortex is said to be the "command headquarters." As shown in humans, during deep thought, the prefrontal cortex has a particular form of synchronous electrical activity, known as theta rhythm. Insightful and self-reflective thinking in humans has been attributed to the prefrontal cortex, as has planning future actions, decision-making, memory, language, and emotional as well as artistic expression (Semendeferi et al., 1997).

Earlier research has reported that the prefrontal cortex occupies about one quarter of the human neocortex, which is an apparent advance on the great apes with a prefrontal cortex occupying only about 14 percent of their neocortex (Falk, 1992). However, Semendeferi et al. (1997) found that this may be incorrect. Using a new technique of three-dimensional reconstruction of images of brain scans (obtained by magnetic resonance imaging), the researchers measured the size of the frontal lobes relative to the rest of the cortex in one of each of the following subjects: chimpanzee, gorilla, orangutan, gibbon, and macaque. They found no evidence of a difference in the size of the frontal lobes relative to the whole brain in any of these subjects. Moreover, the subregions of the frontal lobes were proportionately of the same size in all of the subjects examined.

Although the sample size of this study is too small to draw any final conclusions, the result does alert us to the need for more such measurements and to be very cautious when attributing a distinction between humans and great apes using this anatomical criterion. It is possible that, as evolution progressed, the size of the frontal lobes increased at the same rate as the rest of the brain and so the relative size remained unchanged. While it remains a possibility that primates may be separated from other mammals by the relative size of their frontal lobes, this needs to be investigated. I tend to agree with Semendeferi et al. (1997) that cognitive evolution within the primates does not appear to be directly associated with the prefrontal cortex.

RELATIVE DIFFERENCES IN THE SIZE OF DIFFERENT REGIONS OF THE BRAIN

Brain weight is a gross measurement even when it is adjusted for body weight. Some refinement of approach has been achieved by calculating the volume of the executive brain relative to the hindbrain, or the relative size of the prefrontal cortex with respect to the rest of the brain, but these are still generalizations that bring little satisfaction to the neurobiologist. Using these measures may have some explanatory power for the differences in cognitive capacity between the species within a single taxonomic class but more understanding might be gained by taking into account the fact that brains are comprised of numerous regions, usually connected into systems and often overlapping, each specialized to process certain types of information and/or to control one or more functions and not others. (Note that I prefer not to use the term "modules" for these regions/systems because that term has been used by evolutionary psychologists to imply strong genetically determination, whereas experience and learning may well affect their development; for discussion see Kaplan and Rogers, 2003.)

The question arises therefore whether differences between species in the size of regions or systems has come about by "mosaic evolution," meaning that natural selection has acted selectively on the behavioral capacities of each region/system separately, as suggested by Barton and Harvey (2000), or by coordinated evolution, meaning that natural selection for an increase in the size of one region/system leads to an increase in the size of all of the other regions, as proposed by Finlay and Darlington (1995). I will discuss each of these hypotheses separately. Before doing so, however, it is important to note that localization of functions to specific brain regions is not without its critics, as elegantly outlined by Uttal (2001), and the reductionist approach of tying specific functions to specific locations has very real technological difficulties which are often overlooked (also made clear by Uttal, 2001).

Coordinated Size Change

In 1995, Finlay and Darlington proposed a model that might explain the accelerated expansion of the neocortex in the evolution of mammalian brains, the human brain being at the top of an exponential increase in the size of this important region of the brain.

Recognizing that each species is subjected to the forces of natural selection, leading it to optimize its behavior in a particular environment, Finlay and Darlington (1995) asked what changes might take place in the brain to allow it to develop a specialized ability or behavior, carried out by a specific region or system. In other words, they wondered whether adaptation of a species to perform a special behavior in a particular environment led to the expansion of only the region (or system) of the brain needed for that behavior or whether other areas increased along with it. They reasoned that, amongst the mammals at least, the latter may apply because, when a brain is developing, new neurons form in a particular order that is largely the same in all mammalian species. Hence, an increase in the size of one region of the brain would, they suggested, have to be coordinated with an increase in the size of all regions that develop with and after the required region: all of these regions would have to enlarge along with the one undergoing selection.

Accordingly, natural selection for one specialized ability (e.g., hands to catch prey) would lead to an expanded capacity to perform other specialized functions. There are some examples that seem to support this proposition. The Australian striped possum (*Dactylopsila trivirgata*) has a special adaptation for its mode of feeding in the canopy of the rainforest: it has one digit longer than the others and it can use this digit to get insects out of holes in trees. It also has the largest brain, corrected for its body weight, of all marsupials. Thus, its adaptation of a special digit may have led to an overall increase in brain size, not just an increase in the size of the brain region controlling the digit itself.

If so, acquiring one special ability may enhance the brain's capacity for performing many other special functions. For example, when paws evolved to hands (of primates), not only did the region of the neocortex used to control the forelimbs expand in size but so did the entire neocortex. Hence, along with the specific adaptation to evolve hands and the ability to use them to manipulate objects, came many other cognitive abilities.

Finaly and Darlington measured the sizes of different regions of the brain, relative to total brain size, of a large number of mammalian species living in different environments and they found that, as total brain size increases, the size of the neocortex, as a whole, expands relative to all of the other regions of the brain (Figure 2). The size of the neocortex increases exponentially compared to the other regions of the brain. To cite an example given by Finlay and Darlington, the brain of the smallest shrew is some 20,000 times smaller than the human brain, whereas its neocortex is more than one 100,000 times smaller.

The human neocortex is at the top of the exponential curve, perhaps the result of selection for some specific adaptations such as walking in a more upright posture and use of the hands in making tools. Thus, along with these specific adaptations, we might have acquired the increased brain capacity for thinking, consciousness, etc. We might also apply this hypothesis to answer the question whether primates are special. Since primates have evolved hands and use them for specific purposes, the hypothesis would suggest that this adaptation elevated their cognitive abilities above those of other mammals.

Some non-primate mammals make extensive use of their hands, a prime example being the raccoon, which uses its hands to catch prey. Compared to its nearest relatives, it has a greatly enlarged primary somatosensory cortex and a greater proportion of this cortex is devoted to processing information received from the hands. How is the required enlargement of this specialized area of the brain to perform a particular adaptive function achieved and have other regions of the brain structures expanded along with this one?

The Finlay and Darlington hypothesis should also apply to branches of evolution other than the mammalian trajectory and it would gain credence if it held for the avian trajectory. It is known that avian species that store food and later retrieve it have a larger hippocampus (Figure 1) than species that do not store food (see Chapter 1 by Emery and Clayton). Do these food-storing species have other special abilities that they evolved along with their ability to store using their larger hippocampus? Do owls, which are perfectly adapted to searching for their food in the dark of night using specialized abilities for locating the source of sounds made by their prey, have other special abilities that they acquired with this one?

To summarize, have the constraints of sequential development of different regions of the brain prevented one region of the brain from expanding independently of the others, and so have certain specialized adaptations enabled expansion of the neocortex as a whole, thus freeing neural capacity for other types of higher cognition? This question has yet to be answered.

Mosaic Evolution

Barton and Harvey (2000) have provided empirical evidence against the hypothesis of coordinated size increase and for selection of increased size of specific cortical regions. They showed that the neocortex is five-fold larger in primates than in the insectivora (the order of mammals mostly insect-eating, including shrews, moles and hedgehogs) even after accounting for the weight of the rest of the brain, and that those regions that are linked structurally and functionally have evolved together and independently of other brain regions. In fact, there was some indication that an increase in the size of one neural system might be traded off against another: in primates there was a significant negative correlation between the size of the regions involved in olfaction and those involved in vision, possibly related to the divergent trajectories to nocturnal versus diurnal niches, as the researchers suggested. Hence, particular demands of particular environments involved changes in the size, and presumably the functional capacity, of specific regions of the brain, and that change in size might be an increase or a decrease. The conclusion reached is that mosaic evolution has been an important aspect of adaptive radiation of the mammalian brain. It would now be interesting to see whether the same applies to the rich adaptive radiation of the avian brain.

A subsequent investigation by de Winter and Oxnard (2001) confirmed the conclusion that the relative proportions of functionally linked brain structures (nuclei) vary independently between different orders of mammals, and so follow separate evolutionary trajectories. Hence the hypothesis of mosaic evolution was supported. They analyzed a large number of variables of measured sizes simultaneously, using multidimensional methods, and found that the proportional sizes of different functional systems vary independently between the different orders of mammals, thus showing independent evolution. In addition, the analysis revealed some clusters of unrelated species that occupy similar ecological niches. Within primates there was clustering according to the mode of locomotion used (e.g., those with hindlimb dominated leaping clustered separately from forelimb-dominated brachiation). Thus, convergence of lifestyle (e.g., as in the case of forelimb-dominated climbing and feeding in New World spider monkeys and Old World apes) is matched by similar proportions of subregions of the brain. Similar separations occurred within the bats and insectivores (shrews, moles, etc.). This is evidence that mosaic evolution is caused by adaptation to ecological niche. The authors conclude their paper with the following statement, "Hence, rather than a single evolutionary progression in general 'intelligence' across mammals, different dimensions of 'intelligence' appear to have evolved independently in different lineages" (de Winter and Oxnard, 2001, p. 714). This suggests that not only are primates special, in the sense that they are different, but so are the bats and the insectivores.

A separate study by Clark et al. (2001) drew rather similar conclusions to the one just discussed but suggested a reconciliation between coordinated size change and mosaic evolution. These researchers found that within taxa brain regions maintain fixed proportionality of size irrespective of the total brain size, and this indicates that development follows a common set of rules. At the same time, there are separate cerebrotypes characteristic of each taxon, which suggests that there has been selection for variation in the development processes.

LINKING THE SIZE OF BRAIN REGIONS TO SPECIFIC BEHAVIOR

Only if one thinks of "intelligence" in a unitary way is the overall size of the brain size a matter for consideration. I have mentioned previously that there is little evidence that such considerations have validity, despite that fact that scientists, primatologists especially, often refer to one species as being "more intelligent" that another. Moreover, this line of thinking is used to set humans apart from all other animals: we cite our large brain relative to body weight as direct evidence of our intellectual superiority. In attempts to link function to brain size, it might be fruitful to move away from such generalizations and to begin by considering that each brain region is specialized for behavior adapted to the particular niche in which the species exists. We might therefore look at the size of particular regions (single and linked) within the brain, rather than the whole brain itself, and see whether these measures correlate with specialized skills or types of cognitive capacity.

This approach was taken with successful outcome by Krebs et al. (1996), who measured the size of the hippocampus, which is involved in spatial learning, in birds that store their food compared to birds that do not store their food. They calculated the volume of the hippocampus relative to the rest of the forebrain as well as adjusting for body weight and found that the relative size of the hippocampus is larger in species that store and retrieve food than in species that do not do so. Apparently, the need for the storing bird to remember where it has stored its food has been met by an enlargement of the area of the brain that processes spatial information (see also Harvey and Krebs, 1990; Chapter 2 by Emery and Clayton).

Another example of enlargement of a specific region of the brain for specialized behavior is that of the song nucleus called the high vocal center (Figure 1). DeVoogd et al. (1993) found that the size of this nucleus in different species of songbirds correlates with the species complexity of song. Thus, its size appears to reflect its functional capacity. Even within a species, there may be a relationship between the size of the higher vocal center and the individual's song repertoire. Nottebohm (1989) ranked the songs of canaries according to their complexity and found that this correlated positively with the size of the higher vocal center, although there was a reasonable amount of variation in the data. The variation suggests that other factors are also involved in this relationship.

From these initial studies linking specific functions to the sizes (and structural organization) of specific brain regions, we can move to look at connected regions and then proceed to build a broader picture.

Correlations between Brain Size and Behavior

Others have preferred to tackle the broader questions first and asked what factors might have caused the evolution of a larger neocortex, as a whole, in some species compared to others. Amongst the non-human primates, it has been argued that the means by which animals obtain their food and/or the complexity of their social relationships might have been significant factors selecting for intelligence and hence, by association, determining the size of the neocortex. Since it is possible to obtain rather broad-spectrum comparative data on these two types of behavior, some interesting empirical studies have been published, and will be discussed below. However, such data must necessarily be so generalized that it is clearly wide open for alternative explanations. Also, including "intelligence" as a step in this line of argument, it seems to me, is problematic: it would be possible, and preferable, to avoid this value-laden intermediate step, especially since intelligence cannot be defined or accurately assessed. Others have replaced the term by "behavioral flexibility" but, although there have been attempts to measure this, they have been limited to laboratory tasks, and rather restricted ones at that, as I have discussed previously.

Foraging for Food

Sawaguchi (1992) divided a large number of non-human primates into different groups according to their diet, habitat, and social structure, and measured the volume of the neocortex relative to the volume of the rest of the brain. The results showed that those species that feed primarily on fruit, although they might take some insects and leaves, have higher relative volumes of the neocortex than species that feed predominantly on leaves. In other words, diet might be a factor driving evolution of the neocortex, and one possible explanation for this may be the fact that fruit-eaters have to search for their food, which might be rather sparsely distributed, whereas leaf-eaters find their food more evenly distributed. Also, fruit ripens at only certain times of the year and fruit-eaters must remember when that is. Fruit-eaters, therefore, may rely on well-developed abilities to form and remember spatial and temporal maps of their environment. However, a body of research, some of which I have discussed previously, shows that demands on spatial memory and mental maps lead to changes in the size of the hippocampus, not the neocortex. Therefore, it seems unlikely that this hypothesis is correct.

The same study found that social structure also correlated with the relative size of the neocortex, as discussed in the next section. Of course, all of these relationships do not tell us directly what the causal factors are. We can only speculate and should remember that the influences could be indirect, caused by some other factor associated with eating fruit or being in a larger troop, such as encountering different kinds of predators depending on where food is found and on more or less protection depending on troop size.

It would now be important to add other mammals to such comparisons of foraging behavior and neocortex size. In this case, carnivores would add an important new variable and it would be important to note that their contribution might not be in terms of diet as much as in terms of cooperative strategies in hunting. If so, a comparison of the Canidea with the Felidae should be revealing since they are cooperative versus solitary hunters, respectively (Rogers and Kaplan, 2003).

Furthermore, birds make an extremely interesting case, since they provide a rich spectrum of all modes of feeding and diets. Bennett and Harvey (1985) made a comprehensive study of a large number of avian species, looking for associations between relative sizes of different regions of the brain (the hemispheres, brain stem, optic lobes, and cerebellum), but found very few such associations even though they tested a range of ecological variables. The strongest single effect that they found was larger brain size (hemispheres, optic lobes, cerebellum, and whole brain, but not brain stem) in the altricial species, which reflects the known constraints of altricial versus precocial development (Kaplan and Rogers, 2001; Stark and Ricklefs, 1998). They did not find any association between brain size and diet per se, but brain size was associated with mode of prey capture: species hunting from perch to perch and from perch to ground had larger brain sizes than those hunting by staying on the ground or in the air. There is no obvious explanation for this effect.

Social Intelligence

Group living is characteristic of most mammals and, in general, we consider that those living in larger groups have increased social complexity. This concept is the basis of the "Social Intelligence Hypothesis," first formulated by Jolly (1966) and later developed by Humphrey (1976) and Byrne and Whiten (1988). The hypothesis links the evolution of intelligence to increasing complexity of social interactions. The assumption is that individuals in larger social groups have to remember more faces, vocalizations and behavioral characteristics of their troop members than do those in smaller groups. This, in turn, is based on the assumption that, with increasing group size, each individual interacts socially with a larger pool of conspecifics. While this may be correct for some social groups and some species, there are exceptions. On the one hand, some animals living in large groups still confine most of their interactions to a small subset of individuals in that group and, on the other hand, some animals living in small groups engage in very complex social interactions. For example, there is no evidence that orangutans are less intelligent (in terms of problem solving, insight, learning, or sign language acquisition) than chimpanzees or gorillas, even though they tend to live in smaller groups or, in some cases, are solitary for most of the time (summarized in Kaplan and Rogers, 1999). These exceptions must be kept in mind but they do not invalidate the general picture showing trends for relative neocortex size to increase with increasing group size, as has been shown in several studies of primates (Barton, 1996; Dunbar, 1992, 2000, 2001; Kudo and Dunbar, 2001; Sawaguchi, 1992).

As I have mentioned the Sawaguchi study above, here I will add to it. In his comparative study of neocortex size in primates (discussed above), Sawaguchi (1992) found that the species in which males have many female partners had significantly larger relative neocortex volumes than species in which single male–female partnerships were formed. The size of the neocortex was also influenced by the size of the troop size: the larger the troop size, the larger was the relative size of the neocortex.

Kudo and Dunbar (2001) examined neocortex size and the size of social networks in primates, the latter being assessed in terms of cliques that the animals form when they groom each other. The size of these grooming cliques correlated with troop or group size and also with the relative size of the neocortex. This approach is important because it controls for the number of other animals that interact regularly within the larger group.

Of course, grooming cliques are not the only manifestation of social complexity. Tactical deception and flexibility to exploit strategies outside the main social structure are other relevant variables. In addition, grooming may be replaced by other forms of social communication (e.g., contact calls in primates and other species, and language in humans). Overall, however, size of the neocortex correlates positively with the size of the social group maintained as a coherent entity (Dunbar, 2001). The latter, of course, is not so simple to determine for most species since group sizes and social interactions change with season and also according to circadian rhythms. Levels of predation also impact on group size, which increases to provide safety in numbers, but such forced aggregations may not demand increased levels of social interaction (e.g., grooming and other forms of communication).

Group size and size of the neocortex have also been found to correlate positively in carnivores and some insectivores (Dunbar and Bever, 1998) and cetaceans (Marino, 1996), although Connor (1998) has mentioned that the latter result should be qualified by lack of available evidence on social group size and by fission–fusion of groups in species such as bottlenosed dolphins. To my knowledge, no other similar studies of neocortex size and group size have been published for other mammalian taxa. While comparisons within orders may reveal similar associations to those found in primates, it is already apparent that the relationship does not apply without exceptions across orders; for example, many ungulates live in very large herds but the relative size of their neocortex remains smaller than that of primates, even those living in small social groups.

In birds, the study of Bennett and Harvey (1985), mentioned above, found a significant association between brain size and mating behavior: monogamous species have larger brains than polygynous species. As the researchers point out, this is the opposite relationship to that found in primates (Clutton-Brock and Harvey, 1980). It should be noted, however, that mating patterns do not always depend on group size in birds: many species flock in one season of the year and form pairs in another. This would need to be taken into account in future studies.

Social Learning, Innovation, and Tool Use

A recent investigation by Reader and Laland (2002) tested the hypothesis that large brains should be better in performing three types of behavior: social learning (learning from conspecifics), innovating (novel solutions to problems, including learning that is not social), and using tools. Although it can be said that this hypothesis is, in some ways, an extension of the one concerning social intelligence (discussed above), it is important to note that, since social group size and social learning do not correlate, social learning may be a separate expression of social intelligence (discussed also by Seyfarth and Cheney, 2002). The authors considered the three categories to represent behavioral flexibility. They correlated assessments of these patterns of behavior with the relative size of the "executive brain" (discussed above) and found a positive relationship between the latter and all three of the behavioral categories: the larger the executive brain, the more the non-human primate innovates, uses tools, and learns from conspecifics. Interestingly, social learning and innovation correlated positively, suggesting that they may be linked, in some way. We note too that the association between brain size and frequency of tool using supports another idea claiming that technology and technical ideas might favor expansion of the executive brain (Deaner et al., 2000; Whiten and Bryne, 1997).

Avian species are an excellent comparative test of this hypothesis, since there is now a body of literature reporting tool use in birds, and this approach has been taken in four studies by the same research team (Lefebvre et al., 1997, 1998, 2002; Nicolakakis and Lefebvre, 2000). The first of these looked at innovations in feeding behavior in birds of North America and Britain and found that those species with more such innovations listed (hawks and falcon) have larger relative volumes of the forebrain than lesser innovators (ducks). This finding was repeated on birds of Australia and New Zealand, with one difference between the two samples: there was a high frequency of innovations in Australian parrots but not in New Zealand parrots. Next it was repeated on European species and then the team examined tool using against brain size and found another positive correlation. In fact, tool-using species had a larger neostriatal region of the forebrain (Figure 2) than non-tool users. Amongst the tool-users, those meeting the most strict definition of tool using (i.e., using an object held in the beak, as in the case of breaking open an egg by hitting it with a stone) had a larger brain size than those with tool use such as simply breaking open an egg by dropping it, and other similar examples. These data are entirely similar to the data for primates, indicating parallel evolution, as well as the fact that primates are not special in this respect, as Seyfarth and Cheney (2002) speculated.

One should mention before closing this section that all of these studies have relied on sifting through the reported scientific literature to glean observations of innovation, social learning, or tool use. They are, therefore, all compromised by the accuracy of observation and reporting and by the amount of effort that has been made to observe a particular species. It is possible that, for example, these types of behavior are more likely to be reported when they are observed in species to which humans have more affiliation, apes and parrots, for example. As a result, the associations being investigated may represent a self-fulfilling hypothesis. On the other hand, Lefebvre and colleagues, in particular, have noted this problem and taken stringent procedures to attempt to rule it out.

HEMISPHERIC SPECIALIZATION

Lateralization of the brain usually refers to specialization of the two hemispheres of the brain to carry out different functions, to process different sorts of information and to control different patterns of behavior, but lateralization occurs at subcortical levels as well, as in the case of hormonal control of behavior at hypothalamic level (e.g., Nordeen and Yahr, 1982) and control of the pituitary–adrenal axis (Wittling, 1995). Lateralization was originally thought to be a unique feature of the human brain, underlying the ability of humans to use tools (and so, associated with right handedness) and language. However, three decades ago, lateralization was discovered in non-primates, first by Nottebohm (1971) in the canary (for control of song) and soon after in the chick for visual searching and attack and copulation (Rogers and Anson, 1979), and in rat for affective and other behavior (Denenberg, 1981). These findings hinted strongly that it might be a widespread characteristic of vertebrates. The latter has now been confirmed, since lateralization has been shown in fish, amphibians, reptiles, a number of other avian species, and a number of mammalian species, including primates (summarized in Rogers and Andrew, 2002). Not only is it present in the wide range of vertebrate species studied, but also several forms of lateralization appear in all of these species, as is the case for attacking leftwards under guidance by the left eye and right hemisphere and for preferentially directing prey catching or feeding responses rightwards under right eye and left hemisphere control (summarized in Rogers, 2002).

Lateralization is considered to increase neural capacity and so be a means of enhancing cognition, but there is little direct evidence so far available to support this concept. Indeed, when this explanation was originally applied to lateralization in the human brain, as an explanation of our superiority to other animals, it was no more than an assumption. The discovery of lateralization in animals has opened the way for controlled research on this and other questions, and the results so far indicate that lateralization of the brain does enhance neural capacity. Experiments with weakly and strongly lateralized chicks have shown that lateralization enhances detection of a model predator moving overhead while the chick feeds (Rogers, 2000). Comparison of weakly and strongly lateralized pigeons has shown that stronger lateralization enhances visual discrimination of food objects (Güntürkün, 2000). Also consistent with this, chimpanzees with stronger hand preferences are more efficient at fishing for termites than weakly lateralized ones (McGrew and Marchant, 1997). Hence, it may be a general principle for vertebrates that brain lateralization increases neural capacity, and that it is not special for humans or even for primates.

Despite the now widespread evidence for lateralization in vertebrate brains, as mentioned above, there are those who remain unconvinced by the evidence and lay claim to a unique connection between lateralization and language ability appearing only in *Homo sapiens* and even suggest that a single gene event might have caused it (Crow, 2002). Although the appearance of language in humans might have depended on the existence of brain lateralization, as seems very likely, and that aspect might have been quite discrete from the other forms of lateralization present in non-human vertebrates, it would be foolhardy to deny the existence of a range of other types of lateralization in vertebrates. Indeed, the latter might have been precursors to the evolution of those forms of lateralization essential for language, as perhaps manifested in the primate equivalent, or precursor to, Broca's area (Cantalupo and Hopkins, 2001).

Those who choose to deny the existence of any form of lateralization with a population bias in vertebrates frequently turn to handedness in primates as supporting evidence for their position (Crow, 2002). Indeed, primates, or at least some species of them, may not exhibit population biases for handedness, as indicated by the meta-analysis of McGrew and Marchant (1997) and the recent paper by Palmer (2002) suggesting that the data collected from small samples may be observer-biased, since handedness is more often reported for smaller sample sizes than for larger ones. However, handedness in primates is a special case of lateralization since the hands can be used to reach on either side of the body. Hence, hand use is likely to be freed from lateralized control by the hemispheres, except perhaps in the constrained case of writing (Andrew and Rogers, 2002). Other lateralized functions are likely to be as strongly and consistently lateralized at the population level in primates as they are in other vertebrates. For example, Hook-Costigan and Rogers (1998) have shown consistent right eye preferences in a group of marmosets (20 out of 21 marmosets had a significant right eye preference to view a favorite food through a peep hole, the mean strength of preference being about 75 percent right). Added to this, there is some evidence for structural lateralization of the brain in great apes: the planum temporale region of the cortex is larger in the left than in the right hemisphere (Gannon et al., 1998; Hopkins et al., 1998), and the hemispheres show the same torque as in humans, meaning that the left occipital lobe and the right frontal lobe are larger than their counterparts (Hopkins and Marino, 2000).

If population level lateralization (group bias) is not as common or as strong in non-ape primates as it is lower vertebrates and birds, they must have lost asymmetries that were strongly expressed in their ancestors and regained them only with the appearance of *Homo sapiens.* However, the weight of current evidence points to the presence of hemispheric specialization in non-human primates, as in other vertebrate species (Rogers and Andrew, 2002; Ward and Hopkins, 1993). In fact, consistent group-biased hand preferences are present in some primate species, even if its presence remains controversial in apes (Hook-Costigan and Rogers, 1996; MacNeilage et al., 1987; see also Chapter 10 by Hook).

CORPUS CALLOSUM

Well-developed interhemispheric connections evolved along with the neocortex. In fact, the corpus callosum was an entirely new structure that evolved to transmit information between the cerebral hemispheres of mammals. This tract is not present in reptilian or avian brains, which have a number of much smaller tracts connecting each side of the brain. The size of the corpus callosum, relative to the rest of the brain, is largest in humans. Thus, humans have more neocortex and more connections between the separate neocortical regions of the left and right hemispheres than all other animals. This has relevance to hemispheric specialization since the corpus callosum has an important role in preventing the left and right hemispheres from both carrying out the same functions: that is, from duplicating functions. This appears to be possible because the corpus callosum links regions in one hemisphere to their equivalent regions in the other hemisphere, and lateralization of the hemispheres is generated by one region inhibiting the other (Denenberg, 1981).

However, despite the large size of the corpus callosum in humans, its size relative to the rest of the brain is smaller than in other apes and monkeys. Rilling and Insel (1999b) measured the area of the corpus callosum in magnetic resonance images of 11 species of primates, including humans, and related it to the volume of the brain. They found that, as the brain increases in size (from monkeys to apes), the size of the corpus callosum also increases but not at the same rate as the brain. Hence, per unit of brain tissue, monkeys have more connections between their two hemispheres (via the corpus callosum) than do apes, and, amongst the apes, humans have the least number of such interhemispheric connections. Rilling and Insel suggested that this diminishing ratio of the size of the corpus callosum to brain size might be a reflection of the evolution of lateralization on the grounds that fewer interhemispheric connections might lead to a lesser degree of lateralization. This hypothesis is unlikely to be correct. As explained above, lateralization is widespread among vertebrates and it is as strong in avian species as in humans, despite the fact that birds have very few interhemispheric connections. Moreover, the hypothesis fails to take into account that arborization of the interhemispheric neurons is an important aspect of the effectiveness of the callosal fibers. Glossing over structural details such as this, combined with formulating a functional outcome that goes well outside the data collected is rather too prevalent in this field. However, the predicted association between relative size of the corpus callosum and one type of functional lateralization (hand preference) was later tested and, using the limited data from just one study reporting hand preferences of primates, a significant correlation was found between corpus callosum size relative to neocortex size and hand preference (Hopkins and Rilling, 2000). Nevertheless, this data set did not support the original hypothesis since the species with the highest ratio of corpus callosum to neocortex size were left handed rather than being without handedness. The hypothesis was then modified to suit the data, which is acceptable. The main problem with this study is its failure to take into account the data on hand preferences reported by a number of other researchers and failure to consider that captive and wild primates might express different hand preferences. Data on hand preferences in chimpanzees and orangutans are now known for subjects living in natural and semi-natural conditions, and they are somewhat at odds with the data for captive members of their species (also see Chapter 10 by Hook).

Gazzaniga (2000) sees the corpus callosum as essential to human mental function. In his words, "By having the callosum serve as the great communication link between redundant systems, a pre-existing system could be jettisoned as new functions developed in one hemisphere, while the other hemisphere could continue to perform the previous functions for both half-brains"

(Gazzaniga, 2000, p. 1293). This conceptualization is, of course, the same as the hypothesis for increased neural capacity resulting from lateralization, as mentioned previously. While I do not disagree with this opinion, particularly since there is some evidence to support it, it is now clear that this role of the corpus callosum is not essential for lateralization, since birds, reptiles, and amphibians do not have either a neocortex or a corpus callosum and they have lateralization of the brain (discussed above). In the case of birds, at least, we know that lateralization is as strong as in humans. It seems possible, therefore, that lateralization may be achieved through rather different means in the mammalian than in other taxa but the outcome may be rather similar (Rogers, 2002). The human brain might be at the summit of the primate line but the road to there was not a unique path via lateralization.

EXPERIENCE AND BRAIN SIZE

There is another important factor that we must take into account when we consider brain size. The overall size of the brain is affected by experience and the size of any region of the brain is affected by performance of the behavior associated with that region. Considering the overall brain size first, some years ago Diamond (1988) demonstrated that rats raised in an enriched environment develop a larger brain that those kept in an impoverished environment. The enriched environment was one with other rats present and it contained novel objects and toys, and the impoverished environment was isolation in a standard, sterile laboratory cage. The size of the brain of the rats in the enriched condition increased by expansion of the thickness of the neocortex, and it did so by increasing the number of connections between the neurons and by a remarkable 40 percent increase in the size of the synapses. That is, enrichment caused an increase in the amount of connectivity between neurons in the cortex, and the cognitive capacity of the rats changed along with this. The rats from the enriched environment had superior abilities in solving mazes to find food. These changes occurred after as little as 30 days in the enriched environment and in both young and old rats. Thus, cortex size is not a fixed aspect of an individual but varies with experience.

The size of the rat hippocampus can also be changed by experience in early life, as shown by Verstynen et al. (2001). Rat pups were exposed to a novel environment for 3 min each day over the first 3 weeks of their life. At 8 months of age, this treatment lead to an increase in the volume of the hippocampus in the right hemisphere. This lateralized effect on the hippocampus is not without relevance since the right hippocampus is dominant in spatial learning and rats can perform the spatial task of finding a hidden platform in a water maze when they use the left eye, which provides input to the right hemisphere, but not when they use the right eye (Cowell et al., 1997). Hence, early experience in novel spatial arrangements enhances the size of neural structures used for processing spatial information and it does so quite specifically and has a long-lasting effect.

A similar dependence of size on experience has been found for the hippocampus in the food-storing birds. The opportunity to store food is essential for enlargement of the hippocampus in food-storing marsh tits (Krebs et al., 1996). Following the storing experience, and at all ages, more neurons are formed and fewer are lost by natural attrition. If marsh tits are completely prevented from storing food, the volume of the hippocampus decreases because the hippocampal neurons are not replaced as fast as they are lost.

In fact, the effect of processing spatial information on hippocampal size has even been seen in humans. Maguire et al. (2000) used magnetic resonance imaging to determine the volume of the hippocampus in London taxi drivers and found that the volume of their right posterior hippocampus was larger than the same brain region in age-matched controls. This region of the brain is used for representation of spatial information. The researchers also found a positive correlation between the number of months that individuals had spent driving taxis and the volume of their right posterior hippocampus. Of course, other explanations can be made for this result, such as longer and better performance as a taxi driver in those who happen to have a larger right hippocampus, but the studies in animals would tend to suggest that experience is the operative factor. Interestingly, the opposite relationships were found for the anterior hippocampus: it was smaller in the taxi drivers than in controls and the volume of this region correlated negatively with the months of taxi driving. Apparently, use or disuse of spatial maps leads to reorganization of the hippocampus, and this regional plasticity is a characteristic of the adult brain.

Other studies using rats have shown that development of the corpus callosum is also influenced by experience in early life, as shown in rats (Cowell and Denenberg, 2002; Denenberg, 1981). Handling male rat pups over the first 20 days of neonatal life leads to a larger corpus callosum and this experiential effect interacts with the level of sex hormones circulating in the pup such that handling plus elevated levels of testosterone cause the corpus callosum to become even larger than in rats that are simply handled. Estrogen interacts in the opposite direction, but here these details are not particularly relevant. The point is that the size of this important interhemispheric tract is influenced by experience, and this fact is so often ignored in studies comparing brain sizes.

Experience-dependent growth of the corpus callosum has also been shown in rhesus monkeys. Sánchez et al. (1998) found that the corpus callosum was smaller in rhesus monkeys raised by humans in a nursery for the first year of line than in rhesus monkeys raised in a semi-natural social environment.

These recent findings show us that the size of specific regions of the brain is influenced by experience in early life; social experience and exposure to novelty affect the size of the corpus callosum and the demands of forming spatial memories, and recalling them, affect the size of the hippocampus. Added to this, the sizes of different regions of the brain are modulated by experience throughout life. We can say now that the brain is in constant interaction with the environment and that use or disuse affects its size and neural circuitry. Of course, here we are talking about effects within a species. As far as we know, it is not possible to make one species equivalent to another through experience, even in the case of closely related species. Giving non-storing birds the opportunity to retrieve food strategically hidden inside small holes in artificial trees did not increase the size of their hippocampus (Krebs et al., 1996). The environmental demand had not changed the hippocampus of the non-storing species into that of the storing species. Hence, we can consider large differences between species from an evolutionary point of view, as characteristic of the species but malleable with experience.

ASSUMPTIONS/NEW NEURONS

The concept that "bigger is better" is the basis of most theories about the evolution of the human brain made by anthropologists and many primatologists. While this hypothesis has some validity when one is comparing closely related species, for example, apes and humans, recent knowledge about the avian brain certainly throws the assumption that bigger is always better into doubt. Birds can perform problem-solving tasks and other complex cognitive tasks just as well as can primates, despite the fact that they have very much smaller brains and, of more importance, a lower ratio of brain to body weight: pigeons have abilities surpassing those of humans in tasks requiring matching of rotated symbols, and they are able to form abstract concepts such as "sphericity," "water," "above," or "below" (e.g., Delius, 1985, 1987). In fact, the ability of pigeons to categorize objects in this way is remarkably similar to that of humans (Delius et al., 2000), but these functional similarities must be achieved through different neural processes since the mammalian brain and the avian brain are organized quite differently. There is another major difference between avian and mammalian brains: new neurons can be made readily in the adult avian brain but they are made quite rarely in the adult mammalian brain (Nottebohm, 1989).

No one knows why adult birds retain the ability to make new neurons in adulthood, but Nottebohm (1989) has said that they need to keep their brains light for flying. Hence, birds may vary the sizes of different parts of the brain at different times of the year as required. For example, the sizes of the song nuclei in the forebrain increase during the breeding season when singing is required. Presumably, at the same time, the sizes of other brain regions might shrink so that the increased size of the song nuclei can be accommodated within the cranium. Of course, there might be other means of accommodation such as diminishing the volume of the ventricles in the forebrain or decreasing the extracellular fluid in the gaps between neurons. So far no one has compared the size changes in the song nuclei with other regions of the same brain.

Although adult mammals cannot substantially increase the size of regions of their brains by making new neurons, they can, as mentioned previously, increase the size and number of connections between neurons depending on experience, and this expands the size of the particular brain region. Thus, even in mammals, the size of various brain regions is not fixed and not a result of biological predestination, but rather of the interaction between biological determining events and environmental factors throughout the life span.

The avian brain has solved its cognitive demands in quite a different way than has the mammalian brain, and its small size indicates nothing of its cognitive complexity. It may well be that the size of a particular, specific region of the brain correlates with the complexity of its specific behavioral function, but total brain size does not indicate a great deal about overall cognitive capacity, or "intelligence." In this way and other ways, comparison of the cognition of birds and primates has added greatly to our understanding and brought into sharp focus some of the assumptions that have been made about primate superiority.

CONCLUSION

The body of available evidence seems to suggest that increases in relative brain size go with increased cognitive abilities (sometimes relatively specific ones rather than any sort of overall "intelligence"). This is perhaps evidence that primates are special. However, recent research on birds undermines any complacent belief that that is the end of the matter. Study of the avian brain shows us that there are solutions to achieving higher cognition, of a level equivalent to that of the apes in some avian species, which do not depend on having a neocortex or highly developed interhemispheric connections (i.e., the corpus callosum). This should tell us that there is more than one route to higher cognitive functioning and the alternatives may be less dependent on increasing brain size. Within an order there is a base plan for brain development but experience can modify this, not only during development but also throughout life. In other words, the ground plan is not an absolute. We need to envisage the brain as a system of change in constant interaction with its environment. This move away from a static view of the brain leads us to address new questions and to focus on function (behavioral capacities) rather than, or at least in addition to, focusing merely on measuring the size of the brain, or even its various regions.

REFERENCES

Andrew, R. J. and Rogers, L. J., 2002, The nature of lateralization in tetrapods, in: *Comparative Vertebrate Lateralization*, L. J. Rogers and R. J. Andrew, eds., Cambridge University Press, Cambridge, pp. 94–125.

Barton, R. A., 1996, Neocortex size and behavioural ecology, *Proc. Roy. Soc. Lond.: B.* **263**:173–177.

Barton, R. A. and Harvey, P. H., 2000, Mosaic evolution of brain structure in mammals, *Nature* **405**:1055–1058.

Bayer, S. A. and Altman, J., 1991, *Neocortical Development*, Raven Press, New York.

Bennett, P. M. and Harvey, P. H., 1985, Relative brain size and ecology in birds, *J. Zool., Lond.* **207**:151–169.

Braitenberg, V. and Schüz, A., 1998, *Cortex: Statistics and Geometry of Neuronal Connectivity*, Springer, Berlin.

Byrne, R. W. and Whiten, A., eds., 1988, *Machiavellian Intelligence: Social Expertise and the Evolution of Intellect in Monkeys, Apes and Humans*, Clarendon Press, Oxford.

Cant, J. G. H., 1987, Positional behavior of female Bornean orang-utans (*Pongo pygmaeus*), *Am. J. Primatol.* **12**:71–90.

Cantalupo, C. and Hopkins, W. D., 2001, Asymmetric Broca's area in great apes: A region of the ape brain is uncannily similar to one linked with speech in humans, *Nature* **414**:505.

Chevalier-Skolnikoff, S. and Liska, J., 1993, Tool use of wild and captive elephants, *Anim. Behav.* **46**:209–219.

Clark, D. A., Mitra, P. P., and Wang, S-H., 2001, Scalable architecture in mammalian brains, *Nature* **411**:189–193.

Clutton-Brock, T. H. and Harvey, P. H., 1980, Primates, brains and ecology, *J. Zool., Lond.* **190**:309–323.

Connor, R. C., 1998, Quantifying brain–behavior relations in cetaceans and primates: Reply from R. C. Connor et al., *TREE* **13**:408.

Cowell, P. E. and Denenberg, V. H., 2002, Development of laterality and the role of the corpus callosum in rodents and humans, in: *Comparative Vertebrate Lateralization*, L. J. Rogers and R. J. Andrew, eds., Cambridge University Press, Cambridge, pp. 274–305.

Cowell, P. E., Waters, N. S., and Denenberg, V. H., 1997, The effects of early environment on the development of functional laterality in Morris maze performance, *Laterality* **2**:221–232.

Crow, T. J., ed., 2002, *The Speciation of Modern* Homo sapiens, Oxford University Press, Oxford.

Deacon, T. W., 1997, *The Symbolic Species: The Co-evolution of Language and the Brain*, W. W. Norton & Co., New York.

Deaner, R. O., Nunn, C. L., and van Schaik, C. P., 2000, Comparative tests of primate cognition: Different scaling methods produce different results, *Brain Behav. Evol.* **55**:44–52.

Delius, J. D., 1985, Cognitive processes in pigeons, in: *Cognition, Information Processing and Motivation*, G. D'Ydelvalle, ed., Elsevier, Amsterdam, pp. 3–18.

Delius, J. D., 1987, Sapient sauropids and hollering hominids, in: *Geneses of Language*, W. Koch, ed., Brockmeyer, Bochum.

Delius, J. D., Jitsumori, M., and Siemann, M., 2000, Stimulus equivalencies through discrimination reversals, in: *The Evolution of Cognition*, C. Heyes and l. Huber, eds., The MIT Press, Cambridge, MA, pp. 103–122.

Denenberg, V. H., 1981, Hemispheric laterality in animals and the effects of experience, *Behav. Brain Sci.* **4**:1–49.

DeVoogd, T. J., Krebs, J. R., Healy, S. D., and Purvis, A., 1993, Relations between song repertoire size and the volume of brain nuclei related to song: Comparative evolutionary analyses amongst oscine birds, *Proc. R. Soc. Lond. B* **254**:75–82.

Diamond, M. C., 1988, *Enriching Heredity: The Impact of the Environment on the Anatomy of the Brain*, The Free Press, New York.

Dunbar, R. I. M., 1992, Neocortex size as a constraint on group size in primates, *J. Hum. Evol.* **20**:469–493.

Dunbar, R. I. M., 2000, Causal reasoning, mental rehearsal, and the evolution of primate cognition, in: *The Evolution of Cognition*, C. Heyes and L. Huber, eds., The MIT Press, Cambridge, pp. 205–219.

Dunbar, R. I. M., 2001, Brains on two legs: Group size and the evolution of intelligence, in: *Tree of Origin: What Primate Behavior Can Tell us about Human Social Evolution*, F. B. de Waal, ed., Harvard University Press, Cambridge, pp. 174–191.

Dunbar, R. I. M. and Bever, J., 1998, Neocortex size predicts group size in carnivores and some insectivores, *Ethology* **104**:695–708.

Eccles, J. C., 1989, *Evolution of the Brain: Creation of the Self*, Routledge, London.

Falk, D., 1992, *Brain Dance*, Henry Holt, New York.

Finlay, B. L., and Darlington, R. B., 1995, Linked regularities in the development and evolution of mammalian brains, *Science* **268**:1578–1584.

Galdikas, B. M. F., 1982, Orang-utan tool use at Tajung Putting Reserve, Central Indonesian Borneo (Kalimantan Tengah), *J. Hum. Evol.* **10**:19–33.

Gannon, P. J., Holloway, R. L., Broadfieldd, D. C., and Braun, A. R., 1998, Asymmetry of chimpanzee planum temporale: Humanlike pattern of Wernicke's brain language area homolog, *Science* **2279**:220–222.

Gazzaniga, M. S., 2000, Cerebral specialization and interhemispheric communication. Does the corpus callosum enable the human condition? *Brain* **123**:1293–1326.

Gould, S. J., 1981, *The Mismeasure of Man*, Norton & Co., New York.

Güntürkün, O., 2000, Asymmetry pays: Visual lateralization improves discrimination success in pigeons, *Curr. Biol.* **10**:1079–1081.

Harvey, P. H. and Krebs, J. R., 1990, Comparing brains, *Science* **249**:140–146.

Hook-Costigan, M. A. and Rogers, L.J., 1996, Hand preferences in New World primates, *Int. J. Comp. Psychol.* **9**:173–207.

Hook-Costigan, M. A., and Rogers, L. J., 1998, Eye preferences in common marmosets (*Callithrix jacchus*): Influence of age, stimulus and hand preference, *Laterality* **3**:109–130.

Hopkins, W. D. and Marino, L., 2000, Asymmetries in cerebral width in nonhuman primate brains as revealed by magnetic resonance imaging (MRI), *Neuropsychologia* **38**:493–499.

Hopkins, W. D. and Rilling, J. K., 2000, A comparative MRI study of the relationship between neuroanatomical asymmetry and interhemispheric connectivity in primates: Implications for the evolution of functional asymmetries, *Behav. Neurosci.* **114**:739–748.

Hopkins, W. D., Marino, L., Rilling, J. K., and MacGregor, L.A., 1998, Planum temporale asymmetries in great apes as revealed by magnetic resonance imaging (MRI), *NeuroReport* **9**:2913–2918.

Humphrey, N. K., 1976, The social function of intellect, in: *Growing Points In Ethology*, P. P. G. Bateson and R. A. Hinde, eds., Cambridge University Press, Cambridge, pp. 303–317.

Innocenti, G. M. and Kaas, J. H., 1995, The cortex, *Trends Neurosci.* **18**:371–372.

Jerison, H. J., 1973, *Evolution of Brain and Intelligence*, Academic Press, New York.

Jerison, H. J., 1976, Principles of the evolution of the brain and behavior, in: *Evolution, Brain, and Behavior: Persistent Problems*, R. B. Masterton, W. Hodos, and H. Jerison, eds., Lawerence Erlbaum Associates, Hillsdale, pp. 23–45.

Jerison, H. J., 1994, Evolution of the brain, in: *Neuropsychology*, D. W. Zaidel, ed., Academic Press, San Diego, pp. 53–104.

Jolly, A., 1966, Lemur social behavior and primate intelligence, *Science* **153**:501–507.

Kaas, J. H., 1995, The evolution of the isocortex, *Brain Behav. Evol.* **46**:187–196.

Kamin, L. J., 1977, *The Science and Politics of IQ*, Penguin, Harmondsworth.

Kaplan, G. and Rogers, L. J., 1994, Race and gender fallacies: The paucity of biological determinist explanations of difference, in: *Challenging Racism and Sexism*, E. Tobach and B. Rosoff, eds., The Feminist Press, New York.

Kaplan, G. and Rogers, L. J., 1999, *The Orang-utans*, Allen & Unwin, Sydney.

Kaplan, G. and Rogers, L. J., 2001, *Birds: Their Habits and Skills*, Allen & Unwin, Sydney.

Kaplan, G. and Rogers, L. J., 2003, *Gene Worship: Moving Beyond the Nature/Nurture Debate Over Genes, Brain, and Gender*, Otherpress, New York.

Karten, H. J., 1991, Homology and evolutionary origins of the "neocortex," *Brain Behav. Evol.* **38**:264–272.

Krebs, J. R., Clayton, N. S., Healy, S. D., Cristol, C. A., Patel, S. N., and Jolliffe, A. R., 1996, The ecology of the avian brain: Food-storing memory and hippocampus, *Ibis* **138**:34–36.

Krubitzer, L., 1995, The organization of the neocortex in mammals: Are species differences really so different? *Trends Neurosci.* **18**:408–417.

Krubitzer, L., 1998, What can monotremes tell us about brain evolution? *Phil. Trans. R. Soc. Lond. B* **353**:1127–1146.

Kudo, H. and Dunbar, R. I. M., 2001, Neocortex size and social network size in primates, *Anim. Behav.* **62**:711–722.

Lefebvre, L., Whittle, P., Lascaris, E., and Fikelstein, A., 1997, Feeding innovations and forebrain size in birds, *Anim. Behav.* **53**:549–560.

Lefebvre, L., Gaxiola, A., Dawson, S., Timmermans, S., Rosza, L., and Kabai, P., 1998, Feeding innovations and forebrain size in Australasian birds, *Behaviour* **135**:1077–1097.

Lefebrve, L. N., Nicolakakis, N., and Boire, D., 2002, Tools and brains in birds, *Behaviour* **139**:939–973.

MacNeilage, P. F., Studdert-Kennedy, M. G., and Lindblom, B., 1987, Primate handedness reconsidered, *Behav. Brain Sci.* **11**:748–758.

Maguire, E. A., Gadian, D. G., Johnsrude, I. S., Good, C. D., Ashburner, J., Frackowiak, R. S. J., and Frith, C. D., 2000, Navigation-related structural change in the hippocampi of taxi drivers, *Proc. Natl. Acad. Sci. USA* **97**:4398–4403.

Marino, L., 1996, What dolphins can tell us about primate evolution, *Evol. Anthropol.* **5**:81–86.

Marino, L., 1998, A comparison of encephalization between odontocete cetaceans and anthropoid primates, *Brain Behav. Evol.* **51**:230–238.

Marino, L., 2002, Convergence of complex cognitive abilities in cetaceans and primates, *Brain Behav. Evol.* **59**:21–32.

McGrew, W. C. and Marchant, L. F., 1997, On the other hand: Current issues in and meta-analysis of the behavioral laterality of hand function in nonhuman primates, *Year Book Phys. Anthropol.* **40**:210–232.

McGrew, W. C. and Marchant, L. F., 1999, Laterality of hand use pays off in foraging success for wild chimpanzees, *Primates* **40**:509–513.

Nicolakakis, N. and Lefebvre, L., 2000, Forebrain size and innovation rate in European birds: Feeding, nesting and confounding variables, *Behaviour* **137**:1415–1427.

Noble, W. and Davidson, I., 1996, *Human Evolution, Language and Mind*, Cambridge University Press, Cambridge.

Nordeen, E. J. and Yahr, P., 1982, Hemispheric asymmetries in the behavioural and hormonal effects of sexually differentiating mammalian brain, *Science* **218**:391–394.

Northcutt, R. G. and Kaas, J., 1995, The emergence and evolution of mammalian cortex, *Trends Neurosci.* **18**:373–379.

Nottebohm, F., 1971, Neural lateralization of vocal control in a passerine bird. I. Song, *J. Exp. Zool.* **177**:229–261.

Nottebohm, F., 1989, From bird songs to neurogenesis, *Sci. Am.* Feb.:56–61.

Palmer, A. R., 2002, Chimpanzee right handedness reconsidered: Evaluating the evidence with funnel plots, *Am. J. Phys. Anthropol.* **118**:191–199.

Posthuma, D., De Geus, E. J. C., and Boomsma, D. I., 2001, Perceptual speed and IQ are associated through common genetic factors, *Behav. Genet.* **31**:593–602.

Povinelli, D. and Cant, J. G., 1995, Arboreal clambering and the evolution of self-conception, *Quart. Rev. Biol.* **70**:393–421.

Reader, S. M. and Laland, K. N., 2002, Social intelligence, innovation, and enhanced brain size in primates, *Proc. Natl. Acad. Sci. USA* **99**:4436–4441.

Riddell, W. I. and Corl, K. G., 1977, Comparative investigation of the relationship between cerebral indices and learning abilities, *Brain Behav. Evol.* **14**:385–398.

Rilling, J. K. and Insel, T. R., 1999a, The primate neocortex in comparative perspective using magnetic resonance imaging, *J. Hum. Evol.* **37**:191–223.

Rilling, J. K. and Insel, T. R., 1999b, Differential expansion of neural projection systems in primate brain evolution, *NeuroReport* **10**:1453–1459.

Rogers, L. J., 1997, *Minds of their Own: Thinking and Awareness in Animals*, Allen & Unwin, Sydney.

Rogers, L. J., 2000, Evolution of hemispheric specialisation: Advantages and disadvantages, *Brain Lang.* **73**:236–253.

Rogers, L. J., 2002, Lateralization in vertebrates: Its early evolution, general pattern, and development, in: *Advances in the Study of Behavior*, Vol. 31, P. J. B. Slater, J. S. Rosenblatt, C. T. Snowdon, and T. J. Roper, eds., Academic Press, San Diego.

Rogers, L. J. and Andrew, R. J., eds., 2002, *Comparative Vertebrate Lateralization*, Cambridge University Press, Cambridge.

Rogers, L. J. and Anson, J. M., 1979, Lateralization of function in the chicken forebrain, *Pharm. Biochem. Behav.* **10**:679–686.

Rogers, L. J. and Kaplan, G., 2003, *Spirit of the Wild Dog*, Allen & Unwin, Sydney.

Rumbaugh, D. M., 1997, Competence, cortex, and primate models: A comparative primate perspective, in: *Development of the Prefrontal Cortex: Evolution, Neurobiology, and Behavior*, N. A. Krasnegor, G. R. Lyon, and P. F. Goldman-Rakic, eds., Paul H. Brookes, Baltimore, MD, pp. 117–139.

Rumbaugh, D. M., Beran, M. J., and Hillix, W. A., 2000, Cause–effect reasoning in humans and animals, in: *The Evolution of Cognition*, C. Heyes and l. Huber, eds., The MIT Press, Cambridge, MA, pp. 221–238.

Sánchez, M. M., Hearn, E. F., Do, D., Rilling, J. K., and Herndon, J. G., 1998, Differential rearing affects corpus callosum size and cognitive function of rhesus monkeys, *Brain Res.* **812**:38–49.

Sawaguchi, T., 1992, The size of the neocortex in relation to ecology and social structure in monkeys and apes, *Folia Primatol.* **58**:131–145.

Semendeferi, K., Damasio, H., and Frank, R., 1997, The evolution of the frontal lobes: A volumetric analysis based on three-dimensional reconstructions of magnetic resonance scans of human and ape brains, *J. Hum. Evol.* **32**:375–388.

Seyfarth, R. M. and Cheney, D. L., 2002, What are big brains for? *Proc. Natl. Acad. Sci. USA* **99**:4141–4142.

Stark, J. M. and Ricklefs, R. E., eds., 1998, *Avian Growth and Development*, Oxford University Press, New York.

Uttal, W. R., 2001, *The New Prenology: The Limits of Localizing Cognitive Processes in the Brain*, MIT Press, Cambridge, MA.

Verstynen, T., Tierney, R., Urbanski, T., and Tang, A., 2001, Neonatal novelty exposure modulates hippocampal volumetric asymmetry in the rat, *Dev. Neurosci.* **12**:3019–3022.

Ward, J. P. and Hopkins, W. D., eds., 1993, *Primate Laterality*, Springer-Verlag, N.Y.

Whiten, A. and Bryne, R. W., 1997, *Machiavellian Intelligence II. Extensions and Evaluations*, Cambridge University Press, Cambridge.

Winter, de W. and Oxnard, C. E., 2001, Evolutionary radiations and convergences in the structural organization of mammalian brains, *Nature* **409**:710–714.

Wittling, W., 1995, Brain asymmetry in the control of autonomic-physiologic activity, in: *Brain Asymmetry*, R. J. Davidson and K. Hugdahl, eds., MIT Press, Cambridge, MA, pp. 305–357.

CHAPTER TEN

The Evolution of Lateralized Motor Functions

Michelle A. Hook

Among the most influential and controversial theories on the evolutionary origins of human handedness (a population-level bias for use of the right hand) is the "Postural Origins Theory" proposed by MacNeilage and colleagues (1987). This theory challenges the view that handedness is a characteristic unique to humans, and suggests that the evolutionary precursor of human handedness first emerged in the prosimians, the earliest primates. To overcome problems of postural control when feeding, prosimians evolved a division of function between the hands. Specifically, they evolved a left-hand (right hemisphere) specialization for visually guided reaching and a complementary right-hand (left hemisphere) preference for holding onto a branch, thereby facilitating postural control. According to the postural origins theory, these original specializations lay the foundation for further divisions of function between the hemispheres for controlling performance in manual tasks, including a right-hand preference for the performance of fine motor tasks in humans (MacNeilage et al., 1987). The evolutionary account posited by MacNeilage et al. (1987) was challenged on two fronts. Some researchers maintained that handedness in primates is not analogous to handedness in

Michelle A. Hook • Department of Psychology, Texas A&M University, College Station, TX, USA.
Comparative Vertebrate Cognition, edited by Lesley J. Rogers and Gisela Kaplan. Kluwer Academic/Plenum Publishers, 2004.

humans (Annett, 1987; Corballis, 1987; Kolb and Fantie, 1987; Warren, 1987). Other researchers cited evidence of population-level biases for motor function in non-primate species, and asserted that hemispheric specialization for motor control must have evolved before the emergence of primates (Bradshaw, 1987; Denenberg, 1988).

The controversy generated by the postural origins theory underscores the significant impact that it had on the field of laterality research. First, this theory led to the resurgence of research on laterality of hand use in non-human primates. Over the last 15 years, many studies have investigated whether hand preferences are present in diverse primate populations. Moreover, primate studies began to consider not only whether laterality is present, but also the factors that might affect the distribution of hand preferences in primates, including age (e.g., Hook and Rogers, 2000; Hopkins and Bard, 2000), task demands (e.g., Fagot and Vauclair, 1991; Hook-Costigan and Rogers, 1996; King, 1995), and experience (e.g., Hook-Costigan and Rogers, 1996; Hopkins, 1995; Hopkins et al., 2001). Comparative data on the lateralized hand use of diverse primate species has provided a basis from which we can now consider the mechanisms underlying the evolution of manual specializations.

Motor preferences are present, however, in species whose ancestors diverged very early from the line of evolution that gave rise to primates. Parrots, for example, show footedness for feeding (Friedman and Davis, 1938; Rogers, 1989). Cats also show "pawedness" in a variety of situations. For example, Fabre-Thorpe et al. (1993) found left pawedness in cats on a task that required them to track a moving spot of light. Population-level biases (meaning that a majority of individuals in the population display the same direction of lateralization) referred to by the terms "pawedness," "handedness," and "footedness" are suggestive of hemispheric mediation of task performance. Laterality expressed at an individual level only (meaning that individuals display strong preferences for one direction, but the laterality is not necessarily expressed at a group level) need not be related to hemispheric specializations. As laterality is present at both the population and individual levels in parrots and cats, hemispheric asymmetries for the control of the limbs in the performance of motor tasks may be present in birds and non-primate mammals, and perhaps in their common ancestors.

Given the evidence for population-level asymmetries for motor control in non-primate species, it seems unlikely that primates are unique in terms of their hemispheric specializations. The question that remains, therefore, is whether hand preference biases present in primates are qualitatively different from those expressed by non-primate species. Whereas motor biases expressed by non-primate species may be exclusively task specific, consistent lateralization across tasks requiring particular forms of motor control may emerge in primate species with closer evolutionary links to humans. Humans display right handedness on most manual tasks. As such, it has been suggested that handedness in humans is due to a left hemisphere specialization for motor control. To date, no studies have found consistent lateralization across a variety of manual tasks in primate or non-primate species, but there are similarities between human and non-human laterality that provide clues about how hemispheric specializations may have evolved (Andrew, 2002).

Over almost the last two decades considerable data have been collected on motor asymmetries in a wide range of species, including fish, amphibians, birds, non-primate mammals, and primates (summarized in Rogers, 2002a; Rogers and Andrew, 2002). This chapter addresses whether non-human primates indeed represent a significant transition in the evolution of motor lateralization, as MacNeilage et al. (1987) suggested. Because of the considerable amount of information to be reviewed, this chapter does not address the influence of subject variables such as sex and age on the expression of handedness. The approach taken is to review the current evidence for lateralized motor functions in diverse species throughout the evolutionary continuum, and to ascertain whether there is an explicit point in evolution (i.e., with the evolution of primates) at which hemispheric specialization for motor function may have evolved.

WHOLE-BODY TURNING

Turning biases are recorded when a subject rotates all or a part of its body in one direction. These biases are a useful index of motor asymmetries as they can be assessed in species throughout the evolutionary continuum, including species that evolved as early as fish and amphibians.

Lower Vertebrates: Fish, Amphibians, and Reptiles

Fish show population-level biases when turning. However, the direction of population-level biases appear to be task- and species-specific. In a comparative study of 16 fish species, Bisazza et al. (2000) found that 5 species turned right, 5 turned left, and 6 had no directional preference (at a population level) when required to swim around a barrier to face a simulated predator. Leftward turning biases were reported in two additional species of fish observed when detouring a barrier to approach a group of female fish (*Gambusia nicaraguensis, Poecilia reticulata*, Bisazza et al., 1997b). Moreover, using a similar paradigm, Bisazza et al. (1997a) found that male mosquito fish (*Gambusia holbrooki*) show a leftward turning bias when required to swim around a barrier to approach a simulated predator, but they display a rightward bias when detouring around an opaque barrier. Turning biases in fish, therefore, are contingent on the visual stimulus the fish are approaching and may be dependent on perceptual lateralization (Facchin et al., 1999).

Amphibians also display population biases for whole-body turning. The smooth newt (*Triturus vulgaris*) displays a leftward turning bias during courtship and spermatophore deposition (Green, 1997), although no such population bias is seen in the alpine newt (*Triturus alpestris*; Marzona and Giacoma, 2002). Toads show population biases for "righting"; rolling to an upright position after the body has been overturned. Like turning biases, righting probably requires use of the axial musculature to coordinate movement of the whole body. The South American cane toad (*Bufo marinus*) pivots preferentially to the left side when overturned underwater and does so by applying pressure with its right forelimb (Bisazza et al., 1996). On the ground, the European green toad (*Bufo viridus*) also rolls to the left (Robins et al., 1998). By contrast, *Bufo bufo* (the European common toad) and *B. marinus* roll to the right when overturned on the ground (Robins et al., 1998). The task specificity of righting biases in *B. marinus* may depend in part on the musculature used to control the response. On the ground the hindlimbs are used to pivot the body, and thus asymmetries in this condition may depend primarily on appendicular muscles that control the limbs rather than the axial musculature that is probably used underwater. Nonetheless, the evidence suggests that turning biases in toads are present at a population level and, like fish, are task- and species-specific (for more details, see Malashichev, 2002; Rogers, 2002b).

Unlike fish and amphibians, reptiles do not appear to be lateralized at a group level for whole-body turning. Casey and Sleigh (2001) observed which way leopard geckos (*Eublepharis macularis*) and snapping turtles (*Chelydra serpentina*) turned to enter one of two laterally placed identical enclosures. Neither species displayed a population-level bias for turning. Moreover, individual subjects did not display consistent turning biases across the two testing sessions in which they were observed. Differences in how the fish, amphibians, and reptiles were tested may have influenced their lateralization. The fish and amphibians were tested with perceptual stimuli (e.g., predators) and affective states (e.g., courtship) that may have influenced their behavior, whereas the reptiles were observed in neutral conditions. In the absence of confounding sensory input, geckos and turtles may not display population-level turning biases. It would be interesting to look at the interaction between perceptual cues and rotational biases in reptiles, to determine whether lateral asymmetries are also task dependent in these species.

Birds

Leftward turning biases have been documented for several avian species. Domestic chicks (*Gallus gallus*) and bobwhite quail (*Colinius virginianus*) chicks display strong group biases to turn left in a T-maze (Casey and Sleigh, 2001). Japanese quail chicks (*Corturnix corturnix japonica*), however, demonstrate no group-level turning bias on the same task. Casey and Sleigh (2001) suggest that the differential lateralization of these species of chicks may be related to their prenatal sensory experiences. In the egg, chicks (*G. gallus*) are oriented toward the top of the egg with the right side facing outward. This embryonic position provides more prenatal sensory experience to the right side of the body, and seems to facilitate the development of postnatal behavioral asymmetries (Rogers, 1991). Casey and Lickliter (1998) have shown that changing the prenatal environment disrupts the population-level turning bias displayed by bobwhite quail. It appears that, in birds, even prenatal sensory input plays a large role in the development of subsequent motor asymmetries.

Non-Primate Mammals: Rodents, Dolphins, Cats, and Dogs

Turning biases of rats, mice, and hamsters have been observed in the T-maze (Rodriguez and Afonso, 1993), during spontaneous nocturnal circling (Glick and Cox, 1978), and during swimming (Schmidt et al., 1999). Overall, it appears that rodents express strong individual turning biases that are symmetrically distributed at a population level (Glick and Cox, 1978; Schmidt et al., 1999). Some evidence for slight biases for rightward turning has been reported for hamster pups separated from their litter and rotating in a cylindrical area (Uziel et al., 1996), and strong rightward turning biases were found for rats tested in electrified T-mazes (Castellano et al., 1989). This suggests that the lateralization of turning biases in rodents may be influenced by affective states (e.g., fear), with rightward turning being observed in arousing conditions.

Rotational behaviors in rodents are associated with a hemispheric imbalance of dopaminergic activity within the basal ganglia (Glick and Cox, 1978; Glick et al., 1976; Zimmerberg et al., 1974). Unilateral lesions of the dopaminergic nigrostriatal system resulted in rats circling toward the side of the lesion. This suggest that the intact, contralateral nigrostriatal region had predominantly assumed control of circling behavior. Similarly, unilateral treatment with 6-hydroxydopamine, which selectively destroys dopaminergic neurons, resulted in rats turning toward the lesioned hemisphere, again suggesting contralateral control of the subsequent circling behavior. In normal rats, the dopamine levels in the left and right striata differ by about 10–15 percent (Glick et al., 1974) and this asymmetrical distribution of dopamine in the nigrostriatal system appears to be the primary system involved in the mediation of individual spontaneous circling biases in rats (Glick et al., 1976).

Glick and Shapiro (1985) suggested that circling behavior may, in fact, be a stereotyped form of spatial behavior. Dopamine asymmetries do seem to be correlated with side preferences on some spatial tasks (Zimmerberg et al., 1974), and rats with stronger circling biases seem to be better at discriminating right from left (Zimmerberg et al., 1978). Population biases for the performance of spatial tasks (Adelstein and Crowne, 1991; Cowell et al., 1997), however, suggest that there is not a direct relationship between hemispheric control of spatial behaviors and circling preferences. Bradshaw and Rogers (1993) suggest that spatial behaviors per se may be mediated by asymmetries of dopamine levels in the cortex, whereas individually lateralized behaviors, such as circling, are primarily influenced by asymmetries in the nigrostriatum. Thus, hemispheric specialization for spatial function influences circling behavior, but may not directly mediate this form of motor asymmetry.

Dolphins (*Tursiops truncatus*) spontaneously exhibit stereotypic swimming in circles. Interestingly, at a group level they tend to swim in a leftward direction (Sobel et al., 1994). Moreover, rotational biases of individuals are robust and persist despite attempts to force them to switch their direction of bias.

Pharmacologically induced rotation has also been observed in dogs and cats. When administered apomorphine (0.4 mg/kg, i.v.) dogs begin to run in circles and continue without interruption for about 1 hr (Nymark, 1972). Of 19 dogs observed, 13 ran toward the right, and 5 ran in a leftward direction (Nymark, 1972). Similarly, of nine cats observed during amphetamine-induced rotation, six rotated to the right and three rotated to the left (Glick et al., 1981). These data are not only suggestive of a rightward turning bias in these species, they also suggest that an asymmetry in dopaminergic function may underlie rotation behavior in dogs and cats, as found for rats.

Non-Human Primates

There have been relatively few studies of rotational behavior in non-human primates. In fact, studies thus far have been limited to prosimians and capuchin monkeys. For the prosimians, lemurs (*Microcebus murinus*) and galagos (*Galago moholi*) display left-sided whole-body turning biases (Dodson et al., 1992). Subjects were scored when they made a whole-body rotation (90°–360°) around the long axis of the body. By contrast, Ward and Cantalupo (1997) found no evidence of a group bias for one direction of whole-body turning in small-eared bushbabies (*Otolemur garnettii*). They found that most subjects were ambipreferent for turning in both a quadrupedal and bipedal posture. Westergaard and Suomi (1996a) also found a symmetrical distribution for rotational biases in tufted capuchin monkeys (*Cebus apella*). Whole-body turns were recorded when subjects were calm, and made uninterrupted 180° turns. Like non-primate species, rotational biases appear to be species-specific and may depend on the subject's affective state and sensory conditions.

Summary of Turning Biases

Whole-body turning biases are clearly expressed by species throughout the evolutionary continuum. In fact, rather than primates displaying greater lateralization of function at population and individual levels, they seem less to be biased than early vertebrate species and non-primate mammals, at least as far as we can tell from the species tested so far. Any claim that non-primate species are more lateralized for rotation than primates, however, would be premature because of the paucity of studies in this area. In fact, differences in the lateralization of rotational behavior in primates compared to non-primate species may be due in part to variations between methodological procedures adopted across studies. In fish, lateral biases for turning, at individual and population levels, seem to be contingent on the perceptual stimuli used to induce rotation: if so, summation of turning across stimuli would mask any population bias for turning in these species. Moreover, the lateralization of rotational behavior in birds seems to depend on prenatal sensory experiences, and rats only express population biases for turning in arousing contexts. These data suggest that whole-body turning might be strongly influenced by perceptual lateralization in early vertebrates and potentially in mammals. From an evolutionary perspective, therefore, it seems likely that perceptual biases preceded this form of motor asymmetry even though the behavior clearly involves the motor system, specifically the axial musculature. Rotational asymmetries in fish and early vertebrates, generating biases in the axial musculature, may have been the foundation for further evolution leading to other forms of motor asymmetry. The association between turning and asymmetrical distributions of neurobiological substrates (e.g., dopamine) in rats, cats, and dogs, show definitively that mechanisms underlying turning behavior in mammalian species have been retained across species.

HAND PREFERENCES FOR SIMPLE ACTIONS

Use of the forelimbs, or forepaws, when reaching for and holding food has been studied extensively in non-human primates. Less work has been done on the laterality of forelimb functions in non-primate mammals and particularly few studies have looked for potential handedness in lower vertebrates. This unequal distribution of studies of handedness underscores a qualitative shift in the motor abilities of species throughout the evolutionary continuum. Primates are special in terms of their manual dexterity. The evolution of a pseudo-opposable thumb in non-human primates facilitated use of the hands for food holding and manipulative function. Nevertheless, lower vertebrates and non-primate mammals do use the forelimbs for swiping, reaching and for simple food holding, and toads have population-level lateralization for some of these behaviors.

Lower Vertebrates

European toads (*B. bufo*) use the right forepaw preferentially to remove an elastic balloon wrapped around their head (Bisazza et al., 1997c). They also display right pawedness when required to wipe wet pieces of paper off their snouts (Bisazza et al., 1996). By contrast, toads from two other species (*B. marinus, B. viridius*) display individual lateralization on the "paper wiping task," but neither species displays a group-level bias for this behavior. Therefore, paw preferences appear to be species-specific in anurans.

Birds

Birds use their feet for feeding, and population biases have been documented in these species. Friedman and Davis (1938) reported left footedness for feeding in several species of African parrots. Of 15 species studied, 6 species used the left foot preferentially when manipulating and holding food while one species was right footed (reanalyzed in Rogers, 1980). Australian parrots and cockatoos also display footedness during feeding. Rogers (1981) found that eight of nine species of Australian birds were left footed and one, the crimson rosella, was right footed. On the whole, it appears that most species of parrot display left footedness for feeding activities.

Chickens (*G. gallus*) also display footedness when scratching the ground for food. Rogers and Workman (1993) scored the foot used by six chicks to initiate bouts of scratching. They found a significant tendency for chicks to initiate scratching with the right foot. These observations were replicated by Tommasi and Vallortigara (1999). As noted in Bradshaw and Rogers (1993), this right-foot preference correlates with the specialization of the right eye found in tasks requiring chicks to search for food hidden among pebbles (Andrew et al., 1982; Zappia and Rogers, 1987). Again, there appears to be a strong relationship between motor and perceptual lateralization in avian species.

Non-Primate Mammals: Rodents

Peterson (1934) observed the paw preferences of rats when reaching in a tube to obtain a food reward. He found that individuals displayed strong preferences for one paw, but these preferences were symmetrically distributed at a population level.

Initial studies of mice also suggested that 50 percent of mice are left handed and 50 percent are right handed when reaching into a medially placed cylindrical tube for food (Collins, 1975). In more recent, large-scale replications of Collins' original study, significant, albeit slight, population biases for left pawedness have been found for inbred strains of mice (Signore et al., 1991; Waters and Denenberg, 1994). Overall, most studies suggest that paw preferences on Collins' test of medial reaching are symmetrically distributed in rats and mice with a very weak bias toward left pawedness suggested in some studies.

Paw preferences in mice, however, appear to be task-specific. Waters and Denenberg (1994) tested mice on a task that required them to select one of two hoppers from which to take food. The hoppers were placed laterally so that the mice could access one with the left hand and the other with the right hand (Waters and Denenberg, 1991). By comparing the amount of food consumed from the left and right hoppers, respectively, Waters and Denenberg (1994) determined that 61 percent of mice are right pawed. These same mice were left-pawed when required to reach into a medially placed cylindrical tube. Rogers and Bulman-Fleming (1998) confirmed the dissociation between these two measures of paw preferences in mice. Although they did not find evidence of left or right pawedness on either the medially placed or laterally placed reaching tasks, they found that paw preferences on the separate tests were not correlated. Rogers and Bulman-Fleming (1998) suggest that spatial preferences may influence reaching on the laterally placed reaching task and not influence medial reaching. Thus, at least two separate lateralized subsytems may guide simple reaching behavior in mice.

One of the most interesting findings from studies of laterality in mice is the evidence for heritability of the degree, and possibly the direction of paw preferences. There is strong evidence that the degree of preference for one hand, irrespective of direction, is under genetic control in mice. Collins (1991) was able to produce strongly lateralized and weakly lateralized mice by selectively inbreeding individuals with strong or weak paw preferences. Significant differences between strongly lateralized and weakly lateralized populations persisted, with random breeding, for at least 16 generations (Collins, 1991). Moreover, recent studies suggest that the direction of paw preferences may also be influenced by genetic factors (Biddle et al., 1993; Waters and Denenberg, 1994). Because of substantial knowledge about the genetics of mice, these species are *ideal* models for further understanding the role of genetic factors in the expression of lateralization. Further investigation of the relationship between genes and laterality in mice may provide important clues about the mechanisms underlying the evolution of motor preferences.

Non-Primate Mammals: Cats and Dogs

To elicit use of the paw, rather than the mouth, when feeding, most studies using cats have required them to reach into containers to obtain food. Cats display individual paw preferences on these simple reaching tasks, but the distribution of preferences at the population level is inconsistent between studies. Pike and Maitland (1997) found that 90 percent of cats show paw preferences for reaching, and there was an equal distribution of left- and right-paw preferent cats in the group. Similarly, Tan and Kutlu (1991) found that paw preferences in a large group of 109 cats were symmetrically distributed at a group level, with most cats displaying an individual preference for reaching with one paw. By contrast, other studies have found "pawedness" using simple reaching tests. Cole (1955) found left pawedness in cats trained to reach into a glass tube to obtain a piece of meat. It should be noted that in this study most of the cats were ambipreferent, but more of those with preferences preferred to reach with the left paw. Forward et al. (1962) also report a tendency toward left pawedness in cats observed reaching into containers: 6 were right pawed and 10 were left pawed. When a tendency for pawedness appears in cats, therefore, it appears to be for the left paw.

By contrast, dogs seem to be right pawed. Tan (1987) scored the paw use of 28 dogs while they tried to remove pieces of sticking plaster that covered their eyes. In this group, 16 dogs were right pawed, 5 were left pawed, and 7 did not display a preference. Differences between the dogs and cats may be related to variations in task demands rather than a phylogenetic shift in laterality. On the one hand, placing plaster over both eyes is likely to be stressful for dogs. Stress may influence lateralization. Indeed, right turning biases were found in mice when they were placed in an arousing electrified T-maze (Castellano et al., 1989). Alternatively, removing plaster from the eye may have strong manipulative components and this might cause a shift toward right pawedness (if manipulation is linked to right pawedness).

Non-Human Primates

Many prosimians display left handedness during feeding. Ward et al. (1990) conducted a large-scale study of 194 lemurs from 6 species (*Lemur catta, Lemur coranatus, Lemur macaco, Lemur mongoz., Lemur rubriventer, Lemur fulvus*). They found a bias toward left handedness, particularly in males. Similar biases toward left handedness have been found for black- and white-ruffed lemurs (*Varecia variegata variegata*, Forsythe et al., 1988), mouse lemurs (*M. murinus, Microcebus rufus*, Pohl et al., 2000), lesser bushbabies (*Galago senegalensis*, Sanford et al., 1984), and small-eared bushbabies (*O. garnettii*, Ward and Cantalupo, 1997). Some species do not seem to show handedness at the population level (Mason et al., 1995; Stafford et al., 1993), but the main body of data does support the postural origins hypothesis and suggests that left handedness is characteristic of prosimians.

Despite the prevalence of left handedness in prosimians, hand preferences in New and Old World monkeys seem to be species-specific. Most monkeys display individual hand preferences, but handedness is not consistently expressed in either the platyrrhine or catarrhine infraorders. Among the New World primates, squirrel monkeys (Laska, 1996b) display left handedness in feeding. Marmosets (Box, 1977: de Sousa et al., 2001; Hook and Rogers, 2000; Matoba et al., 1991) and spider monkeys (Laska, 1996a) do not display handedness for reaching for and holding food, whereas tamarins (Diamond and McGrew, 1994; King, 1995), muriquis (Ades et al., 1996), and capuchins (Masataka, 1990; Westergaard et al., 1997; Westergaard and Suomi, 1993a) are right-handed when taking food. It may be argued that New World primates are in transition from left-handed ancestors toward right handedness, as found in humans. As the specific order in which the New World primates evolved is not really known (Ferrari, 1993), however, we can only speculate that handedness in New World primates represents a transitional state.

Similarly, hand preferences in Old World monkeys vary across species. Simple hand use appears to be unlateralized in wild Hanuman langurs (*Presbytis entellus*), with few individuals displaying significant hand preferences (Mittra et al., 1997). By contrast, captive Hanuman langurs (Neves and Dolhinow, 1995), baboons (*Papio papio*, Vauclair and Fagot, 1987), and patas monkeys (*Erythrocebus patas*, Teichroeb, 1999) display individual hand preferences, but symmetrical distributions of hand use biases at a group level. Japanese (*Macaca fuscata*, Kubota, 1990; Tokuda 1969) and rhesus macaques (*Macaca mulatta*, Brookshire and Warren, 1962; Drea et al., 1995) display left handedness, and golden and leaf monkeys (*Rhinopithecus roxellanae* and *Presbytis* sp.) are reported to be right handed (Yuanye et al., 1988).

Between species variation is even evident among closely related macaque species. In a recent comparative study, Westergaard et al. (2001) observed left handedness in rhesus macaques (*M. mulatta*) and adult pig-tailed macaques (*M. nemestrina*), whereas cynomolgus macaques (*M. fascicularis*) did not display a group bias when reaching for food. They suggest that the inconsistencies among closely related species may relate to interspecies differences in temperament. Focusing on the comparisons between rhesus and cynomolgus macaques, Westergaard et al. (2001) suggest that left handedness may be associated with relatively high levels of negative stress in the captive environment. This is not a new idea. Cameron and Rogers (1999) suggested that left handedness may be associated with avoidance behavior in marmosets (*Callithrix jacchus*): right-handed marmosets entered a novel room faster and displayed more exploratory behaviors than left-handed marmosets. Right-handed chimpanzees also approach novel objects faster and display more exploration than their nonright-handed counterparts (Hopkins and Bennett, 1994). Clearly, differences in temperament are associated with the hand preferences of captive primates. Further investigations on temperament, as a potential causal factor underlying species differences in handedness, are warranted.

Studies of hand use in the apes have reported a symmetrical distribution of preferences, or a slight bias for one hand, when holding food or reaching for food. Olson et al. (1990) assessed hand preferences in gorillas (*Gorilla gorilla*), orangutans (*Pongo pygmaeus abelli*), and gibbons (*Hylobates lar*). They found individual lateralization, but no group lateralization when individuals retrieved food from the floor. Similar symmetrical distributions of individual hand preferences during feeding activities have been reported for wild orangutans (*P. pygmaeus pygmaeus*, Rogers and Kaplan, 1996) and other captive gorillas (*G. gorilla*, Annett and Annett, 1991). Olson et al. (1990) found, however, that when subjects were required to adopt a bipedal posture, to get food off wire mesh, the gibbons showed a significant left-hand bias at the population level whereas the gorillas were significantly right-handed. A bipedal posture facilitated the expression of population-level biases. The difference between the direction of population biases in the gorillas and gibbons may be explained by differences in the natural environments of these two species. While the gorilla is primarily terrestrial, the gibbon is an arboreal species. As MacNeilage et al. (1987) suggested, arboreal primates may show left handedness for simple reaching as, in adapting to the postural constraints of the natural environment, they use their right hand to support themselves while reaching. Being freed from postural constraints, terrestrial primates may have evolved a generalized preference for use of the right hand, and hence gorillas display right handedness.

The findings of postural effects on handedness have been replicated in chimpanzees and bonobos. For quadrupedal reaching and food holding, most studies report a symmetrical distribution of individual preferences in captive (Colell et al., 1995; Hopkins and Bard, 2000; Hopkins et al., 1994) and wild (Marchant and McGrew, 1996; Sugiyama et al., 1993; Tonooka and Matsuzawa, 1993) populations. Reaching from a bipedal posuture has been shown to intensify individual preferences (Colell et al., 1995; Hopkins et al., 1993), and in some cases to expose right handedness in populations of chimpanzees (Hopkins, 1993; Hopkins et al., 1993).

This strong effect of posture on hand preferences is not limited to the ape species. The assumption of a bipedal or vertical posture appears to increase right-hand use in bushbabies (Ward and Cantalupo, 1997), lion tamarins (Singer and Schwibbe, 1999), squirrel monkeys (King and Landau, 1993), muriquis (Ades et al., 1996), capuchins (Spinozzi et al., 1998; Westergaard et al., 1997), rhesus macaques (Westergaard et al., 1998), gorillas (Olson et al., 1990), and chimpanzees (Hopkins, 1993; Hopkins et al., 1993). In other species (cotton-top tamarins, bonobos), however, right-hand use decreases with adoption of a vertical posture (de Vleeschouwer et al., 1995; Diamond and McGrew, 1994) or individual preferences are intensified without affecting the distribution of lateralization at the population level (Parr et al., 1997). The effects of posture on hand preferences in primates are still controversial and warrant further investigation. However, the finding of shifts toward right-hand use with the adoption of a vertical posture in many primate species, suggests that the adoption of a vertical posture may have significantly impacted the course of evolution toward human right handedness.

In bipedal postures, primates are able to display coordinated bimanual hand use and on these tasks there seems to be a preponderance of evidence for "handedness." In fact, most studies have found right handedness on bimanual tasks requiring subjects to hold a tube with one hand while extracting food from inside it with the other hand. Of 26 capuchins, 19 preferred to use the right hand to scrape banana from a tube (holding the tube with the left hand; Spinozzi et al., 1998). Similarly, rhesus macaques and chimpanzees display population-level right handedness for extracting peanut butter from tubes (Hopkins, 1995; Westergaard and Suomi, 1996a). Byrne and Byrne (1991) also found significant right handedness in wild gorillas (*G. gorilla beringei*) for the fine manipulation components involved in the bimanual processing of spiny thistle leaves before consuming them. Coordinated bimanual tasks seem to frequently elicit population-level hand use biases. It is acknowledged that some studies have not found handedness for bimanual food processing (e.g., Harrison and Byrne, 2000) but most have, and it is for right handedness.

Summary of Hand Preferences for Simple Actions

Despite the prevalence of left handedness for simple hand use in prosimians, there is no evidence that this form of lateralization was subsequently retained by later evolving monkey and ape species. No clear pattern emerges for handedness in New and Old World monkeys: some species are left handed, some are right handed, and some do not display handedness. Moreover, there is evidence for population-level lateralization for paw use in species whose ancestors evolved before primates. Some species of toads are right handed, most birds are left-handed, and dogs are right-handed when prying sticking plaster off their eyes. The prevalence of individual lateralization across the evolutionary continuum suggests that the potential for lateralization of motor function may be inherent in all (or most) species. Nevertheless, the haphazard appearance of population biases indicates that handedness for simple food holding and reaching tasks are not evolutionary stable characteristics. Instead, it would seem that the expression of hand preferences for such unimanual tasks may be influenced by a variety of contextual, experiential, and genetic factors.

In primates, however, right handedness emerges when tasks require subjects to assume a bipedal posture or demand coordinated bimanual hand use. *Most* species display increased right-hand use when they adopt a vertical posture and reach or hold food and *all* evidence of handedness in bimanual tasks collected thus far are indicative of right handedness. These bimanual hand use data provide the first indication of population-level handedness in primates that may not be species-specific. Whether this form of handedness is representative of the evolutionary predecessor of human handedness for writing, hammering, and other tasks, awaits further investigations with bimanual tasks in a greater diversity of non-human primate (*and* non-primate) species.

In an unpredictable environment, one might benefit from the use of a particular hand at any given time. Therefore, having only one hemisphere able to perform a simple motor task would not be adaptive. For example, what would an exclusively left-handed subject do when required to retrieve a fruit only accessible by use of the right hand? How would an exclusively right-handed subject catch an insect that passes on the left side of the body? In a coordinated bimanual situation, however, it is probably beneficial to develop a systematic routine that maximizes productiveness. A division of function between the hands would be beneficial.

COMPLEX VISUOSPATIAL TASKS

Non-Primate Mammals: Cats

The visuospatial reaching preferences of cats have been studied in a variety of complex tasks. Burgess and Villablanca (1986) observed the paw used by cats when trying to catch a swinging piece of string. Fourteen of the seventeen cats displayed a left-paw preference on this task. When tracking a moving spot of light, approximately 50 percent of cats were ambipreferent (Fabre-Thorpe et al., 1993) but, of the remaining 50 percent of subjects, 17 were left pawed and 6 were right pawed. The number of left-pawed cats significantly outnumbered the number of right-pawed cats. In fact, the left paw was shown to be quicker to respond and more accurate than the right paw when reaching toward a moving target (Lorincz and Fabre-Thorpe, 1996). This left-paw bias when reaching for a moving target may be indicative of specialization of the right hemisphere for processing spatial information. Moreover, it is consistent with the left hand specialization that MacNeilage et al. (1987) suggested evolved in non-human primates for visuospatial reaching.

Interestingly, the strong bias for the left paw in cats decreases with practice on the task (Lorincz and Fabre-Thorpe, 1996). When required to track a moving light, cats initially (first 200 trials) showed a strong preference for the left paw (used 81 percent of the time). Ten of the twelve cats tested displayed a significant individual preference for the left paw, one cat was right pawed and one was ambipreferent. However, after several months of practice on the task, the percentage of left paw use decreased to 71 percent of total trials. After practice, nine cats were left pawed and three were right pawed. Lorincz and Fabre-Thorpe (1996) suggest that the increase in right paw use with practice may be due to the initially incompetent hemisphere acquiring the ability to solve the task. Indeed, the performance advantages initially observed for the left hand (faster reaction time and superior accuracy) also disappeared after practice. These data have implications for other studies of hand use in non-primate and primate species. Left-hand biases for visually guided reaching may be masked by practice effects, and novel tasks may be needed to expose population biases on these tasks (Fagot and Vauclair, 1991). Moreover, in species that rely on accurate visuospatial assessment for survival (i.e., in an arboreal environment), both hemispheres are likely to be capable of performing visuospatial tasks. The absence of strong group biases for reaching in primate and non-primate species may be influenced by not only the familiarity of the specific task, but also by the extensiveness of use of the underlying hemispheric function.

Non-Human Primates

Hook-Costigan et al. (1999) recorded the hand preferences of 11 male and 11 female bushbabies (*O. garnettii*) when capturing live crickets and swimming goldfish. When reaching for crickets, 50 percent of the bushbabies did not display a significant hand preference. Of the 22 subjects, 11 were ambipreferent, 5 were left handed, and 6 were right handed. Thus, the bushbabies' hand preferences on this task were normally distributed and did not differ from that expected by chance. Seven of the twenty-two bushbabies were left handed when reaching for fish, ten were right handed, and only five were ambipreferent. Therefore, in contrast to the cricket catching task, a significant majority of the bushbabies displayed hand preferences when fishing. However, this species did not display handedness for the tasks. It is quite possible that this nocturnal prosimian species does not have hemispheric specialization for visually guided reaching.

The visuospatial reaching preferences of diurnal marmosets, tamarins, squirrel monkeys, and capuchins have also been recorded when reaching for moving objects and live prey. Hook-Costigan and Rogers (1995, 1996) recorded the hand preferences of marmosets reaching for food on a swinging piece of string, and on rotating discs. The marmosets displayed very strong hand preferences, particularly on the string test, but these preferences were bimodally distributed at the group level. Similarly, the hand preferences of tamarins were bimodally distributed on a task requiring them to retrieve food from a rotating disc (King, 1995).

Singer and Schwibbe (1999) recorded the hand preferences of *Callithrix* spp., *Saguinus* spp., and *Leontopithecus* spp. when catching large, mobile crickets. In contrast to the bushbabies, individual marmosets and tamarins displayed strong and significant hand preferences. Moreover, *Leontopithecus* displayed a group bias for right handedness on this task: 7 of 11 subjects were right handed. This right handedness is in contrast to the left handedness found for squirrel monkeys when catching live fish in wading pools (King and Landau, 1993). There have been too few studies of visuospatial reaching in New World primates to determine whether these species have handedness for these tasks, and to postulate about the mechanisms underlying the expression of preferences in marmosets, tamarins, and squirrel monkeys.

Studies of visually guided hand use in Old World primates are also sparse. Kawai (1967) recorded the hand preferences of Japanese macaques (*M. fuscata*) when they were catching sweet potatoes. He found that 50 percent of individuals used one hand exclusively to catch the potatoes, but these preferences were symmetrically distributed at the group level. Baboons, on the other hand, have demonstrated left handedness on a visuospatial alignment task (Fagot and Vauclair, 1988a). Five of six baboons displayed left-hand preferences for precisely aligning two apertures so that they could reach through the hole created and retrieve a food reward. Seven of eight gorillas (*G. gorilla*) were left handed on this same visuospatial alignment task (Fagot and Vauclair, 1988b). Capuchins (*C. apella*), however, displayed strong individual hand preferences, but no handedness when aligning the horizontal apertures (Spinozzi and Truppa, 1999). Perhaps the visuospatial alignment task was not demanding enough for the arboreal capuchin species. As suggested previously, arboreal species may not be lateralized for visuospatial tasks as in their natural environment both sides of the body (and brain) must be capable of precisely estimating the position and distance of a branch that they are jumping toward.

Summary of Complex Visuospatial Tasks

Although simple reaching tasks have some visuospatial requirements, the studies reviewed in this section are dependent on this form of processing. Most of these tasks require the subjects to reach for a moving object. These tasks, therefore, require the continuous analysis of the temporal aspects of the moving stimulus and should be more spatially complex than simple reaching tasks in which subjects reach for static foods. Nonetheless, even on these complex and novel tasks many of the primate species did not display left handedness as hypothesized by MacNeilage et al. (1987).

Cats were initially left handed for complex visually guided reaching, but their left pawedness decreased with practice. These data suggest that preference and performance asymmetries between the hands can be masked by practice, as suggested by Fagot and Vauclair (1991). Perhaps "practice effects" also mask asymmetries of hand use on complex visuospatial tasks in many arboreal primate species. It seems logical that both sides of the brain and body should be capable of complex visuospatial assessments in arboreal primates. Nonetheless, some primate species do display handedness on novel and complex visuospatial tasks. First, primarily terrestrial gorillas and baboons display left handedness on a visuospatial alignment task. The left pawedness of cats, squirrel monkeys, baboons, and gorillas may be indicative of an underlying right hemisphere specialization for visuospatial processing as has been found in chicks (e.g., Andrew, 1988; Vallortigara et al., 1988), rats (Bianki, 1988; Cowell et al. 1997), and humans (e.g., Guiard et al., 1983; Wednt and Risberg, 1994). Lion tamarins, however, are right handed for catching crickets, while nocturnal prosimians and marmosets do not seem to be handed. In the natural habitat, tamarins feed on "hard-to find" foods exploring "crevices and knotholes," "rummaging through palm fronds; jumping rapidly to ground to seek cryptic prey" (Garber, 1993). Perhaps these species are predisposed toward use of a more manipulative strategy in foraging, and their right handedness is a reflection of this type of manual specialization. Overall, there are indications of a tendency toward left handedness, and right hemisphere specialization, for visuospatial tasks in primate and non-primate species. Further exploration with more complex and novel visually guided tasks may unmask the evolutionary history of manual specialization for visuospatial reaching. Notwithstanding, it is clear that any right hemisphere dominance for visuospatial function is not unique to primates.

MANIPULATION AND TOOL USE

Birds

Goldfinches (*Carduelis carduelis*) display right footedness when manipulating doors and latches to retrieve a food reward (Ducker et al., 1986). In fact, all of the individual birds tested displayed significant right-foot preferences on these tasks.

In a recent, ingenious study, Hunt (2000) also found that New Caledonian crows (*Corvus moneduloides*) display right sidedness for making tools. They use leaves from *Pandanus* trees to form "stepped-cut tools," which they insert into the cracks and holes of trees and branches to extract prey. Sculpting of the tools requires precise manipulation of the bill. With the leaf still attached to the tree, the crows precisely position their bill so that they can make a cut (on approximately a 50° angle to the longitudinal axis of the leaf) using one side of the bill only. They then tear the leaf slightly down the longitudinal axis, and make another cut, tear again make another cut, and then remove the leaf (Hunt, 2000). The result is a strip of the leaf with a step-tapered edge. By examining the leaf pieces that the crows left on the trees after making these tools, Hunt (2000) discovered that a significant majority (65–89 percent) of these tools were made with the left side of the leaf. Moreover, with knowledge of how the tools are constructed, Hunt (2000) determined that the crows had used the right sides of their beaks to make the precise step-forming incisions in the leaves, and their right eye to guide this behavior. The accessibility of the leaves from the left or right seemed to slightly influence the crows' behavior, but even when on a tree where the leaves were theoretically more accessible from the right side, the crows made tools from the left side of the leaves in a majority of incidences (65 percent). This unprecedented study provides substantial evidence of tool use in birds that rivals the manufacturing and manipulative skills reported for primates. The population-level right-sided bill use of the crows when manufacturing tools also provides tangible evidence of a left hemisphere specialization for manipulation in birds, whose evolutionary ancestors diverged from the "hominid" evolutionary tree long before primates.

Non-Human Primates

Despite various theories on how tool use may have been the primary impetus behind the evolution of human-like lateralization (Frost, 1980; Kimura, 1979), relatively few studies have looked at limb preferences in tool-using tasks or for manipulation.

Prosimians have not been reported to use tools. Aye-ayes, however, use their third finger, which moves independently of the other fingers, in a manipulative way to extract wood boring larvae from cavities. Milliken (1995) recorded the finger (third finger of the left or right hand) that four aye-ayes used to extract larvae from horizontal and vertical cavities, as well as the efficiency of each of the fingers on these tasks. All of the subjects were right handed on this task. Moreover, when probing horizontal cavities, the right finger was more efficient than the left, extracting more larvae with relatively fewer probes (Milliken, 1995). The aye-aye, therefore, represents a prosimian that has evolved rudimentary manipulative skills in one finger and, as suggested by MacNeilage et al. (1987), the right finger/left hemisphere appears to be specialized for extraction tasks that require fine motor control.

Amongst the New World primates, only capuchins have been reported to use tools, and their hand preferences have been recorded during a variety of tool-using activities, including probing, hammering, sponging, and cutting. Capuchins display individual preferences for probing (Anderson, 1996; Westergaard, 1991), but no group bias. For hammering and sponging, they show a tendency for left (Westergaard and Suomi, 1993c) and right (Westergaard and Suomi, 1993b) handedness, respectively. For cutting, however, strong right handedness was found (Westergaard and Suomi, 1996b). In fact, Westergaard and Suomi (1996b) found that six out of six capuchins used the right hand preferentially when using stones to cut. Moreover, when the capuchins struck stones against hard surfaces (perhaps modifying them for later tool use), nine subjects were right handed, one was left handed, and four were ambipreferent. These data suggest that there is a dissociation between the lateralization displayed on different types of tool-using tasks similar to that found for reaching activities. In terms of handedness, however, only cutting elicits a strong group-level lateralization (right hand) for tool use in capuchins.

The Old World monkeys have been neglected in studies of tool use. There has been only one study of probing in lion-tailed macaques (*Macaca silenus*; Westergaard, 1991). Three of the four subjects were handed on this task; two were left handed and one was right handed. During the "invention" of a stone-throwing strategy to dislodge fruit from a pipe, two of three Japanese macaques (*M. fuscata*) were strongly left handed and one was exclusively right handed (Tokida et al., 1994). On the basis of these small samples very little can be concluded regarding the handedness of Old World primates on tool-using tasks. Further research is warranted.

Even for the great apes, whose propensity for tool use has been documented in captive and wild populations, little is known about hand preferences for manipulation. Boesch (1991) found no evidence of handedness during nut cracking in wild chimpanzees, but individuals were strongly lateralized for this behavior. These data have been replicated in several other studies of wild chimpanzees (McGrew et al., 1999; Sugiyama et al., 1993). Like capuchins, wild chimpanzees display a tendency for right handedness during sponging (Boesch, 1991), and probing results in strong individual preferences but no handedness at a group level (McGrew and Marchant, 1992). Similarly, captive chimpanzees do not display handedness for unimanual probing, though many individual subjects display hand preferences. When a bimanual strategy for probing is used, however, a significant population right-hand bias is found for captive chimpanzees (Hopkins and Rabinowitz, 1997). These data underscore the significance of bimanual compared to unimanual hand use in primate studies.

Summary of Manipulation and Tool Use

When handedness is displayed for tool use or manipulative function, it is indicative of a right-hand specialization. As MacNeilage and colleagues (1987) suggested, therefore, the left hemisphere may have become specialized for the mediation of tasks requiring fine motor manipulation in primates. However, as birds also display right "handedness" for manipulation, it would seem that any specialization of hemispheric function evolved well before the emergence of primates or evolved more than once (in at least primates and birds). Moreover, on many tool-use tasks, primates do not seem to display handedness. Even chimpanzees, with a well-documented propensity for tool use, express handedness only when required to use a bimanual strategy to manipulate objects. These data suggest that hand preferences for manipulation are not simply mediated by the left hemisphere, but rather both hemispheres may be involved in processing them. In unimanual tasks, individual hand preferences may depend on the completion strategy adopted by the subjects. Some subjects may solve probing tasks, for example, using a predominantly visuospatial strategy (focusing on the parameters of the hole to be probed), while others may use a manipulative extraction strategy. Perhaps in a bimanual situation, where both hands are forced to work simultaneously in a coordinated way, the complementary specializations of the hemispheres are expressed. In this situation, many subjects may show the same division of function between the hands; with the right hand manipulating the probe while the left eye (predominantly right hemisphere) perhaps assesses the visuospatial components of the task. Bimanual tasks appear to be a better measure of handedness in chimpanzees, and other primate species. As pointed out previously, it is only on these tasks that strong and consistent hand use biases are found within and across species. The evolution of the capacity to use both hands simultaneously to perform a task, most commonly seen in primates (but also present in some other mammals), may have been an important development in the evolution of manual specialization and these types of tasks may be important for understanding the history of lateralization.

FOOT PREFERENCES IN LOCOMOTION

Birds

Davies and Green (1991) recorded the foot preferences of pigeons during take off and landing from flight. They were able to determine the leading foot using post hoc video analyses. Six of the sixteen pigeons displayed a foot preference for landing, and the preferences of these subjects were bimodally distributed. Davies and Green (1991) suggest that greater foot preference may be observed in pigeons if the landing maneuver was made more stressful. As noted previously, both hemispheres may be capable of performing this well-rehearsed landing behavior in pigeons, and a task with extensive visuospatial demands may be necessary to uncover an inherent specialization of one hemisphere for this function.

Non-Human Primates

Foot preferences in non-human primates are recorded during the initiation of locomotory activities. Most studies of non-human primates have found right leading-limb (hand or foot) preferences for initiating walking. For example, Forsythe and Ward (1987) found that four ruffed lemurs, *V. variegata*, displayed right-limb preferences for initiating quadrupedal locomotion. Group-level biases for right handedness (footedness) have also been reported in captive chimpanzees (*Pan troglodytes*: Heestand, 1986, cited in Hopkins et al., 1993), bonobos (*Pan paniscus:* Hopkins et al., 1993; Hopkins and de Waal, 1995), orangutans (*P. pygmaeus:* Cunningham et al., 1989; Heestand, 1986, cited in Hopkins et al., 1993), and gorillas (*G. gorilla*; Heestand, 1986, cited in Hopkins et al., 1993). Leading-limb preferences have been found in neonatal chimpanzees also: Hopkins et al. (1997) found a population-level bias for the right hand in the initiation of crawling by neonates, although the same chimpanzees did not have foot preferences for initiating stepping (Bard et al., 1990; Hopkins et al., 1997). Therefore, prosimians and apes may be right-limb preferring when initiating locomotion.

One study of wild chimpanzees did not find right handedness during quadrupedal locomotion (Marchant and McGrew, 1996). Marchant and McGrew (1996) suggested that the absence of a group bias in the wild chimpanzees compared to the captive chimpanzees may be due to the increased opportunity for arboreal locomotion in the natural environment. Indeed, gibbons (*Hylobates syndactylus, Hylobates concolor, Hylobates lar*) do not have a group bias for one limb in the initiation of brachiation (Stafford et al., 1990). However, Hook and Rogers (2002) found that marmosets display right footedness for a form of arboreal locomotion. In this study, leading-limb preferences were recorded during the initiation of walking, leaping, and landing. Individual preferences were symmetrically distributed for walking and leaping activities, but there was a significant bias for the right hand to contact a solid substrate first during landing. Clearly differences between the wild and captive chimpanzees are not simply due to changes in the degree of terrestrial versus arboreal locomotion. The dichotomy between the lateralization of captive and wild apes (and generally captive and wild primates) may instead be due to differences in the selective pressures influencing the behavior of these populations. With decreased proximate selective pressures, captive settings may "free the hands" of primate populations (Marchant and McGrew, 1996) recapitulating some of the situational changes that led to the evolution of human handedness and footedness.

Summary of Foot Preferences in Locomotion

Right leading-limb preferences are expressed by many primates in the initiation of locomotion. This suggests that the left hemisphere may be mediating this form of motor function. However, the observation of birds and wild primate populations do not reveal footedness at a population level. As Davies and Green (1991) suggest, the absence of footedness in these studies may be due in part to practice effects masking underlying asymmetries. Indeed, in the natural environment, both hemispheres need to be able to assess a landing substrate's position and distance so that the subject has the flexibility to move in any direction and is not constrained by lateralization, which would be nonadaptive (Marchant and McGrew, 1996). The finding of very subtle differences between the two limbs during landing in marmosets lends support to this hypothesis (Hook and Rogers, 2002). Both hemispheres seem capable of initiating landing in marmosets, but frame-by-frame analyses of these behaviors show a significant bias for the right hand to lead the left by approximately 1/24 of a second. Such subtle differences between the limbs would not be apparent without precise technological analyses (which are not always possible in the natural environment). Overall, on the basis of the predominant observation of right footedness in primates, it seems that a left hemisphere specialization has evolved for leading locomotion which is consistent with that found for humans (e.g., Day and MacNeilage, 1996; Gabbard and Iteya, 1996; Porac, 1996; Searleman, 1980).

PRODUCTION OF EMOTIONAL RESPONSES AND VOCALIZATIONS

Lower Vertebrates: Fish, Amphibians, and Reptiles

B. marinus, the South American cane toad, direct significantly more agonistic tongue strikes at conspecifics in the left visual hemifield (right hemisphere) than in their right (Robins et al., 1998; Vallortigara et al., 1998). The green anole lizard (*Anolis carolinensis)* also displays most aggressive behaviors, including biting and threatening, when viewing a conspecific with their left eye (Deckel, 1995). Because of an almost complete decussation of the optic tract in *Anolis*, and the absence of a corpus callosum in this species, it seems likely that aggressive responses are primarily mediated by the right hemisphere (Deckel, 1995). Similarly, the right hemisphere seems to mediate aggression in free-ranging male tree lizards (*Urosaurus ornatus*). Hews and Worthington (2001) found that male tree lizards display more aggressive behaviors when viewing conspecifics in their left visual field. Thus, it seems that right hemisphere specialization for the production of emotional responses may be characteristic of early vertebrates, as is claimed for higher vertebrates including humans.

The left hemisphere specialization for the production of communicative signals, known to be characteristic to humans, also appears to be present in lower vertebrates. The channel catfish (*Ictalurus punctatus*) displays a preference for use of the right fin when producing pectoral stridulation sounds (Fine et al., 1996). During pectoral stridulation sound production, catfish rub a ridged process on the first pectoral spine against the rough surface of a groove in the pectoral girdle (Fine et al., 1996). From post hoc video recordings, Fine et al. (1996) determined that 10 of 20 catfish displayed a preference for one fin for making the pectoral stridulation sounds, and 9 of the 10 preferred the right fin. Nine times as many fish preferred the right fin over the left. Therefore, the left hemisphere appears to mediate production of a social communication signal in the channel catfish.

Frogs (*Rana pipiens*) also appear to have a left hemisphere specialization for the production of species-specific vocalizations. Bauer (1993) cut the dorsal-ventral extent of frogs (*R. pipiens*) brains, on either the left ($n = 11$), or right side ($n = 11$), between the cerebellum and the tectum. Following recovery, he recorded the number of vocalizations made by each group and compared the totals to intact and sham-operated control groups. The left lesioned group made significantly fewer vocalizations than the right, intact, and sham-operated groups (Bauer, 1993). The quality and duration of the vocalizations emitted from the left lesioned group were also significantly impacted by the experimental manipulation. These data provide evidence for left hemispheric dominance in the production of vocalizations in amphibians that concurs with the direction of lateralization found for catfish.

Birds

Chicks peck more at strangers than at their cagemates when viewing these conspecifics with their left eye, whereas they display random pecking at familiar and unfamiliar chicks viewed with the right eye (Vallortigara, 1992). Chicks also show a preference for fixating a conspecific using the left lateral visual field prior to pecking aggressively (Vallortigara et al., 2001). The left-eye system, therefore, appears to play a role in the initiation of aggressive responses. Pharmocological treatments that elevate attack behaviors, such as testosterone, also elevate the expression of aggressive behavior when chicks view a hand (shaped to simulate an attacking conspecific) with their left eye only, but not when they view the "hand" with their right eye only (Andrew, 1966). These data suggest that the right hemisphere may be dominant for controlling attack behavior in the chick, as found for toads and lizards.

A left hemisphere specialization for the production of species-specific vocalizations has also been well-documented for many bird species. Nottebohm and his colleagues have conducted a series of studies providing evidence of the dominant role of the left hemisphere in the production of song in a variety of songbirds. Their initial studies showed that sectioning a branch of the left hypoglossus nerve (which provides motor innervation to the syringeal muscles on the left side) of the adult male chaffinch (*Fringilla coelebs)* resulted in the loss of most of the components of song, whereas similar sectioning of a branch of the right hypoglossus nerve resulted in the loss of few or none of the song's components (Nottebohm, 1970, 1971, 1972). Left hypoglossal dominance in song production has now been shown in the white-throated sparrow (*Zonotrichia albicollis*, Lemon, 1973), the white-crowned sparrow (*Zonotrichia leucophrys*, Nottebohm and Nottebohm, 1976), and the canary (*Serinus canaria*, Nottebohm, 1977). Hartley and Suthers (1990) have also shown that plugging the left bronchus of canaries results in impaired song production, whereas plugging the right bronchus leaves the birds with normal song production. It seems, therefore, that the branch of the left hypoglossal nerve and the left bronchus are dominant for song control in a variety of bird species. As the left hypoglossal nerve and bronchus largely connect (via more than one neurone) to brain centers in the left hemisphere (Bradshaw and Rogers, 1993), the left hemisphere appears to primarily mediate song production. Nonetheless, not all bird species display left-sided control of song production: branches of both the left and right hypoglossus are involved in vocal production in the orange-winged Amazonian parrot (*Amazona amazonica)*. While a majority of bird species, therefore, are left lateralized for song production, some species are not lateralized or lateralized in the opposite direction.

The predominance for left sidedness in song production in many bird species is also evident in higher brain centers. Nottebohm and colleagues (Nottebohm, 1977; Nottebohm et al., 1976) showed that after lesioning of the high vocal center (HVC) in the left hemisphere, adult canaries where able to produce no more than one of the syllables of song that they had produced preoperatively. Similar lesions in the right hemispheres resulted in much less disruption to song production. After right hemisphere lesions, the canaries were able to sing one third to three fifths of the preoperative song syllables, and some birds were able to produce their entire preoperative song. Lesions of the left robustus archistriatalis (RA) in adult canaries reduces the frequency range of the song fundamentals, whereas lesions of the right side do not. The control of singing seems to be primarily located in the left hemisphere.

Non-Primate Mammals: Rodents

Muricide, or mouse killing, appears to be controlled by the right hemisphere in handled rats. Garbanati et al. (1983) put a mouse in the cage of a rat, for a maximum period of 5 days, and observed whether killing occurred. They found that handled rats (rats that had been removed from their mothers as pups and placed in isolation for 3 min before being returned to their mothers) with an ablated left neocortex displayed higher levels of muricide than their handled counterparts with right neocortex ablations and intact handled rats. Nonhandled rats did not show asymmetrical hemispheric control of muricide. This suggests that in handled rats, exposed to a stressor in early life, the right hemisphere induces the production of muricide behavior, and the left hemisphere may inhibit these responses. In fact, if the corpus callosum is sectioned, and the inhibitory role of the left hemisphere is undermined, levels of muricide behavior are increased in intact handled rats (Denenberg et al., 1986).

Conditioned fear responses with taste aversion also appear to be controlled by the right hemisphere in rats (Denenberg et al., 1980). Denenberg et al. (1980) laced a sweetened solution of milk with lithium chloride: a treatment that should result in an association between the milk solution and nausea and cause subsequent avoidance of the milk solution. The degree to which the handled rats learned this association was dependent on whether they had a right hemisphere cortical ablation, left hemisphere ablation, or were intact. Subjects with an intact right hemisphere only avoided the milk more than those with an intact left hemisphere only or those with an intact brain. Nonhandled rats did not show differences in their levels of learning based on whether the left, right, or both hemispheres were available. Denenberg and his colleagues suggested that this was because nonhandled rats are "very emotional" in general and their high level of reactivity might mask subtle differences between subjects. They suggest that handling reduces emotional reactivity in rats (Denenberg et al., 1980). While the differences between handled and nonhandled warrant further investigation, it seems that the right hemisphere produces emotional responses, and the left hemisphere inhibits them, at least in handled rats.

In contrast to the observations of muricide and taste aversion, responses to the odors of stressed and nonstressed conspecifics appear to be under the control of the left hemisphere in rats. A formalin injection into the hindpaw of rats induces pain and licking of the "injured" limb. Stress delays this licking response. Dantzer et al. (1990) observed the latency to perform this formalin-induced licking response when rats with left, right, bilateral, or sham bulbectomies were exposed to the odor of stressed conspecifics. The odor of the stressed conspecifics delayed licking in the rats with right and sham bulbectomies, but did not effect the licking responses of bilaterally or left-bulbectomized rats. As each of the olfactory bulbs projects to the ipsilateral hemisphere, these results suggest that the left hemisphere mediates the response to stressful olfactory stimuli. These data suggest that the right hemisphere may not mediate all affective responses in rats. The conditions dictating which hemisphere controls affective responses in rats, however, are not clear and warrant further investigation. Understanding of the interactions between early stressors, sensory mode, and hemispheric asymmetry in these species would indeed further our knowledge on the plasticity of lateralized brain systems.

Non-Human Primates

Hemispheric specialization for the production of emotional responses has been found for non-human primates. Hauser and Akre (2001) found that the left side of the mouth opens first during the production of a large variety of facial expressions, both positive and negative, in rhesus macaques (*M. mulatta*). These data confirmed a previous finding for this species (Hauser, 1993). Hook-Costigan and Rogers (1998) also showed that common marmosets (*C. jacchus*) display a larger left hemimouth during the production of fear expressions, with and without vocalization. Then in 2002, Fernàndez-Carriba and colleagues extended these findings to chimpanzees (*P. troglodytes*), showing that they too showed a larger left hemimouth area during the production of positive and negative emotional expressions. Fernàndez-Carriba et al. (2002) also showed that, as found for macaques (Hauser, 1993), chimeras of the left hemiface were judged as more expressive (by human judges) than right hemiface chimeras. Thus, like humans and non-primate species, rhesus macaques (Hauser, 1993; Hauser and Akre, 2001), marmosets (Hook-Costigan and Rogers, 1998), and chimpanzees (Fernàndez-Carriba et al., 2002) have specialization of the right hemisphere for the production of facial expressions.

Right hemisphere dominance for the mediation of affective responses in primates has also been found during agonistic interactions. Gelada baboons (*Theropithecus gelada*) use their left visual field to view their opponent significantly more than their right during fights, threats, and approaches (Casperd and Dunbar, 1996). These data are strikingly similar to those found for toads, lizards, and chicks. The various forms of converging evidence for right hemisphere mediation of emotional responses in primate and non-primate species suggest that this form of hemispheric dominance was retained in the primate species (and by humans), possibly originating in the earliest evolutionary ancestors of all present day species.

The left hemisphere specialization for the production of vocalizations, observed in lower vertebrates, and many bird species, may also have been retained by non-human primate species. Hook-Costigan and Rogers (1998) found that marmosets (*C. jacchus*) displayed a larger right hemimouth than left when producing a social contact call, a twitter vocalization. This finding suggests that whereas the right hemisphere modulates negative emotional responses in marmosets, the left hemisphere may be specialized for the production of social communication signals. The authors hypothesized that the hemispheres may be differentially specialized for the production of negative and positive emotional responses, as the marmosets displayed a larger left hemimouth when producing a fear vocalization. This hypothesis, however, has not been supported by research with other species. Rhesus macaques (Hauser and Akre, 2001) and chimpanzees (Fernàndez-Carriba et al., 2002) show a left hemimouth/right hemisphere bias for the production of positive and negative vocalizations and facial expressions. It may, in fact, be the acoustic structure of the twitter vocalization that leads to differential specialization for the production of this call.

Contact twitter vocalizations are composed of sequences of short notes (0.02–0.04 s) separated by intervals of 0.06–0.08 s (Epple, 1968). The detailed temporal modulation required for the production of this call may underlie the left hemisphere advantage for the production of this vocalization (Hook-Costigan and Rogers, 1998). Indeed, Fitch et al. (1993) reported a left hemisphere bias for the discrimination of temporally modulated tone sequences in male Sprague–Dawley rats, but no bias for the discrimination of simple tones. They suggest that these data are indicative of a left hemisphere specialization for auditory temporal processing (Tallal et al., 1995). Further evidence is needed to support this hypothesis in marmosets since Hook-Costigan and Rogers (1998) used a very small sample of subjects, and investigated only one temporally modulated call. Nonetheless, considerable evidence of a right ear advantage for the perception of vocalizations in non-human primates (*M. fuscata*, Heffner and Heffner, 1984; Petersen et al., 1978, 1984; *M. mulatta*, Hauser and Andersson, 1994) supports the hypothesis that the left hemisphere mediates vocalization in primates, as it does in fish, frogs, birds, and rats.

SUMMARY

In all of the species reviewed, there is evidence that the right hemisphere is specialized for the production of emotional responses, such as attack and avoidance behaviors, while the left hemisphere mediates the production of vocalizations. The consistency of lateralization found across species, from lower vertebrates to primates, suggests that the asymmetrical mediation of communication signals appeared very early in evolution and may have been highly conserved throughout evolution.

It is hypothesized that the left hemisphere specialization found in many species, for the production of communicative vocal signals may be dependent on specialization of this hemisphere for the processing and production of temporally modulated sequences. By contrast, the right hemisphere may process signals in which modulation of the frequency conveys information. Indeed, in humans, the right hemisphere is specialized for the analysis of the pitch, timbre, and intensity of auditory signals (Bogen and Gordon, 1971; Curry, 1968; Gates and Bradshaw, 1977; Ley and Bryden, 1982). In marmosets, the mobbing "*tsik*" vocalization produced in a fear context has high frequency modulation and seems to be controlled by the right hemisphere (Hook-Costigan and Rogers, 1998). The frequency modulation of this call could contain information about the identity of the caller, the type of threat being experienced, and the intensity of the threatening context. Dominance of the right hemisphere in the production of emotional responses may have led to specialization of this hemisphere for production of signal details that communicate information about the emotional state of the caller. The idea that the hemispheres may be divided for the production of temporal versus frequency modulated communication signals in a wide range of species remains to be investigated.

ARE PRIMATES SPECIAL?

Population-level biases for various forms of motor function are evident in fish, amphibians, birds, non-primate mammals, and primates. Birds, in fact, show population-level biases in almost all of the motor tasks reviewed and clearly rival primates in terms of the extensiveness to which they are lateralized. Thus, primates *are not special* in terms of hemispheric specialization for motor functions.

In the absence of a clear transition from no specialization to manual specialization in humans, we need to reassess our hypothesis on how hemispheric specializations evolved for motor functions. On the basis of the literature reviewed, there are two potential explanations. First, hemispheric specializations for manual control may have evolved many times in separate species. This explanation seems unlikely, and inefficient, but cannot be ruled out with the sporadic appearance of population biases. A second, more plausible, explanation is that hemispheric specializations for perceptual and communication (including motor components of communication) functions were present in the earliest evolutionary ancestor of present day species. Hemispheric specializations for perceptual and/or communication functions may have preceded and, possibly, given rise to the evolution of hemispheric specializations for other forms of motor control (Andrew et al., 1982; Bradshaw and Rogers, 1993; Corballis, 2003; Vallortigara et al., 1999).

Corballis (2003) proposed that the left hemisphere specialization for many forms of hand use in humans, expressed as right handedness, may have evolved because of a pre-existing specialization of this hemisphere for the mediation of vocalization. He suggests that right handedness may have arisen from an interaction between manual gestural communication and vocalization in the evolution of language (or intentional communication compared to vocalizations that are claimed to be largely involuntary). According to Corballis (2003), a pre-existing manual gestural communication system (present in non-human primates; de Waal, 1982; Tanner and Byrne, 1996; Tomasello et al., 1997) may have gradually incorporated vocal elements that were mediated by an evolutionarily ancient left hemisphere specialization. The integration of manual and vocal communication may have led to lateralization of the hands for gestural communication. That is, the left hemisphere, already specialized for vocal production and perception, assumed its dominant role in the production of gestures leading to the development of an integrated communication system. This left hemisphere specialization for the production of gestures may have mediated the transition to a dominant role in humans for various forms of manual function (Corballis, 2003). This hypothesis is appealing as it recognizes the longstanding lateralization of communication systems and their potential role in the emergence of hemispheric specializations for manual function. It also underscores the *uniqueness* of primates in the evolution of a motor system that "freed the hands" from a perfunctory locomotory role. The theory neglects, however, the evidence for lateralized motor functions in species whose ancestors diverged very early from the line of evolution that gave rise to primates, and the dominant role of the right hemisphere in the production of facial expressions in non-human primates and humans.

It could be argued that specialization of the right hemisphere, rather than the left, had already evolved for gestural communication, at least in the form of facial expressions. Concurring with the theory proposed by Corballis (2003), lateralization of gestural communication does, at present, appear to be unique to primates, including humans. It remains to be ascertained whether aggressive or other emotional displays are lateralized in other species. For example, it would be interesting to look at potential postural and facial asymmetries in the threatening or submissive displays of dogs and other mammals. Indeed, consistent evidence, in species throughout the evolutionary continuum, for the dominant role of the right hemisphere in the mediation of emotional responses suggests that this hemisphere would control the production of emotional displays in a variety of forms (i.e., facial expressions, aggressive stances).

Rather than right handedness emerging from a specialization of the left hemisphere for gestural communication, it may have evolved because of the specialization for the mediation of temporal sequencing. As previously suggested, the role of the left hemisphere in vocalization may be based on a specialization that mediates the processing and production of temporally modulated communication signals. Integration of gestures with vocal signals in communication may have predisposed these manual forms of communication, involving temporal sequencing, to be controlled by the left hemisphere. Because of the role of the left hemisphere in temporal sequencing, other fine motor tasks that rely on precise timing for specialized manual movements may also have become lateralized to the left hemisphere (right hand). Vocal control and handedness (as seen in humans) may be linked because of a common dependence on left hemispheric mediation of the processing and production of behaviors (and communication signals) with high temporal sequencing demands. Manual lateralization in primates and non-primate species may be found in tasks with high levels of temporal processing. Of course, the degree of temporal processing required to unmask any functional specializations must also depend on a species' behavioral repertoire. Novel and complex tasks, for example, might be needed to unmask specializations in arboreal primates and birds that routinely display locomotory behaviors with substantial temporal demands (i.e., flying or rapidly leaping from one branch to another). Indeed, left hemisphere dominance was found for marmosets when initiating landing (Hook and Rogers, 2002), a task that probably relies heavily on complex temporal processing. Whether motor specializations are expressed, or were even retained by present day species, may depend on the species' behavioral repertoire interacting with task demands.

In non-human species, the perceptual requirements of motor tasks seem to have a large impact on the expression of motor preferences. From the review of the literature, it does seem that there is dissociation between hand preferences expressed on motor tasks with different perceptual requirements. For example, while visuospatial tasks tend to reveal left handedness in primates, right handedness is predominantly expressed on manipulative tasks (perhaps because of the temporal sequencing demands inherent to these types of tasks). This division of function between the hands mirrors the complementary specializations of the hemispheres suggested for the processing and production of emotional responses and vocal communication signals. The right hemisphere seems to process, and control responses in, tasks that depend on a holistic (spatial and frequency) strategy while the left hemisphere mediates analytic processing (temporal sequencing) and the production of highly organized and sequenced responses. There may be a division between the hemispheres, based on holistic versus analytic processing strategies (e.g., Bradshaw and Nettleton, 1981; Bradshaw and Rogers, 1993), which appeared very early in evolution. In early vertebrates and birds, perceptual specializations clearly impact the population-level motor biases expressed. This suggests that hemispheric specialization for perceptual processing preceded the evolution of motor specialization (Vallortigara et al., 1999), but it also underscores the significant role of perceptual functions in the expression of motor biases in other species, potentially including primates.

At this point, it is necessary to consider the possible advantages that brought about the evolution of hemispheric specializations. Hemispheric specialization and individual lateralization would not be advantageous if it constrained an individual's behavioral flexibility. Moreover, if one hemisphere alone mediates a function, this predisposes the individual to a loss of function when that hemisphere is damaged. It has been suggested that lateralization evolved to increase the processing capacity of the cerebral hemispheres, by reducing the redundancy involved in duplication of functional control in the separate hemispheres. This idea needs to be further refined. Perhaps hemispheric specialization increases a species neural capacity by allowing both hemispheres to simultaneously process different components of a task (Rogers, 2000). Working together, the hemispheres would be able to process a complex task in half the time, at least. For example, in a visuospatial task, one hemisphere might process the spatial components while the other mediates the manipulative, temporally modulated processes required for task completion. Processing of such a task may theoretically be slower if one hemisphere sequentially analyzed the separate components involved in the performance.

Unimanual tasks that are currently used may not be effective at dictating the primary perceptual strategy adopted by subjects for performing the task (that we think we are measuring) and, thus, subject and contextual variables might mask potential population asymmetries. Bimanual tasks, on the other hand, may bring about expression of the most efficient way that the hemispheres (and their underlying specializations) can work together to process the task. Using bimanual tasks, we may unmask the division of function between the hands and hemispheres. It would also be interesting to examine whether an experiential history of bimanual processing enhances the lateralization expressed by captive primate (and non-primate) populations.

Primates *are special* with respect to their manual dexterity, their ability to perform coordinated, bimanual manipulations of objects, and their potential use of the hands for gestural communication. Thus, it may be that use of bimanual tasks will reveal motor biases, that are comparable to humans, in these species and not in other non-primate species. This remains to be ascertained. It needs to be emphasized, however, *that other species also represent unique models* for investigating aspects of the evolution of lateralization. Birds, for example, are unique in the relative absence of left–right hemisphere interconnections at the forebrain level (Bradshaw and Rogers, 1993). Because of this anatomical characteristic, birds primarily process information from the left and right visual fields separately. This dissociation between the hemispheres makes birds an ideal model system for understanding how the brain is lateralized with regard to visual asymmetries. Because of the well-developed knowledge of the genetic codes of rats and mice, these species are invaluable to investigations of the molecular processes involved in the evolution and expression of hemispheric specialization. Each species is special and will help us to further understand the advantages and disadvantages of having a lateralized brain. Integration of information from various models will further the understanding of the functional and historical causes for the evolution of hemispheric specialization and the mechanisms underlying complex human neural processes.

ACKNOWLEDGMENTS

Thanks to Steven Schapiro for providing helpful comments on an earlier draft of this manuscript.

REFERENCES

Adelstein, A. and Crowne, D. P., 1991, Visospatial asymmetries and introcular transfer in the split-brain rat, *Behav. Neurosc.* **105**:459–469.

Anderson, J. R., 1996, Chimpanzees and capuchin monkeys: Comparative cognition, in: *Reaching into Thought: The Minds of the Great Apes*, A. E. Russon, K. A. Bard, and S. T. Parker, eds., Cambridge University Press, Cambridge, pp. 23–56.

Ades, C., Talebi-Gomes, M., and Strier, K. B., 1996, Hand preferences in food manipulation by wild muriquis (*Brachyteles arachnoides*): Preliminary data. *Proc. XVI Congr. Int. Primatol. Soc./XIX Conf. Am. Soc. Prim.* 346.

Andrew, R. J., 1966, Precocious adult behaviour in the young chick, *Anim. Behav.* **14**:485–580.

Andrew, R. J., 1988, The development of visual lateralization in the domestic chick, *Behav. Brain Res.* **29**:201–209.

Andrew, R. J., 2002, The earliest origins and subsequent evolution of lateralization, in: *Comparative Vertebrate Lateralization*, L. J. Rogers and R. J. Andrew, eds., Cambridge University Press, Cambridge, pp. 70–93.

Andrew, R. J., Mench, J., and Rainey, C., 1982, Right–left asymmetry of response to visual stimuli in the domestic chick, in: *Analysis of Visual Behavior*, D. J. Ingle, M. A. Goodale, and R. J. W. Mansfield, eds., Cambridge, MIT Press, pp. 197–209.

Annett, M., 1987, Handedness as chance or as a species characteristic, *Behav. Brain Sci.* **10**(2):263–264.

Annett, M. and Annett, J., 1991, Handedness for eating in gorillas, *Cortex.* **27**:269–275.

Bard, K. A., Hopkins, W. D., and Fort, C. L., 1990, Lateral bias in infant chimpanzees (*Pan troglodytes*), *J. Comp. Psychol.* **4**:309–321.

Bauer, R. H., 1993, Lateralization of neural control for vocalization by the frog (*Rana pipiens*), *Psychobiol.* **21**(3):243–248.

Bianki, V. L., 1988, The right and left hemispheres: Cerebral lateralization of function, *Monogr. Neurosci.* **3**. Gordon & Breach, New York.

Biddle, F. G., Coffaro, C. M., Ziehr, J. E., and Eales, B. A., 1993, Genetic variation in paw preference (handedness) in the mouse, *Genome* **36**:935–943.

Bisazza, A., Cantalupo, C., Robins, A., Rogers, L. J., and Vallortigara, G., 1996, Right-pawedness in toads, *Nature* **379**:408.

Bisazza, A., Pignatti, R., and Vallortigara, G., 1997a, Detour tasks reveal task- and stimulus-specific behavioural lateralization in mosquitofish (*Gambusia holbrooki*), *Behav. Brain Res.* **89**:237–242.

Bisazza, A., Pignatti, R., and Vallortigara, G., 1997b, Laterality in detour behaviour: Interspecific variation in poeciliid fish, *Anim. Behav.* **54**:1273–1281.

Bisazza, A., Cantalupo, C., Robins, A., Rogers, L. J., and Vallortigara, G., 1997c, Pawedness and motor asymmetries in toads, *Laterality* **2**(1):49–64.

Bisazza, A., Cantalupo, C., Capocchiano, M., and Vallortigara, G., 2000, Population lateralisation and social behaviour: A study with 16 species of fish, *Laterality* **5**(3):269–284.

Boesch, C., 1991, Handedness in wild chimpanzees, *Int. J. Primatol.* **12**:541–558.

Bogen, J. E., and Gordon, H. W., 1971, Musical tests for functional lateralization with intracorticoid amobarbitol, *Nature* **230**:524–525.

Box, H. O., 1977, Observations on spontaneous hand use in the common marmoset (*Callithrix jacchus*), *Primates* **18**:395–400.

Bradshaw, J. L., 1987, But what about nonprimate asymmetries and nonmanual primate asymmetries? *Behav. Brain Sci.* **10**(2):264–265.

Bradshaw, J. L. and Nettleton, N. C., 1981, *Human Cerebral Asymmetry*, Prentice Hall, Englewoods Cliffs, NJ.

Bradshaw, J. and Rogers, L., 1993, *The Evolution of Lateral Asymmetries, Language, Tool Use, and Intellect*, Academic Press, San Diego.

Brookshire, K. H. and Warren, J. M., 1962, The generality and consistency of handedness in monkeys, *Anim. Behav.* **10**(3–4):222–227.

Burgess, W. J. and Villablanca, J. R., 1986, Recovery of function after neonatal or adult hemispercetomy in cats: II. Limb bias and development, paw usage, locomotion and rehabilitative effects of exercise, *Behav. Brain. Res.* **20**:1–8.

Byrne, R. W. and Byrne, J. M., 1991, Hand preference in the skilled gathering tasks of mountain gorillas (*Gorilla gorilla berengei*), *Cortex* **27**:521–536.

Cameron, R. and Rogers, L. J., 1999, Hand preference of the common marmoset (*Callithrix jacchus*): Problem solving and responses in a novel setting, *J. Comp. Psych.* **113**(2):149–157.

Casey, M. B. and Lickliter, R., 1998, Prenatal visual experience influences the development of turning bias in bobwhite quail chicks (*Colinus virginianus*), *Dev. Psychobiol.* **32**:327–338.

Casey, M. B. and Sleigh, M. J., 2001, Cross-species investigations of prenatal experience, hatching behavior, and postnatal behavioral laterality, *Dev. Psychobiol.* **39**:84–91.

Casperd, J. M. and Dunbar, R. I. M., 1996, Asymmetries in the visual processing of emotional cues during agonistic interactions by gelada baboons, *Behav. Proc.* **37**:57–65.

Castellano, M. A., Diaz-Palarea, M. D., Barroso, J., and Rodriguez, M., 1989, Behavioral lateralization in rats and dopaminergic system: Individual and population laterality, *Behav. Neurosci.* **103**(1):46–53.

Cole, J., 1955, Paw preference in cats related to hand preference in animals and man, *J. Comp. Phys. Psych.* **48**:137–140.

Colell, M., Segarra, M. D., and Pi, J. S., 1995, Hand preferences in chimpanzees (*Pan troglodytes*), bonobos (*Pan paniscus*), and orangutans (*Pongo pygmaeus*) in food-reaching and other daily activities, *Int. J. Prim.* **16**(3):413–434.

Collins, R. L., 1975, When left-handed mice live in right-handed worlds, *Science* **18**:181–184.

Collins, R. L., 1991, Reimpressed selective breeding for lateralization of handedness in mice, *Brain Res.* **564**:194–202.

Corballis, M. C., 1987, Straw monkeys, *Behav. Brain Sci.* **10**(2):269–270.

Corballis, M. C., 2003, From mouth to hand: Gesture, speech, and the evolution of right-handedness, *Behav. Brain Sci.* **26**(2):139–153.

Cowell, P. E., Waters, N. S., and Denenberg, V. H., 1997, The effects of early environment on the development of functional laterality in Morris maze performance, *Laterality* **2**(3/4):221–232.

Cunningham, D., Forsythe, C., and Ward, J. P., 1989, A report of behavioral lateralization in an infant orangutan (*Pongo pygmaeus*), *Primates* **30**:249–253.

Curry, F. K. W. A., 1968, A comparison of performances of a right hemispherectomized subject and 25 normals on four dichotic listening tasks, *Cortex* **4**:144–153.

Dantzer, R., Tazi, A., and Bluthé, R.-M., 1990, Cerebral lateralization of olfactory-mediated affective processes in rats, *Behav. Brain Res.* **40**:53–60.

Davies, M. N. O., and Green, P. R., 1991, Footedness in pigeons, or simply sleight of foot? *Anim. Behav.* **42**:311–312.

Day, L. B., and MacNeilage, P. F., 1996, Postural asymmetries and language lateralization in humans (*Homo sapiens*), *J. Comp. Psych.* **110**(1):88–96.

Deckel, A. W., 1995, Laterality of aggressive responses in *Anolis*, *J. Exp. Zool.* **272**:194–200.

Denenberg, V. H., 1988, Handedness hangups and species snobbery, *Behav. Brain Sci.* **11**(4):721–722.

Denenberg, V. H., Hofmann, M., Garbonati, J. A., Sherman, G. F., Rosen, G. D., and Yutzey, D. A., 1980, Handling in infancy, taste aversion, and brain laterality in rats, *Brain Res.* **200**:123–133.

Denenberg, V. H., Gall, J. S., Berrebi, A., and Yutzey, D. A., 1986, Callosal mediation of cortical inhibition in the lateralized rat brain, *Brain Res.* **397**:327–332.

de Sousa, M. B. C., Xavier, N. S., Da Silva, H. P. A., De Oliveira, M. S., and Yamamoto, M. E., 2001, Hand preference study in marmosets (*Callithrix jacchus*) using food reaching tests, *Primates* **42**(1):57–66.

de Vleeschouwer, K., Van Elsacker, L., and Verheyen, R. F., 1995, Effect of posture on hand preferences during experimental food reaching in bonobos (*Pan paniscus*), *J. Comp. Psych.* **109**(2):203–207.

de Waal, F. M. B., 1982, *Chimpanzee Politics*, Jonathan Cape, London.

Diamond, A. C. and McGrew, W. C., 1994, True handedness in the cotton-top tamarin (*Saguinus oedipus*), *Primates* **35**(1):69–77.

Dodson, D. L., Stafford, D. K., Forsythe, C., Seltzer, C. P., and Ward, J. P., 1992, Laterality in quadrupedal and bipedal prosimians: Reach and whole-body turn in the mouse lemur (*Microcebus murinus*) and the galago (*Galago moholi*), *Am. J. Prim.* **26**:191–202.

Drea, C. M., Wallem, K., Akinbami, M. A., and Mann, D. R., 1995, Neonatal testosterone and handedness in yearling rhesus monkeys (*Macaca mulatta*), *Physio. Behav.* **58**(6):1257–1262.

Ducker, G., Luscher, C., and Schultz, P., 1986, Problemlöseverhalten von Steiglitzen (*Carduelis carduelis*) bei "manipulativen" aufgaben, *Zool. Beitr.* **23**:377–412.

Epple, G., 1968, Comparative studies on vocalization in marmoset monkeys (Hapalidae), *Folia Primatol.* **8**:1–40.

Fabre-Thorpe, M., Fagot, J., Lorincz, E., Levesque, F., and Vauclair, J., 1993, Laterality in cats: Paw preference and performance in a visuo-motor activity, *Cortex* **29**:15–24.

Facchin, L., Bisazza, A., and Vallortigara, G., 1999, What causes lateralization of detour behavior in fish? Evidence for asymmetries in eye use, *Behav. Brain Res.* **103**:229–234.

Fagot, J. and Vauclair, J., 1988a, Handedness and manual specialization in the baboon, *Neuropsychologia* **26**(6):795–804.

Fagot, J. and Vauclair, J., 1988b, Handedness and bimanual coordination in the lowland gorilla, *Brain Behav. Evol.* **32**:89–95.

Fagot, J., and Vauclair, J., 1991, Manual laterality in nonhuman primates: A distinction between handedness and manual specialization, *Psychol. Bull.* **109**:76–89.

Fernández-Carriba, S., Loeches, Á., Morcillo, A., and Hopkins, W. D., 2002, Asymmetry in facial expression of emotions by chimpanzees, *Neuropsychologia* **40**:1523–1533.

Ferrari, S. F., 1993, Ecological differentiation in the callithricidae, in: *Marmosets and Tamarins: Systematics, Behaviour and Ecology*, A. B. Rylands, ed., Oxford University Press, Oxford, pp. 314–328.

Fine, M. L., McElroy, D., Rafi, J., King, C. B., Loesser, K. E., and Newton, S., 1996, Lateralization of pectoral stridulation sound production in the channel catfish, *Physiol. Behav.* **60**(3):753–757.

Fitch, R. H., Brown, C., O'Connor, K., and Tallal, P., 1993, Functional lateralization for auditory temporal processing in rats, *Behav. Neurosci.* **107**:844–850.

Forsythe, C. and Ward, J. P., 1987, The lateralized behavior of the ruffled lemur (*Varecia variegata*), *Am. J. Primat.* **12**:342.

Forsythe, C., Milliken, G. W., Stafford, D. K., and Ward, J. P., 1988, Posturally related variations in the hand preferences of the ruffled lemur (*Varecia variegata variegata*), *J. Comp. Psych.* **102**:248–250.

Forward, E., Warren, J. M., and Hara, K., 1962, The effects of unilateral lesions in sensorimotor cortex on manipulation by cats, *J. Comp. Phys. Psych.* **55**:1130–1135.

Friedman, H. and Davis, M., 1938, "Left-handedness" in parrots, *Auk* **80**:478–480.

Frost, G. T., 1980, Tool behavior and the origins of laterality, *J. Hum. Evol.* **9**:447–459.

Gabbard, C. and Iteya, M., 1996, Foot laterality in children, adolescents, and adults, *Laterality* **1**(3):199–205.

Gates, A. and Bradshaw, J. L., 1977, Music perception and cerebral asymmetries, *Cortex* **13**:390–401.

Garbanati, J. A., Sherman, G. F., Rosen, G. D., Hofmann, N., Yutzey, D. A., and Denenberg, V. H., 1983, Handling in infancy, brain laterality, and muricide in rats, *Behav. Brain Res.* **71**:351–359.

Garber, P. A., 1993, Feeding ecology and behaviour of the genus *Saguinus*, in: *Marmosets and Tamarins; Systematics, Behaviour, and Ecology*, A. B. Rylands, ed., Oxford University Press, Oxford, pp. 273–295.

Glick, S. D. and Cox, R. D., 1978, Nocturnal rotation in normal rats: Correlation with amphetamine-induced rotation and effects of nigrostriatal lesions, *Brain Res.* **150**:149–161.

Glick, S. D. and Shapiro, R. M., 1985, Function and neurochemical mechanisms of cerebral lateralization in rats, in: *Cerebral Lateralization in Nonhuman Species*, S. D. Glick, ed., Academic Press, New York, pp. 158–184.

Glick, S. D., Jerussi, T. P., Waters, D. H., and Green, J. P., 1974, Title, *Biochem. Pharmocol.* **23**:3223.

Glick, S. D., Jerussi, T. P., and Fleisher, L. N., 1976, Turning in circles: The neuropharmacology of rotation, *Life Sci.* **18**:889–896.

Glick, S. D., Weaver, L. M., and Meibach, R. C., 1981, Amphetamine-induced rotation in normal cats, *Brain Res.* **208**:227–229.

Green, A. J., 1997, Asymmetrical turning during spermatophore transfer in the male smooth newt, *Triturus vulgaris*, *Anim. Behav.* **54**:343–348.

Guiard, Y., Diaz, G., and Beaubaton, D., 1983, Left hand advantage in right handers for spatial constant error: Preliminary evidence in a unimanual ballistic aimed movement, *Neuropsychologia* **21**:111–115.

Harrison, K. E. and Byrne, R. W., 2000, Hand preferences in unimanual and bimanual feeding by wild vervet monkeys (*Cercopithecus aethiops*), *J. Comp. Psych.* **114**(1):13–21.

Hartley, R. J. and Suthers, R. A., 1990, Lateralization of syringeal function during song production in the canary, *J. Neurobiol.* **21**:1236–1248.

Hauser, M. D., 1993, Right hemisphere dominance for the production of facial expression in monkeys, *Science* **261**:475–477.

Hauser, M. D. and Akre, K., 2001, Asymmetries in the timing of facial and vocal expressions by rhesus monkeys: Implications for hemispheric specialization, *Anim. Behav.* **61**:391–400.

Hauser, M. D. and Andersson, K., 1994, Left hemisphere dominance for processing vocalizations in adult, but not infant, rhesus monkeys: Field experiments, *Proc. Natl. Acad. Sci. USA.* **91**:3946–3948.

Heffner, H. E. and Heffner, R. S., 1984, Temporal lobe lesions and perception of species-specific vocalizations by macaques, *Science* **226**:75–76.

Hews, D. K. and Worthington, R. A., 2001, Fighting from the right side of the brain: Left visual field preference during aggression in free-ranging male tree lizards (*Urosaurus ornatus*), *Brain Behav. Evol.* **58**:356–361.

Hook, M. A. and Rogers, L. J., 2000, Development of hand preferences in marmosets (*Callithrix jacchus*) and effects of ageing, *J. Comp. Psych.* **114**(3):263–271.

Hook, M. A. and Rogers, L. J., 2002, Leading-limb preferences in marmosets (*Callithrix jacchus*): Walking, leaping, and landing, *Laterality* 7(2):145–162.

Hook-Costigan, M. A. and Rogers, L. J., 1995, Hand, mouth, and eye preferences in the common marmoset (*Callithrix jacchus*), *Folia Primat.* **64**:180–191.

Hook-Costigan, M. A. and Rogers, L. J., 1996, Hand preferences in New World primates, *Int. J. Comp. Psych.* **9**(4):173–207.

Hook-Costigan, M. A. and Rogers, L. J., 1998, Lateralized use of the mouth in production of vocalizations by marmosets, *Neuropsychologia* **36**:1265–1273.

Hook-Costigan, M. A., Rehlander, M., Cantalupo, C., and Ward, J. P., 1999, Hand preferences for visuospatial reaching in the bushbaby (*Otolemur garnettii*), *Am. J. Prim.* **49**(1):63.

Hopkins, W. D., 1993, Posture and reaching in chimpanzees (*Pan*) and orangutans (*Pongo*), *J. Comp. Psych.* **17**:162–168.

Hopkins, W. D., 1995, Hand preferences for a coordinated bimanual task in 110 chimpanzees (*Pan troglodytes*), Cross-sectional analysis, *J. Comp. Psych.* **109**(3):291–297.

Hopkins, W. D. and Bard, K. A., 2000, A longitudinal study of hand preference in chimpanzees (*Pan troglodytes*), *Dev. Psychobiol.* **36**:292–300.

Hopkins, W. D. and Bennett, A. J., 1994, Handedness and approach-avoidance behavior in chimpanzees (*Pan*), *J. Exp. Psych.* **20**(4):413–418.

Hopkins, W. D. and de Waal, F. B. M., 1995, Behavioral laterality in captive bonobos (*Pan paniscus*): Replication and extension, *Int. J. Primatol.* **16**(2):261–276.

Hopkins, W. D. and Rabinowitz, D. M., 1997, Manual specialization and tool-use in captive chimpanzees (*Pan troglodytes*): The effect of unimanual and bimanual strategies on hand preference, *Laterality* **2**(3/4):267–278.

Hopkins, W. D., Bennett, A. J., Bales, S., Lee, J., and Ward, J. P., 1993, Behavioral laterality in captive bonobos (*Pan paniscus*), *J. Comp. Psych.* **107**:403–410.

Hopkins, W. D., Bales, S. A., and Bennett, A. J., 1994, Heritability of hand preference in chimpanzees (*Pan*), *Int. J. Neurosci.* **74**(1–4):17–26.

Hopkins, W. D., Bard, K. A., and Griner, K. M., 1997, Locomotor adaptation and leading limb asymmetries in neonatal chimpanzees (*Pan troglodytes*), *Int. J. Primatol.* **18**(1):105–114.

Hopkins, W. D., Dahl, J. F., and Pilcher, D., 2001, Genetic influence on the expression of hand preferences in chimpanzees (*Pan troglodytes*). Evidence in support of the right-shift theory and developmental instability, *Psych. Sci.* **24**(4):299–303.

Hunt, G. R., 2000, Human-like, population-level specialisation in the manufacture of pandanus tools by New Caledonian crows (*Corvus moneduloides*), *Proc. Roy. Soc. Lond.: B.* **267**:403–413.

Kawai, M., 1967, Catching behavior observed in the Koshima troop—a case of newly acquired behavior, *Primates* **8**:181–186.

Kimura, D., 1979, Neuromotor mechanisms in the evolution of human communication, in: *Neurobiology of Social Communication in Primates*, H. D. Stelkis and M. J. Raleigh, eds., Academic Press, San Diego, pp. 197–219.

King, J. E., 1995, Laterality in hand preferences and reaching accuracy of cotton-top tamarins (*Saguinus oedipus*), *J. Comp. Psych.* **109**(1):34–41.

King, J. E. and Landau, V. I., 1993, Manual preference in varieties of reaching in squirrel monkeys, in: *Primate Laterality: Current Behavioral Evidence of Primate Asymmetries*, J. P. Ward and W. D. Hopkins, eds., Springer-Verlag, New York, pp. 107–123.

Kolb, B. and Fantie, B., 1987, Reaching for the brain, *Behav. Brain Sci.* **10**(2):279–280.

Kubota, K., 1990, Preferred hand use in the Japanese macaque troop, Arashiyama-R, during visually guided reaching for food pellets, *Primates* **31**(2):393–406.

Laska, M., 1996a, A study of correlates of hand preferences in squirrel monkeys (*Saimiri sciureus*), *Primates* **37**(4):457–465.

Laska, M., 1996b, Manual laterality in spider monkeys (*Ateles geoffroyi*) solving visually and tactually guided food-reaching tasks, *Cortex* **32**(4):717–726.

Lemon, R. E., 1973, Nervous control of the syrinx of white-throated sparrows (*Zonotrichia albicollis*), *J. Zool.* **171**:131–140.

Ley, R. G. and Bryden, M. P., 1982, A dissociation of right and left hemisphere effects for recognizing emotional tone and verbal content, *Brain Cogn.* **1**:3–9.

Lorincz, E. and Fabre-Thorpe, M., 1996, Shift of laterality and compared analysis of paw performances in cats during practice of a visuomotor task, *J. Comp. Psych.* **110**(3):307–315.

MacNeilage, P. F., Studdert-Kennedy, M. G., and Lindblom, B., 1987, Primate handedness reconsidered, *Behav. Brain Sci.* **10**:247–303.

Malashichev, Y. B., 2002, Asymmetries in amphibians: A review of morphology and behaviour, *Laterality* 7:197–218.

Marzona, E. and Giacoma, C., 2002, Display lateralisation in the courtship behaviour of the alpine newt (*Tritus alpestris*), *Laterality* 7:285–296.

Marchant, L. F. and McGrew, W. C., 1996, Laterality of limb function in wild chimpanzees of Gombe National Park: Comprehensive study of spontaneous activities, *J. Hum. Evol.* **30**:427–443.

Masataka, N., 1990, Handedness of capuchin monkeys, *Folio. Primatol.* **55**:189–192.

Mason, A. M., Wolfe, L. D., and Johnson, J. C., 1995, Hand preference in the sifaka (*Propithecus verreauxi coquereli*) during feeding in captivity, *Primates* **36**(2):275–280.

Matoba, M., Masataka, N., and Tanioka, Y., 1991, Cross-generational continuity of hand-use preferences in marmosets, *Behaviour* **117**(3–4):281–286.

McGrew, W. C., and Marchant, L. F., 1992, Chimpanzees, tools, and termites: Hand preference or handedness? *Curr. Anthropol.* **33**:114–119.

McGrew, W. C., Marchant, L. F., Wrangham, R. W., and Klein, H., 1999, Manual laterality in anvil use: Wild chimpanzees cracking *Strychnos* fruits, *Laterality* **4**(1):79–88.

Milliken, G. W., 1995, Right hand preference and performance biases in the foraging behavior of the aye-aye, in: *Creatures of the Dark: The Nocturnal Prosimians*, L. Alterman, G. A. Doyle, and M. K. Izard, eds., Plenum Press, New York, pp. 261–292.

Mittra, E. S., Fuentes, A., and McGrew, W. C., 1997, Lack of hand preference in wild Hanuman langurs (*Presbytis entellus*), *Am. J. Phys. Anthropol.* **103**:455–461.

Neves, A. and Dolhinow, P., 1995, Hand preference in a colobine monkey (*Presbytis entellus*), *Am. J. Phys. Anthropol. (Suppl.).* **20**:160.

Nottebohm, F., 1970, Ontogeny of bird song, *Science* **167**:950–956.

Nottebohm, F., 1971, Neural lateralization of vocal control in a Passerine bird. I. Song, *J. Exp. Zool.* **177**:229–261.

Nottebohm, F., 1972, Neural lateralization of vocal control in a Passerine bird. II. Sub-song, calls, and a theory of vocal learning, *J. Exp. Zool.* **179**:35–50.

Nottebohm, F., 1977, Asymmetries in neural control of vocalization in the canary, in: *Lateralization in the Nervous System*, S. Harnad, R. W. Doty, L. Goldstein, J. Jaynes, and G. Krauthamer, eds., Academic Press, New York, pp. 23–44.

Nottebohm, F. and Nottebohm, M. E., 1976, Left hypoglossal dominance in the control of canary and white-crowned sparrow song, *J. Comp. Phys.* **108**:171–192.

Nottebohm, F., Stokes, T. M., and Leonard, C. M., 1976, Central control of song in the canary, *Serinus canaries, J. Comp. Neurol.* **165**:457–486.

Nymark, M., 1972, Apomorphine provoked stereotypy in the dog, *Psychopharmacology* **26**:361–368.

Olson, D. A., Ellis, J. E., and Nadler, R. D., 1990, Hand preferences in captive gorillas, orang-utans and gibbons, *Am. J. Prim.* **20**:83–94.

Parr, L. A., Hopkins, W. D., and de Waal, F. B. M., 1997, Haptic discrimination in capuchin monkeys (*Cebus apella*): Evidence of manual specialization, *Neuropsychologia*, **35**(2):143–152.

Pike, A. V. L. and Maitland, D. P., 1997, Paw preferences in cats (*Felis silvestris catus*) living in a household environment, *Behav. Process* **39**:241–247.

Peterson, G. M., 1934, Mechanisms of handedness in the rat, *Comp. Psych. Mono.* **9**(6):1–67.

Petersen, M., Beecher, M., Zoloth, S., Moody, D., and Stebbins, W., 1978, Neural lateralization of species-specific vocalizations by Japanese macaques (*Macaca fuscata*), *Science* **202**:324–327.

Petersen, M. R., Beecher, M. D., Zoloth, S. R., Green, S., Marler, P. R., Moody, D. B., and Stebbins, W. C., 1984, Neural lateralization of vocalization by Japanese macaques: Communicative significance is more important than acoustic structure, *Behav. Neurosci.* **98**:779–790.

Pohl, K., Wrogemann, D., Radespiel, U., and Zimmermann, E., 2000, Hand preference in the mouse lemur *Microcebus murinus* and *M. rufus* for a set of manual tasks in the foraging context, *Folia Primatol.* **71**(4):238.

Porac, C., 1996, Hand and foot preferences in young and older adults: A comment on Gabbard and Iteya, *Laterality* **1**(3):205–213.

Robins, A., Lippolis, G., Bisazza, A., Vallortigara, G., and Rogers, L. J., 1998, Lateralized agonistic responses and hindlimb use in toads, *Anim. Behav.* **56**:875–881.

Rodriguez, M. and Afonso, D., 1993, Ontogeny of T-Maze behavioral lateralization in rats, *Phys. Behav.* **54**:91–94.

Rogers, L. J., 1980, Lateralisation in the avian brain, *Bird Behav.* **2**:1–12.

Rogers, L. J., 1981, Environmental influences on brain lateralization, *Behav. Brain Sci.* **4**:35–36.

Rogers, L. J., 1989, Laterality in animals, *Int. J. Comp. Psych.* **3**(1):5–25.

Rogers, L. J., 1991, Development of lateralization, in: *Neural and Behavioral Plasticity*, R. J. Andrew, ed., Oxford University Press, Oxford, pp. 507–535.

Rogers, L. J., 2000, Evolution of hemispheric specialization: Advantages and disadvantages, *Brain Lang.* **73**:236–253.

Rogers, L. J., 2002a, Lateralization in vertebrates: Its early evolution, general pattern, and development, in: *Advances in the Study of Behavior*, Vol. 31, P. J. B. Slater, J. S. Rosenblatt, C. T. Snowdon, and T. J. Roper, eds., Academic Press, San Diego, pp. 107–161.

Rogers, L. J., 2002b, Lateralised brain function in anurans: Comparison to lateralization in other vertebrates, *Laterality* 7:219–240.

Rogers, L. J. and Andrew, R. J., eds., 2002, *Comparative Vertebrate Lateralization*, Cambridge University Press, Cambridge.

Rogers, L. J. and Kaplan, G., 1996, Hand preferences and other lateral biases in rehabilitated orang-utans, *Pongo pygmaeus pygmaeus*, *Anim. Behav.* **51**:13–25.

Rogers, L. J. and Workman, L., 1993, Footedness in birds, *Anim. Behav.* **45**:409–411.

Rogers, T. T. and Bulman-Fleming, M. B., 1998, Arousal mediates relations among paw preference, lateral paw preference, and spatial preference in the mouse, *Behav. Brain Res.* **93**:51–62.

Sanford, C., Guin, K., and Ward, J. P., 1984, Posture and laterality in the bushbaby (*Galago senagalensis*), *Brain Behav. Evol.* **25**:217–224.

Schmidt, S. L., Filgueiras, C. C., and Krahe, T. E., 1999, Effects of sex and laterality on the rotatory swimming behavior of normal mice, *Phys. Behav.* **65**(4–5):607–616.

Searleman, A., 1980, Subject variables and cerebral organization for language, *Cortex* **16**:239–254.

Signore, P., Nosten-Bertrand, M., Chaoui, M., Roubertoux, P. L., Marchland, C., and Perez-Diaz, F., 1991, An assessment of handedness in mice, *Phys. Behav.* **49**:701–704.

Singer, S. S. and Schwibbe, M. H., 1999, Right of left, hand or mouth: Genera-specific preferences in marmosets and tamarins, *Behaviour* **136**(1):119–145.

Sobel, N., Supin, A. Y., and Myslobodsky, M. S., 1994, Rotational swimming tendencies in the dolphin (*Tursiops truncatus*), *Behav. Brain Res.* **65**:41–45.

Spinozzi, G. and Truppa, V., 1999, Hand preferences in different tasks by tufted capuchins (*Cebus apella*), *Int. J. Primatol.* **20**(6):827–849.

Spinozzi, G., Castorina, M. G., and Truppa, V., 1998, Hand preferences in unimanual and coordinated-bimanual tasks by tufted capuchin monkeys (*Cebus apella*), *J. Comp. Psych.* **112**(2):183–191.

Stafford, D. K., Milliken, G. W., and Ward, J. P., 1990, Lateral bias in feeding and brachiation in *Hylobates*, *Primates* **31**:407–414.

Stafford, D. K., Milliken, G. W., and Ward, J. P., 1993, Patterns of hand and mouth lateral biases in bamboo leaf shoot feeding and simple food reaching in the gentle lemur (*Hapalemur griseus*), *Am. J. Prim.* **29**:195–207.

Sugiyama, Y., Fushimi, T., Sakura, O., and Matsuzawa, T., 1993, Hand preference and tool use in wild chimpanzees, *Primates* **34**(2):151–159.

Tallal, P., Miller, S., and Fitch, R. H., 1995, Neurobiological basis for speech: A case for the preeminence of temporal processing, *Irish J. Psychol.* **16**(3):194–219.

Tan, Ü., 1987, Paw preferences in dogs, *Int. J. Neurosci.* **32**:825–829.

Tan, Ü. and Kutlu, N., 1991, The distribution of right-, left-, and mixed pawed male and female cats: The role of a female right-shift factor in handedness, *Int. J. Neurosci.* **59**:219–229.

Tanner, J. E. and Byrne, R. W., 1996, Representation of action through iconic gesture in a captive lowland gorilla, *Curr. Anthropol.* **37**:162–173.

Teichroeb, J. A., 1999, A review of hand preferences in non-human primates: Some data from patas monkeys, *Lab. Prim. Newslett.* **38**(1):10–13.

Tokida, E., Tanaka, I., Takefushi, H., and Hagiwara, T., 1994, Tool-using in Japanese macaques: Use of stones to obtain fruit from a pipe, *Anim. Behav.* **47**:1023–1030.

Tokuda, K., 1969, On the handedness of Japanese monkeys, *Primates* **10**:41–46.

Tommasi, L., and Vallortigara, G., 1999, Footedness in binocular and monocular chicks, *Laterality* **4**(1):89–95.

Tomasello, M., Call, J., Warren, J., Frost, G. T., Carpenter, M., and Nagell, K., 1997, The ontogeny of chimpanzee gestural signals: A comparison across groups and generations, *Evol. Comm.* **1**:223–259.

Tonooka, R. and Matsuzawa, T., 1993, Hand preferences of captive chimpanzees (*Pan troglodytes*) in simple reaching for food, *Int. J. Prim.* **16**(1):17–35.

Uziel, D., Lopes-Conceicao, M. C., Luiz, R. R., and Lent, R., 1996, Lateralization of rotational behavior in developing and adult hamsters, *Behav. Brain Res.* **75**:169–177.

Vallortigara, G., 1992, Right hemisphere advantage for social recognition in the chick, *Neuropsychologia* **9**:761–768.

Vallortigara, G., Zanforlin, M., and Cailotto, M., 1988, Right–left asymmetry in position learning of male chicks, *Behav. Brain Res.* **27**:189–191.

Vallortigara, G., Rogers, L. J., Bisazza, A., Lippolis, G., and Robbins, A., 1998, Complementary right and left hemifield use for predatory and agonistic behavior in toads, *Neuroreport* **9**:3341–3344.

Vallortigara, G., Rogers, L. J., and Bisazza, A., 1999, Possible evolutionary origins of cognitive brain lateralization, *Brain Res. Rev.* **30**:164–175.

Vallortigara, G., Cozzutti, C., Tommasi, L., and Rogers, L., 2001, How birds use their eyes: Opposite left–right specialization for the lateral and frontal visual hemifield in the domestic chick, *Curr. Biol.* **11**:29–33.

Vauclair, J. and Fagot, J., 1987, Spontaneous hand usage and handedness in a troop of baboons, *Cortex* **23**:265–274.

Ward, J. P. and Cantalupo, C., 1997, Origins and functions of laterality: Interactions of motoric systems, *Laterality* **2**(3/4):279–303.

Ward, J. P., Milliken, G. W., Dodson, D. L., Stafford, D. K., and Wallace, M., 1990, Handedness as a function of sex and age in a large population of lemur, *J. Comp. Psych.* **104**:167–173.

Warren, J. M., 1987, Primate handedness: Inadequate analysis, invalid conclusions, *Behav. Brain Sci.* **10**(2):288–289.

Waters, N. S. and Denenberg, V. H., 1991, A measure of lateral paw preference in the mouse, *Physio. Behav.* **50**:853–856.

Waters, N. S. and Denenberg, V. H., 1994, Analysis of two measures of paw preference in a large population of inbred mice, *Behav. Brain Res.* **63**:195–204.

Wendt, P. E. and Risberg, J., 1994, Cortical activation during visuospatial processing: Relation between hemispheric asymmetry of blood flow and performance, *Brain Cogn.* **24**:87–103.

Westergaard, G. C., 1991, Hand preference in the use and manufacture of tools by tufted capuchin (*Cebus apella*) and lion tailed macaque (*Macaca silenus*) monkeys, *J. Comp. Psych.* **105**:172–176.

Westergaard, G. C. and Suomi, S. J., 1993a, Lateral bias for rotational behavior in tufted capuchin monkeys (*Cebus apella*), *J. Comp. Psych.* **110**:199–202.

Westgaard, G. C. and Suomi, S. J., 1993b, Hand preference in capuchin monkeys varies with age, *Primates* **34**:295–299.

Westgaard, G. C. and Suomi, S. J., 1993c, Hand preference in the use of nut-cracking tools by tufted capuchin monkeys (*Cebus apella*), *Folia Primatol.* **61**:38–42.

Westergaard, G. C. and Suomi, S. J., 1996a, Hand preference for a bimanual task in tufted capuchins (*Cebus apella*) and rhesus macaques (*Macaca mulatta*), *J. Comp. Psych.* **110**(4):406–411.

Westergaard, G. C. and Suomi, S. J., 1996b, Hand preference for stone artifact production and tool-use by monkeys: Possible implications for the evolution of right-handedness in homonids, *J. Hum. Evol.* **30**:291–298.

Westergaard, G. C., Kuhn, H. E., Lundquist, A. L., and Suomi, S. J., 1997, Posture and reaching in tufted capuchins (*Cebus apella*), *Laterality* **2**(1):65–74.

Westergaard, G. C., Kuhn, H. E., and Suomi, S. J., 1998, Bipedal posture and hand preference in humans and other primates, *J. Comp. Psych.* **112**:55–64.

Westergaard, G. C., Lussier, I. D., and Higley, J. D., 2001, Between-species variation in the development of hand preference among macaques, *Neuropsychologia* **39**:1373–1378.

Yuanye, M., Yunfen, T., and Ziyun, D., 1988, The hand preference of *Presbytis* (PR), *Rhinopithecus* (RH), and *Hylobates* (HY) in picking up food, *Acta Anthropol Sinica.* 7:177–181.

Zappia, J. V. and Rogers, L. J., 1987, Sex differences and reversal of brain asymmetry by testosterone in chickens, *Behav. Brain Res.* **23**:261–267.

Zimmerberg, B., Glick, S. D., and Jerussi, T. P., 1974, Neurochemical correlate of a spatial preference in rats, *Science* **185**:623–625.

Zimmerberg, B., Stumpf, A. J., and Glick, S. D., 1978, Cerebral asymmetry and left–right discrimination, *Brain Res.* **140**:194–196.

Epilogue

Gisela Kaplan and Lesley Rogers

Many of us who have worked with great apes have had experiences of profound and long-lasting impact. Part of that impact, one suspects, has to do with moments in which the line between our species and theirs becomes blurred, sometimes unsettling but often astonishingly liberating. Indeed, in 1996, at the International Primatology Meeting in Madison, Wisconsin, a 4-hr session on recollection of "epiphanies" with apes was held to explore how these special events had shaped decisions, careers, attitudes, and research. The room was charged with memories and emotion. Speaker after speaker "confessed" their personal ties with apes and admitted to the ethical quandary and deliberations such feelings imposed on thoughts and research about animals.

Primates will always be very special to us, not just to those of us who work with them, because, in between the clean sheets of science, we slumber in the knowledge of a common evolutionary bed. It is only with a primate that we can hold hands and look into each other's eyes. This encapsules a sense of relatedness and, no doubt, at a deeper level, it gives us some reassurance that we, as human primates, are not an isolated species.

One may be forgiven, perhaps, that along with the emotive content (not rare when working with great apes) comes a fudge factor that may color our science and our views of what we perceive primates to be, what they can do, and what we infer or would like to think we know about them. At the same time, the knowledge of our very relatedness may indeed be one of the greatest motivators for long-term commitments to science and to apes and, indeed, to primates in general. It is not unusual for otherwise sober scientists and scholars to throw caution to the wind and proclaim to the world how unique apes are but, in the cold light of day, some such declarations of partiality cannot pass the test of proof. This book has been an attempt to deal with some of the thorny and exciting research questions asked of and in primatology today, and not only within primatology.

A healthy skepticism, mixed with a dose of passion for the discipline we profess, and a broad ethological perspective allows us to see where we have been misguided, where we have failed, but also where we have succeeded and made enormous progress in our understanding of primates. This book has, sometimes more than surreptitiously, examined our own attitudes to studies of primates, exposed primatocentric views and shown that individual traits, even in complex cognition, in communication and social learning, and in perceptual systems, may be found elsewhere amongst other orders.

We may have been unable to answer the question whether primates are superior to non-primates in an evolutionary or cognitive sense because this is work in progress. It is important, however, to continue asking this question at least within the four parameters that Tinbergen insisted we need to investigate: the phylogeny, ontogeny, evolution, and function of an organism.

The exciting new work of the last 20 years or so has also placed great emphasis on cognitive abilities and such emphasis, we feel, has not only enriched our understanding of primates in general, but it has raised moral questions of a kind that are new and very important in the current status of apes, lesser apes, and all other primates. We know of the endangered status of many of them in the wild, and we know of the fragility of their survival in their habitats. We also know why this is so. Their habitats are being destroyed and their unique attributes, especially of rain forests, cannot be restored. This has certainly been the case of the habitat we know best, Borneo (Pearce, 1990; Sutton et al., 1983). Like many other observers, we have watched helplessly over the last 15 years as the situation for orangutans has worsened dramatically (Kaplan and Rogers, 2000; Smits, 1998). Human greed and exploitation of primates for wide-ranging human benefit, and population pressures extol an ever greater price on primate survival (Rijksen and Meijaard, 1999). Those keen to seek their protection or rehabilitation in the wild have yet another strong motivator to understand primate behavior. But we also need to recognize that primates are not alone in this struggle for survival. Of course, we recognize that primates, as umbrella species and as public "flagships," can be of substantial service to their entire habitat in raising awareness and exerting pressure where needed.

It was not too long ago that perhaps about 80 percent of all rehabilitation and reintroduction projects more or less failed (be this for birds or primates) because the most dedicated and devoted teams of researchers and wildlife managers involved in the process too often lacked any formal training or understanding of animal behavior. So alarming were these figures of failure that, in 2001, the prestigious journal *Nature* published an article emphasizing the importance of behavior in rehabilitation and complaining about widespread ignorance about the significance of knowledge of the behavior of the species supposedly being protected from extinction (Knight, 2001).

And herein lay another important task that we had set ourselves: consistently throughout this book, our emphasis has been on behavior, attempting to understand more fully how primates and non-primates learn and develop, what their behavioral repertoire actually is, and how they would predictably respond in given situations. The value of studies of higher cognitive functions, in an applied sense, is that we have moved on from relegating anything that we do not understand to "instinct." The very moment we concede that primates learn many of the behavior patterns that they need for survival, we have also directly or inadvertently helped set a different tone and standard for rehabilitation and captive breeding programs.

The contributors to this volume have pointed out, often rather courageously when announced within the heartland of primatological studies, that primates are not "special" in many of the attributes in which we once thought they were. No doubt, this is a debate that continues to raise evolutionary questions about what traits have been preserved and which have newly emerged in different orders and for what reason. Evolution and coevolutionary events may have resulted in similar behavioral solutions on the basis of very different biological parameters.

The 21st century has been dubbed the century of brain and mind. There is no doubt, as amply demonstrated by the cutting edge primatological research discussed in this volume, that primatology is a thriving field and a very active participant and facilitator in the brain/mind debates also of other species, and so it is and will be a partner in some of the very important discoveries of this century.

REFERENCES

Kaplan, G. and Rogers, L. J., 2000, *The Orang-Utans.* Perseus, Cambridge, MA.

Knight, J., 2001, If they could talk to the animals, *Nature* **414**:246–247.

Pearce, F., 1990, Hit and run in Sarawak. Under the floodlights of 24-hour forestry, the world's oldest rainforest is disappearing before the eyes of its inhabitants. Who will call a halt? *New Scientist* (12 May):24–27.

Rijksen, H. D. and Meijaard, E., 1999, *Our Vanishing Relative. The Status of Wild Orangutans at the Close of the Twentieth Century*, Kluwer Academic Publishers, Dordrecht, Netherlands.

Smits, W., 1998, Diary Aug. 2–11, 1998, web: http://www.redcube.nl./bos/di980804.htm.

Sutton, S. L., Whitmore, T. C., and Chadwick, A. C., eds. 1983, *Tropical Rain Forest Ecology and Management*, Blackwell Scientific Publications, Oxford.

About the Editors

Both authors are professors in the Centre for Neuroscience and Animal Behaviour, School of Biological, Biomedical and Molecular Sciences at the University of New England, Armidale, NSW, Australia. Together they have spent nearly two decades carrying out joint research on orangutans in the field and on marmosets in the laboratory, resulting in books entitled *The Orang-Utans of Borneo* (1994) and *The Orangutans: Their Evolution, Behavior, and Future* (1999, 2000, 2001). As co-authors they have published many scientific papers and book chapters together and, in addition, have published a number of other books on animal and human behavior, including the highly acclaimed *Birds. Their Habits and Skills* (2001), *Songs, Roars and Rituals. Communication in Mammals, Birds, and Other Animals* (2000, 2002), and most recently *Spirit of the Wild Dog* (2003) and *Gene Worship: Moving Beyond the Nature/Nurture Debate Over Genes, Brain, and Gender* (2003). In addition, they have followed their separate research interests on other species (see below).

Lesley Rogers has a Doctor of Philosophy and a Doctor of Science from the University of Sussex and is an elected Fellow of the Australian Academy of Science. Her area of research is neuroscience and animal behavior, and has covered a wide range of species from lower vertebrates to apes. She has a leading international reputation for her research on brain development and behavior, with a particular emphasis on lateralization of brain function. In addition to the publications listed above, she has authored *The Development of Brain and Behaviour in the Chicken* (1995), *Minds of Their Own; Thinking and Awareness in Animals* (1998), *Sexing the Brain* (1999, 2001), and with John Bradshaw *The Evolution of Lateral Asymmetries, Language, Tool Use, and Intellect* (1993). She has also edited with Richard Andrew *Comparative Vertebrate Lateralization* (2002).

Gisela Kaplan has a Doctor of Philosophy from Monash University, Melbourne, and is now a research professor at the University of New England, in a dual position in Biological Sciences and in Education. She specializes in research on animal communication and cognition and has a strong postgraduate teaching role in animal welfare. A prolific writer of over 150 research articles and 15 books (some highly acclaimed books are in social science), she is also regularly invited to contribute to wildlife magazines, to media science programs, and to community wildlife education, as well as to prestigious handbooks on animal behavior (on apes and birds) internationally. In her spare time, she runs a rehabilitation station for native Australian wildlife, especially birds of prey. Her latest book, not mentioned above, is the *Biology and Behaviour of the Australian Magpie* (2004).

INDEX

Abstract cognition, 83–4, 201, 316, *see also* Visual cognition, *see also* Spatial memory
Acoustic concealment, 197
Acoustic localization, 173–4, *see also* Communication
Acuity, 84, *see also* Visual cognition, *see also* Visual fields
Adaptation, 40, 72, 74, 85, 97, 99, 102, 121, 293, 300–4
Affect, *see* Emotion, *see also* Fear Expression
Affiliative behaviour, 8
Age, 30, 62, 66, 73, 80, 103–4, 119, 127, 161, 176, 235–6, 238, 242, 245, 314, 326–7
Aggression, 6, 34, 158, 192, 207, 211–12, 349–50
Alarm calls, 27–8, 190, 193–8, of birds, 28, 194–5, 198, 204, of primates, 28, 193–8, *see also* Referential communication, *see also* Communication, *see also* Vocalizations
Allocortex, *see* paleocortex,
Amazona amazonica, see Parrot, orange-winged Amazonian
American Sign Language (ASL), *see* Signing, in apes
Amodal completion, 58–68
Amphibians, 310, 314, 327–9, 348–9, 355
Anolis carolinensis, see Lizard, Anole
Anthropocentric view, 4, 231
Anthropoid primate, 290, 292
Anticipation, 33, 35, 256, 264
Apes, 4–5, 7, 9, 27–9, 34, 37, 41, 73, 98, 104–6, 110–18, 121, 123–6, 157, 189, 190, 197, 199–215, 227–58, 276, 278, 280–1, 292, 294–6, 299, 303, 310, 312–13, 316, 318, 337–8, 345, 347, *see also* Great apes,
Aphelocoma californica, see Jay, Western scrub
Aphelocoma coerulescens, see Jay, scrub
Aphelocoma ultramarina, see Jay, Mexican
Archistriatum, 9–10
Artificial fruit test, 13, 14, 111
Artificial predator, 28, 154
Artistic expression, 299
Associative learning, 23, 42, 76
Assumptions, 39, 64, 97, 99, 103, 105, 148, 163–4, 289, 316–17
Asymmetry, *see* Lateralization, *see also* Hemispheric specialization
Ateles geoffroyi, see Monkey, spider
Attention, 15–18, 76, 101, 108, 141, 158, 236, 240, 264–72, 275, 277–9, 281–2, directing, 264, 275–8, 281, following, 271–5, 281, global and local, 65, reading, 265–71, 278, 281, states, 269–71, 281, *see also* Gaze following, *see also* Visual perspectives of others, *see also* Visual perception
Attribute hypothesis, 176
Attribution, 4, 17, 42, 45, of mental states, 42, 45, 142, of special character, 142
Audience effect, 28, 194–5, 265, 266, 278
Audition, 179–80
Autonoetic consciousness, 31
Awareness, 199, 204, of self, 41–3, 234, 294, of others, 42, *see also* Consciousness

Baboon, 60, 63, 107, 198–9, 341–2, gelada, 197, 353, hamadryas, 121, 249
Bat, 100, 172, 174, 176, 303, echo-locating, 175, 179
Bear, 173, 250
Bee-eater, 235
Bees, 81
Begging, 104, 115, 199, 232–3, 236–8, 240, 243–7, 269, 276, *see also* Gestures
Behavioral flexibility, *see* Intelligence
Binocular interaction, 65–7, *see also* Visual cognition
Biological motion, 70, 72, *see also* Visual cognition, *see also* Visual scenes
Bipedalism, 85, 337–9
Birds, 3, 6, 11–16, 18, 20–1, 24, 28, 38, 40, 42, 45, 100, 142, 157, 160, 178–9, 200, 204, 208, 292, 308–9, 314, 327, 343, 349–51, 354, 358, compared to primates, 3–4, 45, 57–85, 172, 195, 204, 290, 298, 308–9, 314–16, cognitive abilities, 36, 40, 42, 84, 204–5, 210, 257, 316, social learning, 40, 148, communication, 192, 198–200, 204, footedness in, 326, 333, 339, 346–7, song learning, 40, turning bias and, 329, 332, *see also* Caching, *see also* Imitative learning, *see also* Tool use, *see also* Pilfering, *see also* Brain of birds, *see also* Lateralization

Body size, 208, compared to brain size, 5, 10, 290–304
Bonobo, 29, 122, 192, 337–8, 347, Kanzi, 29, 203
Brain asymmetry, *see* Lateralization
Brain size, 5–7, 289–305, 313, 317–18, across species, 3–4, 143, 229, 292, 297, 301–6, 317, of primates, 290–4, 295, 309, 313, evolution of, 292–304, behavior and, 293, 304–10, experience and, 314–16, of elephants, 228, mode of prey capture and, 306, tool use and, 308–9, *see also* Encephalization
Brain stem, 306
Brain structure, 4, size differences in different regions, 299–305, functionally linked structures, 303, species comparisons, *see also* Birds, *see also* Primates
Brain volume, *see* Brain size
Brain weight, *see* Brain size
Brain, of birds, 3–4, 8, 43, 77, 205, 292, 297, 306, 308, 312, 318, of elephants, 228, of mammals, 205, of primates, 3–4, 77, 290–2, 297, 313, organization of, 4, 77, forebrain, 45, neostriatum, 45
Brainstem ratio, 6–7, 9
Broca's area, 311
Broken-wing display, 16
Budgerigar, 115
Bufo bufo, see Toad, European common
Bufo marinus, see Toad, South American
Bufo viridus, see Toad, European green
Burial rite, in elephants, 230
Bushbaby, 106, 338, 340–1, lesser, 335, small-eared, 331, 335

Caching, of food, 6, 9, 11–13, 17–22, 32, 35–6, 39, 42–4, 74–5, *see also* Pilfering
Callithrix jacchus, see Marmoset
Callitrichidae, see Tamarin, *see also* Marmoset
Camarhynchus pallidus, see Finch, woodpecker
Canary, 304, 310, 350–1
Canids, 102–3, 206–12, 215, 277, 306
Canis familiaris, see Dog
Captivity, 15, 100, 104, 237, 245, 293, 337, 358
Capra hircus, see Goat
Capuchin, 14, 107, 108–11, 121, 126, 159, 336, 338, 341–2, 344–5, white-faced, 121, tufted, 331
Carassius, see Goldfish
Carduelis carduelis, see Goldfinch
Caribou, 102–3
Carnivore, 34, 103, 205–8, 210–12, 306, 308
Carnivorous behavior, 190, 205–6
Cartesian philosophy, 231–5
Cat, 64, 70, 174, 178, 179, 252, 326, 329–32, 334–5, 339, 340, 342
Catarrhine, 336
Catching, 341, *see also* Hand preference
Catfish, 349
Cebus apella, see Capuchin, tufted
Cebus capucinus, see Capuchin, White faced
Cebus, 107, 121, 126, *see also* Capuchin
Centre localization, 81, 83
Cercopithecus aethiops, see Monkey, vervet
Cerebellum, 306
Cetaceans, 105, 118, 127, 157, 192, 213, 292, 308
Chaffinch, 350
Chantek, *see* Orangutan
Charadius melodus, see Plover, piping
Charadius Wilsonia, see Plover, Wilson's
Cheetah, 105
Chelydra serpentina, see Turtle, snapping
Chick, *see* Chicken
Chicken, 28, 61–2, 64, 66–71, 73–7, 79–83, 172, 177, 195–6, 198–9, 215, 265, 310, 311, 329, 333, 349–50
Chimpanzee, 3–5, 9, 14–15, 23, 25, 28, 30, 33–5, 37, 60, 103, 109–13, 120–6, 162–3, 189–90, 197, 199–201, 203, 205–7, 209–14, 228–40, 245–52, 263–82, 299, 307, 313, 337–8, 352–3, Washoe, 203, 228, 345, 347
Cingulate gyrus, 178
Circling behavior, 330, 331, *see also* Lateralization, *see also* Hemispheric specialization
Clambering, 295
Co-evolutionary events, 204
Cognition, *see* Cognitive ability, *see also* Higher cognition, *see also* Visual cognition, *see also* Cognitive capacity, *see also* Cognitive gap
Cognitive ability, 8, 36–7, 40, 45, 127, 189, 214, 294, 301, 302, *see also* Higher cognition
Cognitive capacity, 45, 289, 295, and brain size, 6, 299, 300–4, 314, 317, and Theory of Mind, 16
Cognitive gap, 112
Cognitive map, 163
Colinus virginianus, see Quail, Bobwhite
Collicular system, 57, *see also* Visual pathways
Command headquarters, *see* Prefrontal cortex
Communication, 101, 189–215, 349, 355, complex communication, 17, 118, 190, 198, 205–15, in primates, 113, 116–18, 121–7, 171–81, 190, 195, 196, 203, 211–12, 266, 270, 277, 281, in birds, 117–18, in dogs, 266, 277, in other animals, 171, 180, 197, 207–11, intentional, 200, 277–8, *see also* Audience effect, *see also* Gestures, *see also* Isolation calls, *see also* Referential communication, *see also* Separation calls, *see also* Symbolic communication, *see also* Vocal signaling

Complex communication, *see* Communication
Complex stimuli, 59, 60
Computer models, 234
Concept of centre, *see* Centre localization
Conceptual skills, 197
Conditioning procedures, 62, 282
Conformity, 146, 152–6
Consciousness, 227–57, 298, 301, *see also* Awareness
Conservation, 372–3
Contact call, 121, 230, 307, 353
Contour perception, *see* Visual cognition
Cooperation, 208–10, 264, 266, 306, 358
Coordinated evolution, 300–1
Coral reef fish, 163
Cord-pulling, 253
Corpus callosum, 312–16, 318, 349, effects of experience on, 315–16, 351
Cortex, 290, 293, 296, 302, 314
Corvids, 4–13, 16–18, 21, 23–4, 39, 41, 45, 204, 215, *see also* Jays, *see also* Crows, *see also* Ravens, *see also* Magpies
Corvus corax, see Raven
Corvus corone, see Crow, carrion
Corvus frugilegus, see Rook
Corvus macrorhynchos, see Crow, jungle
Corvus monedula, see Jackjaw
Corvus moneduloides, see Crow, New Caledonian
Coturnix japonica, see Quail, Japanese
Courtship, 72, 249, *see also* Social attachment
Coyote, 197, 207
Creativity, 264, 282
Critical frequency fusion, *see* Flicker fusion
Crow, 4, 25–7, 34, 41–3, 249, carrion, 5, 9, 10, jungle, 41–3, New Caledonian, 26–7, 34, 204, 343, *see also* Corvids
Crowing, 172
Cues, 63, 69, 70, 74, 77–8, 80, 115–16, 146, 154, 174, 193, 198, 243, 264, 268–70, 272–5, 278–82, 329, *see also* Communication, *see also* Gestures, *see also* Pointing, *see also* Visual perception
Culture, 101, 118–27, 158, 162–3
Cyanocitta cristate, see Jay, Northern blue

Dactylopsila trivirgata, see Possum, Australian striped
Darwin, Charles, 232, 238
Deception, 16–17, 43, 232, 235, 264, 307, signal, 197–9
Decision-making, 299
Declarative-like representation, 75
Delayed matching-to-sample (DMS), 31
Delayed non-matching-to-sample (DNMS), 31, 39, 76
Delayed reinforcement, 33
Demonstrator, *see* Teaching
Dendrons, 298
Depth stratification, *see* Visual scenes
Detours, 271
Devalued food, 75
Development, 62, 80, 101, 103, 113, 116, 156, 171, 175–6, 192, 203–5, 236–7, 265, 300, 303, 306, 315, 329, 398
Diet, 97, neocortex size and, 305–6
Discrimination task, 60, 229–30, 243, two-choice, 296
Discrimination, of stimuli, *see* Stimuli
Dishabituation, *see* Habituation/dishabituation
Distress call, 177, *see also* Communication, *see also* Isolation calls, *see also* Separation calls, *see also* Vocalizations
Dog, 15, 177, 191, 197–201, 205–7, 209–11, 252, 263–82, 329–32, 334–5, 339
Dolphin, 15, 103, 105, 115–16, 190, 197, 200, 215, 282, 292, 329–30, bottle-nosed, 117, 174, 235, 308
Dominant/subordinate, 268–9
Dopamine, 8, 330–1, circling behavior and, 330–1, *see also* Lateralization
Dove, ring, 73
Drosophila melanogaster, see Fruit fly
Duck, 309

Eagle alarm call, 194–5, *see also* Alarm calls, *see also* Vocal signaling, *see also* Communication
Echolocation, 175
Ecological niche, 303
Ecological strategies, 105, 113–14, 214, 294–5
Ecological validity, 4, 41, 45, 100, 105, 113, 116, 236, 293–4
Ecological/ethological approach (EEA), 38, 41, 45
Ecology, 38–9, 100, 162, 191
Elephant, 116, 213, 215, 228–31, 235, 238–48, 251–7, 294, African savannah, 103, 230, Asian, 115, 229, 240–6, 252–7
Elephas maximus, see Elephant, Asian
Embedded food, 118
Emotion, 189–92, 299, 348–54, 356
Emulation, 109, 142
Encephalization, 292, 296, in primates, 5, 296, in birds, 5, evolution of, 292, *see also* Brain size
Enculturalization, 113, 281
Enrichment, 314, *see also* Experience
Episodic memory, 5, 30–3, 35
Equus caballus, see Horse
Erythrocebus patas, see Monkey, Patas
Escape route, 155
Ethological approach, 38, 40, 45

Ethological validity, 4, 32, 116
Ethologically based learning tasks, *see* Ethological approach
Eublepharis macularis, see Lizard, gecko
Evolution, 4, 45, 203–5, 214, 289–307, 309, 311, 325, 331, 339, 346, 348, 371, 373, of hemispheric specialization, 358–9
Executive brain, 297, 299, 309
Experience, 14, 20–1, 23, 28, 30–2, 42, 44, 58, 61, 102, 144, 146, 175–6, 244, 279, 300, 314–18, 326, 354, brain size and, 314–16, 351–2, increased size of hippocampus and, 314–16
Exploratory behavior, 108
Extraction of insects, 122
Extraction of seeds, 122,
Eye contact, 115, 228, 237, 240, 246, 281
Eye gaze, *see* Gaze following
Eye preferences, 311, 348
Eyes, *see* Visual perception

Falcon, 209, 309
Fear expression, 352–3, 356, *see also* Hemispheric specialization, fear responses and
Feeding behavior, 212–15, 309–10
Felids, 103, 206, 306
Finch, woodpecker, 26
Fish, 80, 141–64, 290, 292, 310, 327–33, 354
Fixed action pattern, 17
Flexibility, 115, 126, 253, 264, 290, 294, 296, 305, 307–9, 348, 357, *see also* Intelligence
Flicker fusion, 63
Food calling, 194–8, *see also* Vocal signaling, *see also* Communication
Food sharing, 126
Food-caching, *see* Caching
Food-storing birds, 74, 78, 302, 315–16, *see also* Caching
Footedness, 326, 333–43, 346–8, *see also* Pawedness, *see also* Hand preference
Foraging, 84, 104–5, 113–14, 118, 121, 124, 144–54, 157–64, 206–8, 210, 338, and neocortex size, 305–6, *see also* Diet
Forebrain, 6–10, 45, 297, 304, 317
Fovea, 84
Fox, 207, bat-eared, 103, Artic, 198
French grunt, 163
Frequency-dependent learning, 145, 147, 152
Fringilla coelebs, see Chaffinch
Frog, 349, 354
Frontal lobes, 299, 312
Fruit fly, 248–9

Galago moholi, see Galago
Galago senegalensis, see Bushbaby, lesser
Galago, 292, 331
Gallus gallus, see Chicken
Gambrusia holbrooki, see Mosquito fish
Gaze following, 4, 15, 199–200, 264, 271–3, 279, in primates, 115, 199–200, 234–5, 278, in dogs, 272–80, *see also* Cues, *see also* Attention
General intelligence, *see* Intelligence
Genetically determined behavior, 248
Geniculo-striate system, 57, *see also* Visual pathways
Geometric information, 77–80, 84–5, module, 78, 80, sense of space, 78, 84–5
Geometric regularity, *see* Visual scenes
Geometric spatial cognition, *see* Visual cognition
Gestures, 111, 118, 123, 200–2, 213, 234–8, 243, 247, 266, 269–71, 275–8, 281, 355–6, *see also* Communication, *see also* Pointing, *see also* Begging
Gibbon, 9, 296, 299, 337, 347
Global visual strategy, 65, 67, 70, *see also* Visual cognition
Go/no-go procedure, 60
Goat, 15
Goldfinch, 343
Goldfish, 144, 341
Gorilla gorilla, see Gorilla
Gorilla, 3, 9, 14, 32, 97, 111, 123, 189, 201, 212–13, 299, 307, 337–8, 342, 347, Koko, 203
Grackle, 117, carib, 115
Gracula religiosa, see Mynah bird
Great apes, 73, 110–27, 189–90, 197, 201–15, 234, 257, 280, 294, 299, 312, 345, 371–2, *see also* Bonobo, *see also* Gorilla, *see also* Orangutan
Group complexity, 212–15
Group coordination, 190, 205–15
Group size, 40, 207, 306–8
Group stability, 212–15
Guinea pig, 174, 177–8
Guppy, 143–64, 328
Gymnorhina tibicen, see Magpie, Australian
Gymnorhinus cyanocephalus, see Jay, Pinyon

Habitat, 103, 305
Habituation/dishabituation, 61
Haemulon flavolineatum, see French grunt
Hamster, 329
Hand preference, 310–11, 313, 325–59, and avoidance behavior, in humans, 325–7, 339, 355, in primates, 311, 335–8, 341–6, postural effects on, 335–6
Handedness, *see* Hand preference
Hands, *see* Primates, hands of
Hapalemur griseus, see Lemurs
Hare, arctic, 102–3
Hawk, 209–10, 309

Hedgehog, 292
Hemispheres, 306, 310–13, 325, 339–40
Hemispheric specialization, 310–12, 325–6, 351–8, circling behavior in rodents and, 330, corpus callosum and, 312–14, emotional responses and, 352–4, 356, fear responses and, 329–30, 352–3, 356, for motor control, 326–7, 330, 339, muricide and, 351, song production and, 350–1, species-specific vocalizations and, 353–4, evolution and, 310–12, visuospatial processing and, 340, taste aversion and, 351–2, temporal sequencing and, 356–7, visual cognition and, 66–9, 341, *see also* Lateralization, *see also* Hand preference
Heritable characteristics, 174, 176, 334
Hidden food, 274, 277–9, 281
Hidden objects, *see* Visual cognition
Hierarchies, *see* Social organization
High vocal centre, 297, 304, 317, 351
Higher cognition, 5–46, 117, 142, 190–1, 196, 199, 204–5, 209, 214–15, 292–8, 302, 318, in birds, 17, 28, 36, 45, 127, 298, in primates, 28, 36–8, 45, 103–4, 110–17, 124–7, in elephants, 228–31, *see also* Intelligence
Hindbrain, 297, 299
Hippocampus, 297, 302, spatial learning and; 304–6, in birds, 302, food-storing and, 302–4, 315, experience and, 314–16
Homo sapiens, see Human
Honest/dishonest signaling, *see* Deception
Horse, 15
Human centered, 298
Human superiority, 98, 213, 304, 310–11
Human, 9, 60–4, 66, 70, 73, 78, 80, 82, 98, 106, 109–10, 148, 160, 180, 197, 200, 203, 228, 232, 234, 238, 273, 280, 293, 295–6, 298–301, 310–17, 325–7, 342, 348–9, 355
Hunting, 190, 205–15, 306, *see also* Killing
Hylobates lar, see Gibbon
Hyperstriatum accessorium (HA), 8, 10, 65, *see also* Brain, of birds
Hyperstriatum dorsale (HD), 8, 10, *see also* Brain, of birds
Hyperstriatum ventrale (HV), 8, 10, *see also* Brain, of birds

Iambus call, 179, *see also* Isolation calls, *see also* Separation calls, *see also* Vocalizations
Ictalurus punctatus, see Catfish
Illusion, *see* Visual scenes
Imitation, 9, 12–13, 99–101, 104, 106, 107, 114–20, 142, 157, 160, in birds, 12, 127, in fish, in primates, 4, 99, 106–20, 124, 159, 235, in elephants, 243, *see also* Vocal mimicry
Imitative learning, 12, in apes, 110–18, in birds, 12, 117–18, *see also* Imitation
Imo, *see* Macaque, Japanese
Imprinting, 61–2, 67, 70, 73, 76
Individual learning, in primates, 120
Infant cries, 177–81
Inference, 264
Information ecology, 102
Inheritance, *see* Heritable characteristics
Innovation, 4, 101, 142, 158–64, 308–10, by birds, 27, by fish, 143–4, 148–57, by primates, 4, 23, 158–62, in feeding, 6, *see also* Social learning
Insectivora, 302–3, 308
Insight, 227–57, 299, in primates, 4, 23, 158, 210, in birds, 24
Intelligence, 6, 37–8, 45, 159, 191, 201–4, 211, 227–30, 263–4, 289–98, 302–7, 318, in birds, 45, 317, in elephants, 228–31, in fish, 159, 163, in primates, 37–8, 45, 201–4, 209, 211–12, 290–6, 305, 318, *see also* Complex cognition, *see also* Higher cognition
Intentional behavior, 209
Intentional communication, 195, 198–201, *see also* Communication
Intentional manipulation, *see* Deception
IQ, 293, *see also* Intelligence
Isocortex, *see* neocortex
Isolation calls, detection and discrimination of, 173–81, in primates, 171–81, in mammals, 171–81 in birds, neural mechanisms of, 175–81, *see also* Communication, *see also* Vocalisations

Jackjaw, 9, 10
Jay, 4, 10, blue, 198–9, Florida scrub, 40, Mexican, 11, 39–40, northern blue, 26, pinyon, 11, 39–40, scrub, 11, 32, 36, 39, 42–3, 74–5, western scrub, 12–13, 18, 20–1, 39–40, white-throated magpie, 100, *see also* Corvids

Kanzi, *see* Bonobo
Kea, 14, *see also* Parrots
Killing, 208, 210, in primates, 122, 204–6, 210
Knowledge-based approach, 279
Koko, *see* Gorilla

Landmarks, 82–4,
Language abilities, 119, 190, 196, 201–5, 232, 295, 299, 311, in birds, 29, 196, 204–5, in primates, 28, 196, 201–4, in humans, 80, 197, 201–2, 204, 310–11, 355

Langurs, Hanuman, 336
Lateral visual field, 84, *see also* Visual fields
Lateralization, 310–15, 325–59, functional asymmetry of dopamine, 330, in birds, 66, 83, 311, 333, in mice, 179, 180, 479, of circling behavior, 330, emotional states and, 330, of motor functions, 328, 329, spatial learning and, 315, visuospatial perception and, 83, 328, *see also* Hemispheric specialization, *see also* Hand preference
Learning transfer, 296
Learning, 32, 36, 101, 192, 249, 279, 282, 296, in birds, 17, 29, in primates, 28–30, 236–7, 246–9, 273, 309, in elephants, 229, 252–7, 294, in dogs, 249, *see also* Social learning
Lemur catta, see Lemur, ring-tailed
Lemurs, 106, 335, ring-tailed, 194–5, 335, black and white ruffed, 335, 347, dwarf, 97, 173, mouse, 97, 331, 335
Leontopithecus rosalia, see Tamarin, golden lion
Limb preferences, *see* Hand preference, *see* Footedness, *see* Pawedness
Linguistic abilities, *see* Language abilities
Lizard, green Anole, 348–9, tree, 349, leopard gecko, 328–9
Local enhancement, 99, *see also* Social learning
Local visual strategy, 65–7, *see also* Visual cognition
Locational cues, 193
Locomotion, 97, 295, 302, 347, 357
Long call, 180, *see also* Vocalizations, *see also* Communication
Lorises, 106
Lower vertebrates, 327, 332, *see also* Fish, *see also* Reptiles, *see also* Amphibians

Macaca, see Macaque
Macaca fascicularis, see Macaque, cynomolgus
Macaca fuscata, see Macaque, Japanese
Macaca nemestrina, see Macaque, pig-tailed
Macaca rhesus, see Macaque, rhesus
Macaca silenus, see Macaque, lion-tailed
Macaque, 9, 64–5, 70, 75, 124–5, 148, 176, 178, 296, 299, cynomolgus, 336, japanese, 60, 119–21, 126, 158–60, 162, 193, 336, 341, 345, lion-tailed, 345, pig-tailed, 76, 109, 193, 336, rhesus, 175–6, 178, 193, 297, 336, 338, 352–3
Machiavellian Intelligence, 37
Magnetic resonance imaging, (MRI), 298–9, 312–13, 315
Magpie, Australian, 194, 200, 204, European, 4, 9–10, 62, 73, *see also* Corvids
Maladaptive traditions, 162
Mammals, 76–7, 105, 171–4, 178, 192–215, 290, 292–3, 296, 299–303, 306, 310, 312, 317, 327, 329, 331–2
Manipulative skills, 124–5, 345–6
Manipulative tasks, 108, 110–11, 357
Manorina melanophrys, see Noisy miner
Manual specializations, 326
Marine mammals, 99–100, 192, 213, 215, 282
Marmoset, 13, 104, 109, 111, 175–7, 235, 311, 336–7, 341, 347–8, 352–3, *see also* Tamarin
Marmot, 195
Marsh tit, 26, 315
Marsupial, 102, 172, 301
Matching of rotated symbols, 316
Matching-to-sample, 60, 76, *see also* Delayed matching-to-sample
Maternal care, *see* Parental care
Maze solving, 162, 314–15
Meaningful signals, 191–7, *see also* Communication, *see also* Vocal signalling, *see also* Referential communication
Meat eating, *see* Carnivorous behavior, *see also* Carnivore
Meat sharing, in primates, 122
Medial geniculate body, (MGB), 179
Meer cat, 195, 214
Melopsittacus undulates, see Parakeet
Memory, 6, 31–2, 290, 299, in birds, 74–6, 82–5, in primates, 85, 108, 268–9, in elephants, 294, *see also* Spatial memory, *see also* Episodic memory, *see also* Semantic memory
Mental abilities, similarities and differences between species, 3–4, 45, 107–8, *see also* Intelligence, *see also*, Complex cognition, *see also* Cognitive ability
Mental attribution, 42, 45
Mental maps, 35, 306, *see also* Spatial memory, *see also* Spatial maps
Mental states, 234, in primates, 4, 14, 45, in birds, 16–17, 42
Mental time travel, 5, 29, 30, 32, 35, 45
Mice, *see* Rodents
Microcebus murinus, see Lemur, mouse
Midbrain, 178
Mimicry, *see* Imitation, *see* Vocal mimicry
Minnow, 144
Mirror guided self-recognition (MSR), *see* Recognition, of self
Monkeys, 73, 97, 104–14, 118–27, 158–60, 175–6, 193–4, 200, 204, 209–15, 251, 294, 296, 312–13, 336, 338–9, colobine, 209, Diana, 195, golden, 336, howler, 97, leaf, 336, patas, 336, rhesus, 316, spider, 126, 193, 303, 336, squirrel, 84, 175–6, 178, 180, 292, 336, 338, 341–2, vervet, 27–8, 194–6, *see also* Marmoset, *see also* Macaque, *see also* Tamarin

Monotremes, 296
Morgan's cannon, 282
Morgan's dog, 249–51
Mosaic evolution, 300, 302–3
Mosquito fish, 328
Mothers, 179–81
Motion patterns, 70–1, 85, *see also* Visual cognition, *see also* Visual scenes
Motivation, 116, and signaling, 192
Motor asymmetries, *see* Lateralization
Motor cortex, 178
Motor functions, hemispheric control of, 326–7, evolution of, 310–12, *see also* Hand preference
Motor imitation, 12, 114–18, *see also* Imitation
Motor preferences, *see also* Hand preference, *see also* Footedness, *see also* Pawedness
Mouse, 173, 179, 329, 333–5, *see also* Rodents
Muricide, 351
Muriquis, 126, 336, 338
Musk ox, 102–3
Mynah bird, 12, 62

Naturalistic tests, 4, 12, 17, 38–9, 40–1, 115, *see also* Ecological validity, *see also* Ethological validity
Neocortex, 4, 6, 8, 77, 85, 190, 205, 291, 296–302, 305–8, 318, 351, group size and, 307–8, higher cognition and, 77, 190, 205, 318, evolution of, 296–9, 305, experience and, 314, in primates, 85, 205, 305, 313, size of, 6, 305–6, 314
Neopallium, 229
Neophilia, 118
Neophobia, 118
Neophron percnopterus, see Vulture
Neostriatum caudolaterale (NC), 8
Neostriatum, 8, 10, 45, 205, 298, and tool use, 309
Nestor notabilis, see Kea
Neural capacity, 310, 311, 314, 357
Neuroanatomy, in primates, in birds, *see also* Brain structure, *see also* Brain size
Neurons, 293, 302, 314–17
Newt, alpine, 328, smooth, 328
Nigrostriatum, 330
Noise, 173
Noisy miner, 194
Nongeometric information, 80
Non-vocal communication, 199–201, *see also* Communication, *see also* Gestures
Nucifraga columbiana, see Nutcracker, Clark's
Nut cracking, 25, 35, 122, 162
Nutcracker, Clark's, 36, 39, 83, 84, *see also* Corvids

Object choice, 279–80, 282
Object perception and cognition *see* Visual cognition
Object permanence, 62, 73–7, *see also* Visual cognition, *see also* Visual scenes
Object-object relations, 112
Observational learning, *see* Social learning
Occam's Razor, 235
Occipital lobe, 312
Olfaction, 101, 116, 302
Omeothermism, 85
Operant learning tasks, 293
Opiates, 177
Optic lobes, 306
Orangutan, 3, 9, 103, 111, 122–3, 125, 173, 189–99, 200, 203, 206–7, 295, 299, 307, 313, 337, 347, 372, Chantek, 203
Oryctolagus cuniculus, see Rabbit, European
Otocyon megalotis, see Fox, bat-eared
Otolemur garnettii, see Busbaby, small eared
Owl, barn, 65–6

Paleocortex, 297
Paleostriatum augmentatum, 10
Paleostriatum primitivum, 10
Pan paniscus, see Bonobo
Pan troglodytes, see Chimpanzee
Papio papio, see Baboon
Parakeet, 42–3, 62, 198–9
Parental care, 85, 101, 118, 172, 179–81
Parrot, 4, 6, 12, 16, 29, 45, 62, 73, 115, 192, 204, 211, 215, 309–10, 326, 333, African gray, 12, 16, 29, 41–3, 115, 118, 204, Alex, 204, orange-winged Amazonian, 350
Partial occlusion of objects, *see* Visual cognition
Parus palustris, see Marsh tit
Passer domesticus, see Sparrow
Pawedness, 326, 332–5, 339–40, 342, *see also* Footedness, *see also* Hand preference
Perception, 107
Periaqueductal gray, (PAG), 178
Personality, 149–59, 160
Petter's Rule, 68, 69
Phoxinus phoxinus, see Minnow
Piagetian object permanence, *see* Object permanence
Pica pica, see Magpie, European
Pig, 15, 174
Pigeon, 59, 60, 62–3, 65–7, 70–2, 76–7, 80, 84, 100, 115, 251, 311, 316, 346
Pilfering, 17–18, 20–1, 42, *see also* Caching
Pinniped, 172, 174
Planning, 30, 32, 34–6, 190, 295, 299
Plant processing, 103
Planum temporale, 312
Plasticity, 192

Platyrrhine, 336
Plover, piping, 16
Plover, Wilson's, 16–17
Poecilia reticulata, see Guppy
Pointing, 199–200, 234, 270, 273–9, 282, *see also* Communication, *see also* Referential communication, *see also* Gestures, *see also* Begging, *see also* Cues
Point-light displays, 71–2, *see also* Visual cognition, *see also* Visual scenes
Pongo pygmaeus, see Orangutan
Porpoise, 292
Possum, Australian striped, 301
Postural Origins Theory, 325–6, 335, *see also* Hand preference, *see also* Footedness, *see also* Pawedness
Potato washing, 119, 148, 158–9, 162
Povinelli's experiments, 232–41, 244, 247–8, 269, 295
Practice effects, 342
Prairie dog, 193, 214
Precocial species, 172
Predatory skills, 103, 105
Prefrontal cortex, 4, 205, 298–300, *see also* Neocortex
Pre-linguistic, 197
Presbytis entellus, see Langurs, Hanuman
Prey capture, 103
Primates, 3–4, 74–5, 97–9, 103–27, 142, 156–60, 189–215, 303, 325, 336, 358, advanced cognitive abilities, 4, 8, 29–30, 36–7, 40, 45, 103–6, 110–14, 142, 289, 294–5, brain size, 290–2, 298, 302, 305–7, communication in, 4, 171–81, 189–215, compared to birds, 3–4, 45, 57–85, 309, compared to fish, 142–3, 148, 156–64, hand preferences of, 332, 335–8, 344–5, hands of, 301–2, lateralization and, 310–13, 326–7, 340, 352–4, mental abilities, 45, 110–12, neocortex of, 4, 305–8, posture and, 338, prefrontal cortex, 4, rotational biases in, 331, *see also* Monkeys *see also* Great apes, *see also* Apes
Primatocentrism, 5, 15, 36, 45, 156, 372
Problem solving, 159, 248–51, 316, in primates, 23, 108, 110–11, 159, 250, 264, 268, 294–5, in birds, 24, 249, 251, 316, in fish, 159, in elephants, 253–7, *see also* Insight
Prosimian, 106, 174–6, 292, 325, 331, 335–6, 338, 341–2
Proto-grammar, 204
Psittacid, see Parrot
Psittacus erythacus, see Parrot, African gray
Pteronotus parnellii, see Bat, echo-locating
Pygmy chimpanzee, *see* Bonobo, *see also* Chimpanzee
Pyschons, 298

Quail, bobwhite, 172, 329, Japanese, 13, 115, 172, 329

Rabbit, European, 101
Raccoon, 302
Rana pipiens, see Frog
Rat, naked mole, 102
Rat, 13, 35, 78, 80, 100, 115–16, 160, 177, 252, 310, 314–15, 329–33, 342, 354, black, 102
Rattus norvegicus, see Rat
Rattus rattus, see Rat, black
Raven, 4, 13, 17–18, 24, 115–18, 127, 204, 231, 251–2, 257, *see also* Corvids
Receptive fields, 65, *see also* Visual cognition
Recognition, of self, 41–2, 229, 232, 235, of others, 172–81, 193
Referential communication, 5, 27, 189–97, 204, 214, 278–9, *see also* Alarm calls, *see also* Communication, *see also* Vocalizations
Rehabilitation, 373
Religion, and attitudes, 231
Representing, objects, 72–7, skills, 197, *see also* Visual cognition, *see also* Visual scenes
Reptile, 171, 178, 290, 292, 296, 310, 312, 314, 327, 329
Response facilitation, 107
Rhinopithecus roxellanae, see Monkey, golden
Rodents, 59, 100, 172–3, 179, 192, 329, 333–5, 351–2
Rook, 9–10, 24
Rotational behavior, *see* Circling
Rules, social, 212–13
Rules, of grammar, 204

Saguinus oedipus, see Tamarin, cotton-top
Seal, 105, 173, 282
Searching behavior, 82
Self-awareness, 295
Semantic memory, 30, *see also* Memory
Semantics, 191, 196, *see also* Language abilities
Separation call, 171–2, 176–7, *see also* Communication, *see also* Vocalizations
Serinus canaria, see Canary
Sex differences, 103, 150–1, 161–2
Sex hormones, 31
Shape information-3D, 72–3, *see also* Visual cognition, *see also* Visual scenes
Shrew, 301
Signing, in apes, 28, 201–3
Simple relational tasks, 110–11
Social attachment, 8, 265

Social cognition, in birds, 15–16, 21, 39, in primates, 40, 104–5, 114–16, 124–7, 264, 280, 282, in dogs, 282
Social communication, 307–8, 349
Social facilitation, 99, *see also* Social learning
Social information, 114–19, 263–82
Social intelligence, 212, 306–9, hypothesis, 264, 307
Social knowledge, 264, 275
Social learning, 9, 11, 12, 40, 45, 98–106, 109, 117, 127, 141–3, 158–64, 192, 308–10, in primates, 4, 40, 76, 97–8, 103, 109–10, 113–27, 158–9, 210–11, 281, 309, in birds, 18, 21, 26, 39–40, 127, 142, in fish, 142–64, in elephants, 243, *see also* Learning
Social organization, 161, 191, 205–15, 266, 305–6, *see also* Group complexity
Social problems, 264
Social release hypothesis, 154
Social tolerance, 115, 124, 158, social tolerance hypothesis, 158
Social transmission, 146
Socially mediated learning, *see* Social learning
Somatosensory cortex, 302
Song complexity, *see also* High vocal centre
Song nucleus, *see* High vocal center
Song repertoire, 304, *see also* High vocal center
Songbirds, 12, 304, 310, 350–1
Sound spectrogram, 173
Sparrow, 15, white-crowned, 350, white-throated, 350
Spatial behavior, 330, 334
Spatial localization, *see* Spatial memory
Spatial maps, 305, 315
Spatial memory, 6, 9, 11, 78, 316, spatial learning abilities, in birds, 18, 20, 39, 74, 78, 80, 82–5, in primates, 78, spatial localization, 81, 83–4, 315, and size of hippocampus, 39, 304–6, 314, *see also* Memory
Spatial navigation, 35, 78, 80–5
Spatial task, *see* Visual cognition
Spermophilus beecheyi, see Squirrel, ground
Squirrel, 214, ground, 195, gray, 173
Starling, 13, 118, 194
Stereotypy, 295
Stimuli, discrimination of, 57–65, 70–1, *see also* Visual cognition, *see also* Visual scenes
Stimuli, moving, 63, 70, 342, *see also* Visual cognition, *see also* Visual scenes, *see also* Visuospatial tasks
Stimulus enhancement and local enhancement, 9, 99, 141–2, 146–7, 154, 157, 160, *see also* Social learning
Stone handling, 120
Streptopeia risoria, see Dove, ring
Striatum, 297
String-drawing task, 252, 257
Structure from motion, 70, 72, *see also* Visual cognition, *see also* Visual scenes
Sturnus vulgaris, see Starling
Sus scrofa, see Pig
Symbolic communication, 5, 27, 28, 45, *see also* Communication
Symbolic representation, 28, 112, 119, 123, 203
Synaptic connections, 293
Syntax, 191, 196, *see also* Language abilities

Tactical deception, *see* Deception
Tamarin, 175, 336, 341–2, Cotton top, 61, 180, 338, 341, Golden lion, 103, 338, 341, *see also* Marmoset
Tarsier, 97
Taxonomic relatedness, 98
Teaching, 104, 109, 119, 160, in fish, 146–55, 160
Tectofugal pathway, *see* Visual pathways
Telencephalon, 39, 57, *see also* Forebrain
Temperamental differences, 337
Temporal domain, 69, *see also* Visual cognition
Temporal integration, 84
Temporal maps, 305
Termite fishing, 122, 162, 311
Thalamofugal pathway, *see* Visual pathways
Thalassoma bifasciatian, see Wrasse, blue head
Theory of mind (ToM), 4, 14–16, 21, 23, 37, 43–4, 227–58, *see also* Deception
Theropithecus gelada, see Baboon, Gelada
Theta rhythm, 299
Toad, European common, 328, 332, European green, 328, 332, South American cane, 328, 332, 348
Tool use, 25, 34–5, 45, 160, 163, 295, 301, 308–10, in birds, 24–7, 34–5, 204, 343–6, in primates, 4, 23, 25, 34–5, 45, 108, 110–15, 120–4, 235, 249, 264, 295, 344–6, in elephants, 230, 294
Traditions, 101, 119–64, among monkeys, 119–27, among great apes, 121–7, among birds, 127, among fish, 143–8, 155–7, 160–4
Trainability, 282
Transfer ability, 296, *see also* Learning transfer
Transfer learning, *see* Learning transfer
Trial and error learning, 23, 26, 141, 210, 228, 231, 248–9, 251, 253, 256–7
Triturus vulgaris, see Newt, smooth
Triturus alpestris, see Newt, alpine
Troop size, *see* Group size
Turning bias, 327–32, *see also* Lateralization
Tursiops truncates, see Dolphin, bottle-nosed

Turtle, snapping, 328–9
Two-action model, 108, 111

Ultrasounds, 179
Ungulate, 214, 301, 308
Urosaurus ornatus, see Lizard, tree

Varecia variegata variegata, see Lemurs, Black and white ruffed
Velocity threshold, 84
Vision, 38, 206, 242–3, 269–71, 302
Visual cognition, 57–85, abstract cognition, 29, 83–4, binocular and monocular vision, 83, 84, contour perception, 63–8, hidden objects, 73–4, in birds, 18, 57–85, in primates, 57–85, 267–8, landmarks, 83–4, partly occluded objects, 58–68, 72–4, recognition, representation and localization of objects, 58, 67–85, in elephants, 242, *see also* Cognitive map, *see also* Visual representation
Visual differences, 38–9, 57–65, 77–9, 84–5
Visual fields, 15, 38–9, 84, 199
Visual orientation, in birds, 75
Visual pathways, 57–66, tectofugal, 57–66, thalamofugal, 57–66
Visual perception, 269–71, in birds, 84, in dogs, 270–1, in elephants, 240, in primates, 269, 271, role of the eyes in, 269–70, 278
Visual perspectives of others, 235–48, 264, 269–71, 278
Visual representation, 57–85, in birds, 57–85, in primates, 57–85, in elephants, 242, rules, 68, *see also* Petter's Rule
Visual scenes, 66, depth stratification, 68–9, relative size, 68, illusion, 68, geometric regularity, 69, 82–5, motion, 85, *see also* Visual cognition
Visual Wulst, 65, *see also* Visual pathways
Visuospatial perception, in birds, 58–65, 82–5, in arboreal and terrestrial primates, 58–85, 346, in cats, 64, 70, and lateralization, 342–3, 346
Visuospatial reaching, 339–42
Visuospatial tasks, 339–43, 357–8
Vocal apparatus, 202, 204
Vocal development, in primates, 197, 201–4, in birds, 204
Vocal mimicry, 12, 29, in primates, 115–17, in birds, 117, 204
Vocal repertoire, 197, 213
Vocal signaling, 171–81, 191, 193, 196–9, 213, in primates, 85, 115–17, 172–81, 213, 275, in birds, 85, in dogs, 278, in elephants, 230, *see also* Deception
Vocalizations, 12, 122–3, 171–81, 190–2, 196–9, 213, 230, 273, 348–56, production of, 123, 175–81, effect of size on, 173, 176
Vulture, 25

Warblers, 32
Washoe, *see* Chimpanzee
Whale, 105, 190, 292, killer, 103
Wolf, 197, 205, 208–9, 213, 215, 275
Working memory, 77
Wrasse, blue head, 162

Zonotrichia albicollis, see Sparrow, white-throated
Zonotrichia leucophrys, see Sparrow, white-crowned